HONEY PLANTS OF NORTH AMERICA

(NORTH OF MEXICO)

A GUIDE TO THE BEST LOCATIONS FOR BEEKEEPING IN THE UNITED STATES

WRITTEN FOR THE A. I. ROOT COMPANY
BY
JOHN H. LOVELL
Author of "The Flower and the Bee: Plant Life and Pollination," and Biological Editor of the "A B C and X Y Z of Bee Culture"

ILLUSTRATED FROM PHOTOGRAPHS BY THE AUTHOR

MEDINA, OHIO
THE A. I. ROOT COMPANY
1926

CONTENTS

PREFACE

"The foundation of success in beekeeping," says W. Z. Hutchinson, "is the location. Unless a man has a proper location he had better relinquish beekeeping as a specialty, or else seek the right location." No amount of labor and skill will produce results where there is no source of supply. No one should enter extensively upon the production of honey without first inquiring in regard to the flora, on which he must depend for a marketable surplus. There is a small number of states in which commercial beekeeping is profitable throughout their entire extent; but in most cases there are only a few sections well adapted to this industry, while the remainder of the territory will support a much smaller number of colonies. The importance to American beekeepers of a thorough knowledge of the honey plants of this country can not, therefore, be overestimated. The need of more information has long been generally recognized; and more or less complete lists and bulletins have been published by a number of states of the chief nectar-yielding plants found within their borders, while others have similar lists in preparation. But the honey floras of the majority of the states still remain very imperfectly known, and there is evidently a demand for a book which will cover the entire United States and Canada.

But it is a list of plants actually valuable to bee culture that is wanted, not merely a list of species which secrete nectar. There are thousands of different kinds of nectariferous flowers which are so rare that they are of no practical importance to the beekeeper. Cultivated herbs and trees introduced from foreign countries, which there is no reason to suppose will ever become common, should not be included in the list. Also flowers adapted to insects with very long tongues, which produce nectar inaccessible to honeybees, should be omitted. In selecting the honey plants described in the following pages the technical as well as the practical point of view has received careful investigation.

Flowers valuable to bee culture for pollen only have been strictly separated from those valuable chiefly for nectar. Many beekeepers have failed to observe that there are many blossoms which are nectarless, and this fact has been the cause of numberless mistakes. If a plant was common and frequently visited by bees either for pollen or honey-dew, many beekeepers have been prone to conclude that it yielded "some honey," no matter how improbable such an assumption might be ecologically. The grasses, sedges, rushes, and the millions of acres of cultivated grains must of necessity depend on wind-pollination; for, even if they yielded nectar, the number of flower-visiting insects is wholly insufficient to pollinate them. It should not be concluded in the case of a doubtful species that it yields nectar, until the nectar has actually been observed and the nectary located in the flower under the microscope.

The nomenclature of the seventh edition of *Gray's New Manual of Botany* has been given the preference, which has permitted the continued use of many old familiar Latin names. In the case of a number of varieties in the southern states, which have been raised to the rank of species of doubtful validity, I have retained the older names. The plants have been arranged alphabetically according to their English names, as this arrangement is more convenient for reference than the grouping of the species in families, and renders it unnecessary constantly to consult an index. While no effort has been spared to make the descriptions strictly scientific, technical terms have been avoided so far as possible. Where a plant has more than one common name, cross-references have been given.

The figures of flowers, which are usually of natural size, are nearly all original, and were photographed by the author from specimens often obtained with much difficulty and expense. For the benefit of photographers it may be stated that panchromatic plates and a Wratten ray filter, No. 3, were used to preserve the proper color values in monochrome; and that to obtain details U. S. diaphragm 128 and a long exposure were employed. All of the work was done indoors with a vertical camera stand. With the aid of the illustrations the identification of many species should be comparatively easy. The two figures of Coccidae are from Comstock's *Manual for the Study of Insects,* and are used with the permission of Professor Comstock. The maps, showing the geographical distribution of alfalfa, cotton, the citrus fruits, and other plants, are of great interest and are worth many pages of description. They are taken from the *Geography of the World's Agriculture,* by V. C. Finch and O. E. Baker. The maps relating to the United States were prepared by Mr. Baker, Agriculturist to the Department of Agriculture, and are reproduced here with his approval. In a few cases they are reduced in size, or given only in part.

To the publications and personal letters of the following distinguished apiarists, M. C. Richter, L. F. Scholl, H. B. Parks, H. A. Scullen, Frank Stirling, R. B. Willson, and F. C. Pellett, the author is indebted for important information in regard to many honey plants. Much of the material in Chapters IV and V is taken from the author's book, *The Flower and the Bee: Plant Life and Pollination,* and his technical papers on the ecology and pollination of flowers, as well as many unpublished notes, have been frequently quoted. A very voluminous literature, relating to the geology, physical features, agriculture, floras, and bee culture of the various states, has been very carefully examined.

In the preparation of Part IV, Mr. E. R. Root, President of The A. I. Root Company, has taken a very active interest. During the preparation of this volume Mr. Root traveled over 30,000 miles in the western, southwestern, and southeastern states, during which time he was in constant correspondence with the author, and has in a long series of letters communicated to him the results of his observations and numerous interviews with many successful beekeepers. The entire manuscript has, indeed, been read by Mr. Root, and to a great extent he is responsible for the descriptions of the different kinds of honey. To Mr. Frank C. Pellett, Assistant Editor of the *American Bee Journal,* and author of *American Honey Plants,* I am also indebted for much valuable information. Mr. Pellett has read the descriptions of Illinois, Missouri, and Iowa, of which latter state he was for many years Bee Inspector. The author wishes to place on record his sincere appreciation of the assistance and encouragement received from Mr. H. G. Rowe, Managing Editor of *Gleanings in Bee Culture,* which has been an important factor in the preparation of the book. Without the generous co-operation of hundreds of beekeepers the descriptions of the different states could never have been written. They have assisted me not only by describing their personal experience in beekeeping, but in many instances have gathered additional data and have furnished specimens of the honey plants for identification.

Finally there remains the great pleasure of acknowledging the assistance of Dr. E. F. Phillips, formerly In Charge of Bee Culture at Washington, D. C., now

Professor of Bee Culture at Cornell University, Ithaca, N. Y., who has read the descriptions of the entire forty-eight states, and has given the author the benefit of his investigations and extensive knowledge of beekeeping in every part of the Union. Information of especial value has been received in the case of New York, Indiana, New Mexico, and the Pacific Coast States.

Waldoboro, Maine. JOHN H. LOVELL.

PART I

NECTAR: ITS SECRETION, PROTECTION, AND DISTRIBUTION

Temperature and proper conditions of soil and atmosphere, or what is commonly called "the season," have a thousand times more bearing on the surplus than the amount of bloom or the number of colonies in a single apiary.

—E. W. Alexander.

CHAPTER I

THE STRUCTURE AND ROLE OF THE FLOWER

The Organs of the Flower — The Role or Function of the Flower Is to Secure Pollination — The Advantages of Cross-Fertilization — Plants Multiply Readily by Vegetative or Assexual Methods — Sex was Developed to Secure Variation — The Importance of Variation in the Evolution of Plant Life.

A knowledge of the structure of the flower, of its role or purpose, of the advantages of cross-fertilization, and of the object of sex among plants should form a part of the education of everyone. Undoubtedly the influence of flowers upon the development of the human race has been both profound and far-reaching. We can not imagine what this world would have been without them, or estimate the enjoyment that would have been lost, or the power for good that would have been forever missing. But the ministry of beauty is not the only service rendered by flowers, for they are also of practical importance as the sources of honey and the producers of fruits, vegetables, and cereals. The beekeeper, fruit-grower, and farmer who understand their structure and economy will be protected from many costly errors, and will be able to obtain results that otherwise would be impossible.

Select some common insect-pollinated flower, as a rose, buttercup, or basswood blossom, for examination, and notice that it consists of many small thread-like organs surrounded by two whorls of leaves (Fig. 1). The outer circle of leaves is green and forms the calyx or cup, each leaf of which is called a sepal

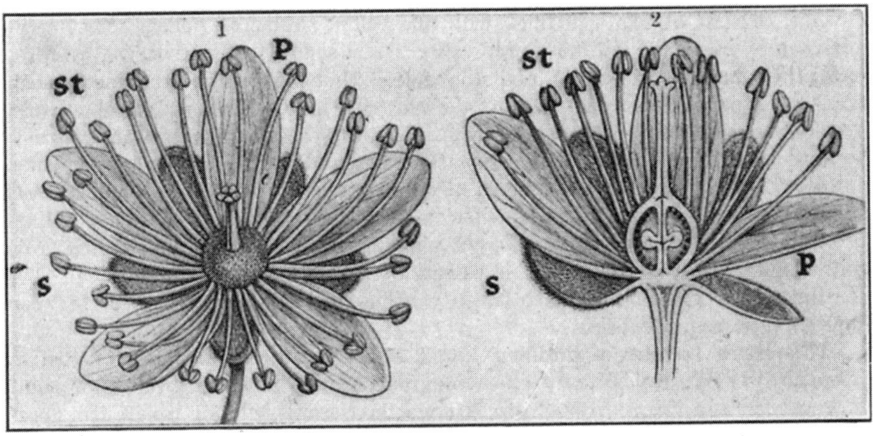

Fig. 1.—Diagram of basswood flower (Tilia); 1, entire flower; 2, lengthwise section of ovary. The outer cycle of five small leaves is the calyx, each member is a sepal (s); the inner cycle of five larger leaves is the corolla, each member is a petal (p); the numerous pinlike organs forming the third cycle are the stamens (st); the little terminal knobs are the anthers and contain the pollen; the thread-like stems are the filaments; in the center of the flower is the pistil; the globular base is the ovary in which are formed the seeds; the stem above is the style bearing the stigma.—(After Kerner.)

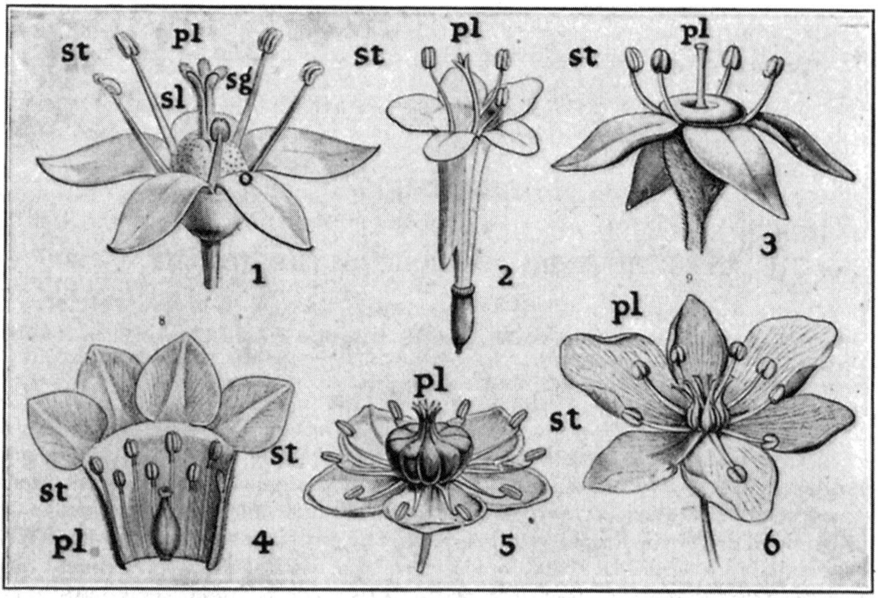

Fig. 2.—The stamens (st) and pistils (pl) of six common flowers; sg, stigma; sl, style; o, ovary. 1, Common Lilac (Syringa vulgaris). 2. Garden Valerian (Valeriana officinalis). 3. Cornelian Cornel (Cornus Mas). 4. Mezereum (Daphne Mezereum). 5. Common Pokeweed (Phytolacca decandra). 6. Flowering Rush (Butomus umbellatus).—(After Kerner.)

(s). The sepals are not always green but may be bright-colored, as in the fuchsia, larkspur, and columbine. The calyx protects the inner, more delicate organs, especially in the bud, when it alone is visible. The second circle of leaves is larger and usually bright-colored—red in the rose, yellow in the buttercup, and blue in the violet—this is the corolla, or little crown, and each leaf is a petal (p). The brilliant hues of the corolla are designed to attract the attention of insects. Next to the corolla stand the stamens (s), threadlike stems bearing the anthers, each of which consists of two sacs filled with fine yellow grains of pollen. The pistils (pl) occupy the center of the flower (Fig. 2). Usually a pistil consists of three parts: the ovary (o), style (sl), and stigma (sg); but the style is sometimes absent, when the stigma becomes sessile, or stemless. The base of the pistil is the ovary, or seed-case, containing the nascent ovules (seeds) which at first are unfertilized; the style is a porous stalk rising from the ovary and bearing at its upper end a glutinous receptive surface or stigma. Pollination is the transfer of pollen from the anthers to the stigma. If the pollen is from the same flower it is self-pollination; but if from a flower on a different plant it is cross-pollination. Pollen from another flower on the same plant is essentially the same as self-pollination. The offspring from cross-pollination between varieties of the same species, or between different species, are hybrids.

The stigma secretes a glutinous liquid which causes the grains of pollen to adhere to its surface. If the proper conditions exist the pollen grain soon sends out a slender tube which grows down through the porous style, by which it is nourished, until it comes to an ovule in the ovary. It enters the ovule by a little orifice (the micropyle, or little gate), and there passes from the end of the tube a male cell or sperm, which fuses with an egg cell in the ovule. This is fertilization. Fertilization may be defined as the union of two specialized cells of unequal size, the smaller being the sperm and the larger the egg. Pollination and fertilization are often confused, but it is clear that they are very different functions. Fertilization does not always occur immediately after pollination; the flowers of the witch hazel are pollinated in autumn, but fertilization does not occur until the following spring; and in the case of pine trees fertilization may not take place until two years after pollination.

But there are comparatively few flowers which, like the buttercup and rose, have all the members present, separate, and regular in form. Many flowers have a part of the organs wanting. The wind-pollinated cat-tails, pondweeds, grasses, and sedges have neither calyx nor corolla, or they are represented only by scales and bristles. Wind-pollinated flowers are small, green or dull-colored, nectarless, and usually scentless. Many hard-wood trees have the stamens and pistils in separate flowers. In Jack-in-the-pulpit the flower is reduced to a single pistil or stamen.

Flowers may be modified also in another way by the fusion or consolidation of the organs. Tubular calyces and corollas, as in the morning-glory and phlox, are stronger and better adapted to protect the nectar than those which consist of separate sepals and petals. In the pulse family the stamens are commonly fused into "brotherhoods." But fusion is nowhere so common as among the pistils. So long as they are separate, pollen must be placed on the stigma of each; but when they are consolidated, one application may serve to fertilize all the ovules. Clearly a flower like the buttercup, where all the parts are separate units, is more primitive than an orchid, where they have coalesced to such an extent that it is difficult to determine their number. In primitive families, again, like the buttercups, mallows, roses, and saxifrages, the flowers, as a rule, have regular forms, the petals are all alike, and are usually white or yellow; but in many families, which have developed more recently, the flowers have assumed strange, bizarre forms, as in the sweet pea, snap-dragon, and sage, and frequently are red, purple, or blue in color. The purpose of these singlar forms is to compel an insect to pursue a fixed path in its visits and thus effect pollination. Such flowers are the youngest and latest to appear in the plant world. The larkspur is of later origin than the buttercup, and the clover blossom than the rose.

What is the role or function of these varied forms, mechanisms, colors, and odors? Their purpose is to secure pollination. This is the significance of the flower. Self-fertilization is better than sterility, but crossing is an advantage. Crossed plants grow more vigorously, bloom earlier and more profusely, and produce more seed capsules. Inbred races tend to disappear. While hybrids between different species often show decreased fertility, there are thousands of cases in which they multiply readily by seed. They show greater variation than the parent forms, have more and larger leaves, stronger stems, better root-systems, bloom sooner, and the forms are larger, better scented, more intense in color, and more persistent. Among the higher plants (Angiosperms) pollination is effected chiefly by the agency of the wind, insects, and birds.

All attempts to explain the structure of the flower failed until the discovery of sex in plants, and, incredible as it may seem, this has been universally recognized for less than one hundred years. In 1694 Camerarius of Tübingen put the question to plants themselves by experiment, and learned that seed, "the most perfect gift of nature," was not produced unless pollen was placed on the stigma. Later James Logan, Governor of Pennsylvania, helped to prove the truth of this theory by experiments with Indian corn.

The lowest and simplest plants consist of one cell, and are usually very small in size. They have no reproductive organs, and multiply by dividing into two equal parts, each of which is like the parent plant. The rapidity with which self-division takes place is almost incredible. Cohn computed that the descendants of a single hay bacillus in 24 hours would exceed four septillions. In a single night many square miles may be reddened by a one-celled red alga (*Protococcus nivalis*). No trace of sexual reproduction can be discovered in many common fungi, as the toadstool, mushroom, and puffball. The higher plants, although they have sex organs, may readily be reproduced vegetatively by buds, rootstocks, runners, stolons, offsets, suckers, bulbs, and tubers. "It is probably true," says Coulter, "that more plant individuals are produced asexually than sexually. Sex is not essential to reproduction. It was the last method of reproduction attained among plants, and it did not replace the older methods but was added to them."*

Since sex is not necessary for reproduction, what then is its significance? It secures greater variation. When plants multiply vegetatively, or asexually, they as a rule vary very little. When a florist obtains a new flower, or the horticulturist a new fruit, he increases the number of his plants by budding or by cuttings. Asexual reproduction is much more likely to give new plants resembling the parent form than sexual. But plants in a state of nature must vary or they are likely to perish. If the conditions under which they live change, and they can not change to meet the new conditions, they will become extinct. Neither can a plant migrate into new territory with different conditions, unless it can adapt itself to them. But in the case of plants which possess sex, when two individuals are crossed, the offspring will inherit the characters of both, and will of necessity be more variable than either. They will also be more vigorous than the parents, since beneficial characters are more likely to survive than those which are injurious, and the progeny will inherit more of them.

The development of sex made cross-fertilization possible, and greater variation and vigor in the offspring hastened the evolution of new species, which were able to adapt themselves to every part of the earth and to every condition of life. Lotsy, in his *Evolution by Means of Hybridization,* regards hybridization as the sole cause of the appearance of new species; and in the opinion of Jeffrey hybridism has played a large role in the evolution of flowering plants. It is worthy of note that, in the great work of pollination, plant-life to-day has found a most efficient helper in the beekeeper.

* The reader who desires to pursue further the study of the origin and significance of sex in plants will find the two following little books very helpful: Coulter, John M., "The Evolution of Sex in Plants." Coulter, John M. and N. C., "Plant Genetics."

CHAPTER II

NORTH AMERICAN HONEY PLANTS

Definition of a Honey Plant — Species Widely Different — Relation of Honey Plants to Soil — Influence of Latitude on Blooming-Time — Relative Importance of the Honey Plants — Great Opportunity for Growth of Bee Culture — Variety of Honeys.

A honey plant may be defined as a common plant which secretes nectar accessible to honeybees in quantities sufficiently large to be of importance to beekeepers. There are in North America, north of Mexico, more than 15,000 kinds of flowering plants, so that, if a list of all plants secreting nectar were desirable, it would be easy to enumerate thousands of species; but very few of them would be of much practical importance to bee culture. Many are rare, others grow in the deep recesses of forests and swamps, while others yield so little nectar that honeybees pass them by unheeded. The buttercups and daisies, which are so conspicuous in June, receive little attention from bees. Potatoes, which are grown by the million acres, secrete no nectar and afford very little pollen, and are seldom visited by insects. Tomatoes are likewise nectarless. The hardhack and the meadowsweet, two of the commonest of shrubs, are almost wholly neglected by honeybees. The bunchberry covers large areas in open woodlands; but the small flowers contain so little nectar that bumblebees ignore them entirely, and honeybees visit them only occasionally. They are left to flies, beetles, and the smaller bees. Then there are many flowers which are adapted to moths, butterflies, bumblebees, and humming-birds, which have the nectar so deeply concealed that it can not be reached by honeybees. Thus the number of species which deserve to be classed as honey plants is comparatively small, not much exceeding 200 in America, north of Mexico, according to present records; while the number which yield a surplus annually is much less.

The only characteristics that the honey plants have in common are that they secrete nectar accessible to honeybees, more or less freely, and are common in one or more localities. Otherwise they exhibit great differences. They belong to widely separated families; include herbs, shrubs and trees, native and exotic species, and wild and cultivated plants. Some are merely local in occurrence; others extend over a large area; and still others occur in several locations far apart. The pulse family, or Leguminosae, contains the most important species, as the acacias, clovers, alfalfa, sweet clover, and locust, while they are wholly absent from the great family of orchids. After the pulse family, the Compositae contain the largest number of honey plants, as the goldenrods, asters, thoroughworts, sunflowers, and thistles.

Honey plants are likewise very variable in the preference they exhibit for different soils. The tupelo, willows, and buttonbush grow in wet swamps; Spanish needles, sneezeweed, and snowvine in marsh lands; the smooth sumac prefers a rocky soil; the mountain sages live on the arid foothills of California, while sourwood flourishes in the mountains of the Alleghanies; the cacti are dwellers in the desert; the gallberries avoid a limestone soil, while the clovers will not grow as well elsewhere. The spikeweed and the alkali-weed thrive in an alkaline soil;

the Rocky Mountain honey plant in a dry saline soil; the black mangrove and salt-marsh goldenrod in a soil and atmosphere impregnated with salt, while fireweed springs up in profusion on burnt land.

The effect of climate on the period of blooming is well shown by the average dates of the beginning and ending of the blooming period of the principal honey plants. In Mississippi and Louisiana white clover and alsike clover bloom from April 15 to June 5-15; in Maryland and Virginia from May 20 to July 5; in New England from June 10 to July 20. Sweet clover in Alabama and Mississippi blooms from May 15 to the end of August. In New England it blooms from July 10-20 to August 20-25. Fruit trees in Texas and Alabama begin blooming April 10-15, and end May 1-10; in New England they bloom from May 15-30 to June 1-15. Alfalfa in Kansas, Oklahoma, and Nevada begins blooming May 20, and ends September 1-30; in Nebraska, Wyoming, and Colorado it blooms from June 15 to September 1. Basswood blooms in Kentucky and Tennessee from June 15 to June 30; in New England from July 10 to July 25. Goldenrod and aster, which throughout the eastern United States bloom from midsummer until late autumn, are less affected by climate. The time of blooming of the plants of more local distribution will be given in the alphabetical list of honey plants.

The four most important honey plants in the United States are sweet clover, alfalfa, white clover, and alsike clover. They furnish not far from 70 per cent. of the surplus honey which enters the wholesale market. The other major honey plants, many of which have a restricted distribution, contribute from one to four per cent. Besides the surplus honey which is sold in the wholesale market, on which the honey-buyer's estimates are based, there is a large amount of honey produced which never enters this market. As a commercial product, basswood has almost entirely disappeared, but the reports of many beekeepers show that this honey is still obtained, but usually mixed with clover. The major honey plants are comparatively few in number. Among trees are orange, tulip tree, the tupelos, sourwood, black locust, mesquite, sumac, and catsclaw; among shrubs, the mountain sages, raspberry, and gallberry; and among herbaceous plants, in addition to those named above, cotton, buckwheat, goldenrod, aster, Spanish needles, and heartsease.

According to E. R. Root, from 80 to 90 per cent. of the nectar available in the honey plants in the eastern half of this country is lost from want of bees to gather it; while in the western half as much as 70 per cent. is gathered from irrigated alfalfa and sweet clover. In Iowa alone the annual product of 10,000,000 pounds, it has been estimated, could be increased to 60,000,000 pounds. A large part of the nectar of tulip tree and other early-blooming plants is lost because the colonies of honeybees, at the beginning of the season, are not strong enough to store all of it. Many orchards of fruit trees and plantations of berry bushes are imperfectly pollinated, because there are not bees enough in the vicinity to make use of the floral food.

Moreover, the area of cultivated honey plants which to-day furnish the larger part of the commercial surplus is steadily increasing. Fruit trees, as the apple, pear, orange, and many small fruits; buckwheat, cotton, and a vast acreage of plants cultivated for fodder, as alfalfa, sweet clover, alsike clover, and crimson clover, are yearly becoming of greater value. The possibilities of this artificial pasturage are only partially recognized. Sweet clover alone will soon add many million pounds to the present output. It is rapidly spreading over North and South Dakota, Nebraska, and in the Rocky Mountain districts, providing new bee territory which for many years will not be overstocked with bees. In 1923 the Kansas State Board of Agriculture reported 48,891 acres grown in 92 counties in Kansas.

Honeys vary greatly in color, density, flavor, and in the length of time they remain liquid. After ten years the honey from white tupelo is still liquid, and the honey from the mountain sages of California is also very slow to granulate. Honeys containing a high percentage of dextrose, or grape sugar, are prone to granulate, while those low in dextrose and high in levulose, or fruit sugar, remain liquid for a long time. On the other hand, the honey of blue curls granulates in the cells before they are sealed; and, in general, a honey, unless heated, granulates within six months after it is extracted. The endless variety of flavors and colors of the different honeys will be described in the list of honey plants.

CHAPTER III

THE SECRETION OF NECTAR

The Structure of Nectaries — The Physiology of Nectar Secretion — External Factors Influencing Nectar Secretion — The Relation of Soils to the Distribution and Vigor of Honey Plants — Rainfall and Honey Production — Humidity — Temperature and Nectar Secretion — Light and the Making of Sugar — Altitude — Extra-Floral Nectaries.

The secretion of nectar is at once one of the most remarkable and baffling functions of plant-life. Foliage leaves are factories in which are made carbohydrates, or sugars and starches, foods required for the growth of plants. But in nectar secretion the plant parts with the food that it has just manufactured. The two functions are antithetical, or directly opposed to each other. While the quantity of nectar secreted by a single nectary is small, thousands of tons are yielded by flowers in a single season. It is a reasonable and logical conclusion that nectar secretion renders a valuable service of some kind to plants. But before the discovery of its use in attracting insects in the pollination of flowers, the early botanists were greatly puzzled by the problem and offered a variety of curious explanations. Pontedera, an Italian botanist, made the ovary absorb nectar for the benefit of the seed. Linnaeus was of the opinion that its significance was not yet known. It was Kolreuter who was the first to point out that the purpose of nectar was to attract insects, which were necessary for the pollination of flowers.

The older flower biologists describe the forms and positions of nectaries at great length, and a few of them, as Caspary, Behrens, and Bonnier, investigated their structure; but of the relation of soil, temperature, and rainfall to the secretion of nectar they knew little. They would have been not a little astonished to learn of the vagaries of alfalfa, white clover, buckwheat, and cotton in yielding nectar in different localities and in different seasons. Most of our information on this subject has been derived from the observations of beekeepers, and will be found recorded in the files of the bee journals. A large amount of data has been accumulated, but the influence of the various factors is still very imperfectly understood.

THE STRUCTURE OF NECTARIES

Nectaries may be divided into floral and extra-floral; the floral occur on all the organs of the flower, as sepals, petals, stamens, and pistils; the extra-floral are found on leaves, stems, and the involucral bracts. They are rare on the calyx, but the concave sepals of basswood secrete nectar in large drops. Nectaries occur in great variety on petals, as in the columbines, buttercups, love-in-the-mist, monkshood, aconite, hellebore, and honeysuckle. Stamens frequently bear nectaries, and sometimes whole stamens are changed into nectaries. In a great many flowers nectar is secreted by a fleshy ring or disc at the top or base of the ovary, as in umbelliferous plants, the cornel, sumac, currant, apple, pear, plum, raspberry, and strawberry.

A typical nectary consists of a little spherical mass of small cells, covered with a single layer of slightly larger cells called the epidermis. The common white clover is a good illustration. The nectar gland is on the inner side at the bottom of the tube or cylinder formed by the cohesion of the filaments of the stamens, at the base of the free stamen. According to Cook, it is a little mound composed of 6 to 8 layers of small cells, which are easily distinguished from the larger cells of the adjacent tissues. The epidermis consists of a single layer of cells. The nectar-secreting tissue often forms a ring around the base of the ovary, or it may line the base of the calyx tube, or again it may develop as a cushion on top of the ovary. Sometimes it is at the bottom of a little cup or pit, as in the buttercups. In basswood, nectar is yielded by many little pegs projecting above the nectary tissue. In the mustard family and in the catalpa there are numerous pegs or knobs on the nectary. In the bindweeds the nectary consists of five scales at the base of the ovary. In the day lily the nectary is a long narrow pore, branched at its lower end and lined with large oblong cells. Nectaries are often white, but they may be of a deep orange color. It would require an entire chapter to describe the many remarkable ways in which nectaries are protected from rain and undesirable visitors. For example, on the petals of the common garden flower love-in-the-mist, there is a little bowl with a scale-like cover which the honeybee raises when it seeks the nectar, and which falls back into place after the departure of the bee.

THE PHYSIOLOGY OF NECTAR SECRETION

The passage of nectar through the epidermis of the nectary in the case of many flowers offers little difficulty. In Spiraea, thornbush, petunia, pumpkin, and many other blossoms there are numerous stomata (openings in the epidermis, protected by guard cells) through which the nectar comes to the surface. Beneath each stoma, or opening, there is a small chamber in which the nectar accumulates before it passes out of the nectary. The stomata may be distributed uniformly over the surface of the nectary, or they may be collected together in special spots. In the willows the nectaries have the form of pegs with a single large pore at the truncated end, through which escapes the colorless nectar. There are also nectaries which are covered with papillae, or single, elongated, thin-walled cells. These cells are not cutinized but consist wholly of cellulose, through which water and substances soluble in it readily pass. Nectar diffuses through the walls of the papillae and collects as drops upon them because of the outward pressure exerted within the distended cells of the nectary.

But not all nectaries have the outer cell-walls readily permeable by nectar. In many flowers the epidermis may be composed of two or three layers of cells, there may be no stomata, and the outer cell-walls may be cutinized. Cutin is a fatty substance which may almost entirely prevent the escape of water. The epidermis of aquatic plants contains little cutin; and for this reason, when exposed to the air, the water quickly escapes and they dry up in a few hours. On the other hand, plants growing in a dry situation usually have the epiderims heavily cutinized. When the time comes for the secretion of nectar the outer layers of such nectaries may swell and disintegrate, forming a gummy or gelatinous substance. With the destruction of the cutinized cells, the remaining cells of the epidermis, composed of cellulose, become readily permeable to nectar or other solutes. As sugar is readily derived from cellulose, it may be formed during the breaking down of the outer cells on the exterior of the nectary.

If the cells of the nectary are distended with water, the passage of nectar through the epidermis offers no great difficulty. In the absence of cutin, and where there is only one layer of cells, it occurs by diffusion; and when cutin is present, the epidermis is rendered permeable by the breaking-down of the outer cell-walls. But there is another membrane, lying within the cell, which it is much more difficult to pass. The cell-walls of all living plant cells are lined with a thin, delicate membrane of protoplasm, known as the ectoplasm, which allows foods to enter the cell, but will not permit them to pass out and be wasted. In what way this living plasma membrane acts can not be fully explained, but it evidently performs a most useful service. As soon as it dies it loses the power of selection, and nectar or cell sap can then readily escape from the cell. It acts only so long as it is living. How, then, does nectar pass through this living membrane and come to the surface of the nectary?

In order that a nectary may secrete nectar, its cells must be distended with water, or a solution of water and sugar. The nectar must exert a strong pressure outward, distending the elastic walls of the cells. If, owing to dry weather, there is not sufficient moisture in the soil, the leaves and young stems of plants will wilt or droop. They are rigid only when the cells, of which they are composed, are distended or swollen with water, or turgid, as it is termed by botanists. As has been pointed out by Sachs, the nectaries of plants secrete nectar only when the parts concerned are turgid. If the cells are flaccid no nectar is secreted. Since growth takes place only when the plant cells are turgid, flaccid cells are largely inactive. The advent of dry weather has often cut short the honey-flow. Plants with leaves and flowers drooping for want of water will not yield nectar. The tugor pressure in nectar cells may amount to as much as five atmospheres. Secretion in all probability begins by the forcing of nectar through the ectoplasm and epiderims, which are perhaps modified by the pressure. There would, of course, be less resistance on the free or outer side of the nectary than on the inner side, which is in contact with other plant tissues. As soon as the nectar comes to the external surface the water evaporates and a solution is formed more dense than that within the cells of the nectary; more nectar might then be drawn outward by osmosis. If nectar formed more rapidly than it escaped by osmosis, tugor pressure might manifest itself repeatedly. Secretion by floral nectaries ceases with pollination or soon after.

The quantity of nectar secreted differs greatly with different species of plants, wholly independent of external conditions. Nectaries are not all equally efficient. In some species small drops appear which are scarcely visible to the naked eye; in others the nectar forms a thin layer, appearing like a coat of varnish; in still others the drops flow together and collect in cups or tubes, which may be filled to overflowing. There is in the flower of the tropical orchid *Coryanthes* a collecting cup, into which nectar gradually trickles until it amounts to nearly an ounce. External conditions do not explain why two species yield very unequally under exactly the same conditions. Why is alfalfa a good honey plant only in the West, while white sweet clover yields well everywhere? Why is alsike clover a more reliable honey plant than white clover, or the bushy goldenrod than the early goldenrod, when both grow side by side and bloom at the same time? These differences depend upon the plant itself, or upon the chemical processes taking place within it.

It should not be forgotten that there are many kinds of glands besides nectaries. Water glands excrete large quantities of water, and many intermediate stages occur between them and nectaries. There are also resin glands, chalk glands, salt glands, slime glands, and the digestive glands of insectivorous plants. Internal glands occur in the leaves of the orange, Eucalyptus, and many other plants, which may not discharge their contents to the surface. Very little is known about the way in which each kind of gland forms the substance peculiar to it. Their origin is not clear, and their functions are often useless.

EXTERNAL FACTORS INFLUENCING NECTAR SECRETION

The principal external factors influencing nectar secretion are the soil, moisture and dryness, heat and cold, light, and altitude. The effects produced by these factors depend in a large degree on locality. White clover secretes nectar better on the limestone soils of southern Minnesota than on those of Alabama because in the former region the temperature is more moderate; but cotton and alfalfa yield well in very hot climates. Few farmers realize that very slight differences in the weather may determine the success or failure of their crops. According to the *Year Book of the Department of Agriculture* for 1920: "An extra quarter of an inch of rain may add thousands of bushels to the corn-planter's harvest; a few degrees lower temperature may put a lot of extra money into the potato-grower's pocket. The way the wind blows is sometimes more important than the cost of farm labor."

THE RELATION OF SOILS TO THE DISTRIBUTION AND VIGOR OF HONEY PLANTS

The soils of the United States differ greatly in chemical composition. All outcropping rocks, as granite, gneiss, trap, serpentine, lava, felspar, slate, shale, limestone, and sandstone, as the result of weathering, disintegrate and form soils. In many regions they have remained where they were formed, but often they have been carried by streams into bottom-lands and deltas, or blown by the wind, as the loess of the central states, or transported by ice, as glacial drift. Plants do not all flourish equally well in the same soil, and the distribution of the vegetation of this country has been greatly influenced by the characteristics of the soils. Let us take the State of New York for illustration: The southern portion of the state is poor in lime, but the sandy soil is well adapted to growing buckwheat. North of this section there is a limestone belt which extends from Buffalo nearly to the Hudson River, on which the clovers thrive and secrete nectar abundantly. Other limestone areas are in Jefferson, St. Lawrence, and Franklin counties in the northern part of the state. It is of great practical importance for beekeepers to know the limits of these areas. An apiary on the margin of one of these limestone tracts might store a large surplus, while another apiary ten miles away on a non-limestone soil might give very poor results. On the sandy soils of the Adirondack region it would be useless to look for clover, and raspberry is the most important honey plant. A beekeeper should be familiar with the soils of his state, and should obtain and study the maps prepared by the Bureau of Soils at Washington.

Plants growing in soils to which they are adapted are more vigorous and produce more nectar than in soils in which they do not flourish. In England, says Sladen, sainfoin yields the largest amount of honey and the best seed in regions where the subsoil is chalk. White clover and sweet clover give the largest surplus on a soil rich in lime, while gallberry and the blueberries grow only in acid soils. Marigold in Texas succeeds best on the black prairies, and is absent from sandy and gravelly ridges. Spikeweed covers thousands of acres of the alkaline plains of the upper San Joaquin, California, where it yields honey by the carload. Buckwheat yields well on a sandy soil, and phacelia on a clay soil. The black mangrove produces a great amount of nectar in soils containing over one per cent. of salt. The chemical composition of soils is closely related to the distribution of honey plants and the surplus of honey.

Soils do not act directly on the nectaries. There is no evidence that any chemical stimulates nectar-secretion directly; but it is well established that vigorous plants will yield more nectar than stunted, poorly developed individuals. Kenoyer found by experiment that vigorous buckwheat plants yielded twice as much nectar as did the weak ones in the same bed. At Medina, Ohio, spider-flower (*Cleome spinosa*) and figwort yielded nectar profusely when growing in deep, rich soil, but only sparingly on a poor soil. Alfalfa, according to Hunter, raspberry, according to Hutchinson, and willow-herb, according to Sladen, secrete the largest amount of nectar under conditions which give the most vigorous growth. The more vigorous the plant, the more active are all its functions and the more sugar it will make. But unless rainfall and temperature are favorable, a suitable soil does not necessarily ensure a honey crop. Sladen gives a most interesting example in England, showing the dependence of nectar secretion on climate, even when the soil is suitable. Heather, in order to yield well, requires not only a peaty surface soil but also a suitable subsoil. In one region near the coast a whinstone subsoil is considered of no value for honey, although the heather grows well and flowers profusely on it. But in the West, where the climate is milder and more moist, a whinstone subsoil gives excellent results. It is even more strikingly shown that a good soil alone does not insure nectar secretion in the case of honey plants which bloom several times during the season, but fail to yield nectar during one or more periods of blooming. Giant sainfoin in England blooms first about June 10, and yields a crop of honey if the weather has been suitable. It blooms again in July, yielding the main honey crop. There is a third period of bloom in middle August, when it yields no honey. In the San Joaquin Valley, California, there are four crops of alfalfa. The first and last crops yield little if any honey; but the second and third, says Richter, if proper climatic conditions prevail, give a great abundance of nectar. A plant may make a fairly good growth on an unsuitable soil, the bad effects of which may appear only in the flowers and fruit, as often happens in the alkali lands of the western states.

It is most important that beekeepers should know that, as regards their adaptation to soils, most plants may be divided into two great groups: plants which grow best in lime or limestone soils, sometimes called calciphiles (lime-lovers) or calcicoles (lime-dwellers); and plants which avoid lime soils and succeed best in acid or sour soil, calciphobes (lime-avoiders). They are common on sandy soils, and hence are sometimes called silicicoles (sand-dwellers). A study of the relation of plants to soils led Unger to divide them into those which grow on limestone only; those which prefer limestone but will grow on other soils; those which grow on sand only; and those which prefer sand but will grow on other soils.

The value of lime in growing crops is recognized in many aphorisms; as, "Lime is the basis of good husbandry," and "A lime country is a rich country." All of the legumes, as the clovers, alfalfa, sweet clover, and sainfoin grow better on limed soils because they form more tubercles on the roots, and the plants are richer in nitrogen than those grown on unlimed soil, as acid is injurious to the nitrogen-fixing bacteria. Kerner found that when plants accustomed to a lime soil were grown in a soil devoid of lime they presented a miserable appearance, with scanty flowers which ripened only a few seeds; while, on the other hand, plants accustomed to sand, when grown in a soil containing lime, soon withered and died without flowering at all.

It is well established that the clovers are most abundant, grow most vigorously, and, other conditions being favorable, yield the most nectar on soils rich in lime. As a rule they are excellent honey plants on the glacial soils of the northern United States. Other extensive areas underlaid by limestone are the Great Appalachian Valley, the Nashville Basin of Tennessee, the Blue Grass Region of Kentucky, and the Ozark Uplands. It is estimated that these limestone areas comprise 68,000,000 acres. Owing to the leaching out of all the soluble salts by heavy rains, most soils in the eastern United States require an application of lime for a good growth of the clovers.

While lime is removed by frequent rains from the soils of the Eastern States as rapidly as it is formed, in the semi-arid Western States where the average annual rainfall ranges from 20 to 10 inches or less, lime is steadily accumulating in the soil as the rocks decay. But many other alkaline salts, which are intensely injurious to vegetation, also remain in the soil, as black alkali and white alkali. Black alkali (sodium carbonate) is commonly known as salsoda, or washing soda. It is corrosive in its action on plants, and the presence of one-tenth of one per cent. will practically prevent the cultivation of all crops, and this salt is therefore greatly dreaded by farmers. White alkali includes a group of salts, as common salt (sodium chloride), Glauber's salt (sodium sulphate) and baking soda (sodium bicarbonate), which form a white incrustation on the surface of the soil. While white alkali is injurious to vegetation, it is not corrosive, but crops will not endure, as a rule, more than one-half of one per cent. of these salts. Thus the soils of the eastern states and the semi-arid western states differ profoundly, and there is a great difference in the native vegetation.

When irrigation is practiced the soil is very fertile and produces enormous crops. If the land has never previously been irrigated, the presence of harmful salts is seldom noticed except in low spots where water has accumulated. But after a few years of irrigation, especially in localities where water has been used excessively, alkali salts may appear at the surface of land that was formerly quite free from them. Capillary attraction draws the alkali to the surface, and in a single season has been known to transform a flourishing alfalfa field into a barren alkali flat. Probably a million acres have been lost on account of alkali.

All of the legumes are very sensitive to the presence of black alkali. Many western farmers have complained that after a few years their alfalfa deteriorated and yielded less nectar. This has resulted from a rise of alkali in the soil. If there is an appreciable quantity of black alkali in the soil the cultivation of alfalfa should not be attempted, but it has succeeded with 0.4 per cent. of white alkali present. The rise of alkali in the soil is injurious to the development of roots and their functions, causing a check in nutrition and the formation of less sugar. The roots of sugar beets are smaller and contain a smaller percentage of sugar than in soils free from alkali. Sugar cane also contains a less amount of sugar. Fruit is seriously injured in sweetness and flavor. Cotton produces a smaller number of flowers, and nectar secretion is impaired or fails. The black locust, plane tree, some species of *Eucalyptus,* and the golden willow will endure a moderate amount of alkali.

There are a few honey plants which are confined to soils rich in common salt, as the black mangrove, the salt-marsh goldenrod, and the sea-grape. The sea-grape grows on the beaches of southern Florida, and on many of the islands, where it is exposed to the full force of the salt spray driven by storms. It is much less valuable for nectar than the black mangrove. Asparagus also will thrive in a soil containing considerable salt. On the other hand, certain families avoid saline conditions, as the rose family. The heaths, likewise, which are acid-soil plants, are absent from such localities.

The influence of lime soils on the distribution of plants has been known for centuries, but the effects of acid soils have been recognized only during the past ten years. A plant might survive in an acid soil, but in general it was supposed to be injurious to plant growth. "During the last five years, however," said Wherry a few years ago, "newly developed methods of interpreting and determining acidity have been applied in several widely separated regions — Sweden, Denmark, the northeastern United States (and subsequently in India and England), with the same result in all cases — recognition of the great significance of the acidity of the soil in controlling the growth and distribution of native plants."

There is an immense area of acid soils in the eastern states. For example, as the result of the study of 1474 samples of soil taken from many counties in Pennsylvania, 72 per cent. of the soil areas were found to be acid. Of the river-bottom soils, 82 per cent. were acid. In Maryland 1500 soils have been tested; 75 showed very strong acidity; 150 strong acidity; 405 medium acidity; 270 were slightly acid; 210 were very slightly acid, and 390 were not acid. Peat bogs, swamps, the majority of river bottoms, the sandy soils of the great pine barrens of the eastern and southern states, and of many mountainous regions, as in the Adirondacks, all have acid soils. The soils of the southern states are more generally acid than those of the northern section, owing to more rapid decomposition of the mineral compounds and the heavy rainfall. Wherry found that "the greatest number of species, as well as individuals, occur in soils lying just on the acid side of the neutral point." Practically all the members of the heath family (Ericaceae), according to Wherry, are acid-soil species. It will be remembered that this family includes many well-known honey plants, as the sweet pepperbush, Labrador tea, *Andromeda,* fetterbush, sourwood, bearberry, and the heathers, as well as the blueberries, cranberries, and huckleberries. In Europe heather (*Calluna vulgaris*) covers immense areas of acid soils, but on lime soils it makes a weak growth and soon perishes.

The knowledge that many plants will succeed only in acid soil is of the greatest value to florists, horticulturists, beekeepers, and to all who cultivate plants. Undoubtedly scores of attempts to grow our native herbs and shrubs have failed because the soil used was not suitable. But it is not sufficient that a soil be acid or alkaline: there must be the right degree of acidity or alkalinity, for determining which with great accuracy there are now provided two well-established methods. Both in Europe and America investigators are giving much attention to the reaction of plants to acid soils.

RAINFALL AND HONEY PRODUCTION

The effect of rain on the production of honey is threefold, according as it influences the growth of the plants, the development of the flowers, and the secretion of nectar. In the eastern states, with their heavy rainfall, there is a varied and abundant flora and a great number of colonies of bees. In the semi-arid regions of the West bee culture is largely dependent on plants grown under irrigation. The wild honey plants are very erratic yielders in this section, varying with the amount of rainfall. Unless there are twenty inches of rain during the winter the mountain sages of southern California will fail to yield a good crop of honey. In 1884, one of the wettest years in the county of San Diego, there was a rainfall of over 27 inches, and in various parts of the county 60 inches fell; and a surplus of 500 to 700 pounds of honey per colony was quite common. In southern Texas, in the Rio Grande Plain, in 1917 and 1918, the honey plants suffered so severely from drought that not far from 50 per cent. of the colonies of bees perished. In this region plant growth responds very quickly to rainfall; thus the palo verde, or green-barked acacia, a desert tree, puts forth leaves in early spring but drops them in the dry weather following; but if there are late rains it puts forth foliage a second time. Stunted plants yield much less nectar than those which are well developed. The time of the rainfall may also be as important as the amount.

The development of the flowers of many plants is very closely connected with a preceding rain. This is well illustrated by the thorny chaparral of southern Texas, which consists of 70 to 80 species of small shrubs and trees. In this semi-arid region a rain at any time during the summer will cause a number of honey plants to bloom in a few days. The yellow rain-lily is the first to respond, blooming in about 60 hours after rain (not valuable for nectar). White brush (*Aloysia ligustrina*) follows in about five days, and it is of interest to note that the flowers are reported to yield more nectar during the rain-induced period of bloom than during the regular blooming time in spring. Besides white brush, soapbush (*Guaiacum angustifolium*), coma (*Bumelia lyciodies*), Brazilwood (*Condalia obovata*), and mesquite, bloom twice or more times in a single season if there are several rains. In Australia flatweed (*Hypochaeris radicata*), a plant resembling the dandelion, sends up flowers at all times of the year after showers following a period of dry weather.

But too much rain may induce a rank vegetative growth which is detrimental to the development of flowers. Much rain will prevent the blooming of mesquite a second time. Several years ago mesquite in Texas twice started "to tag," and heavy rains prevented the buds from maturing. In 1921, in Arizona, according to Vorhies, after an unusually dry winter, there was a fair flow from mesquite. But in July and August the rainfall was exceptionally large, the precipitation at Tucson being 6.2 inches in July alone. A second blooming was expected, but the superabundant moisture induced a heavy vegetative growth and the trees scarcely blossomed at all a second time.

As has been already pointed out, nectaries secrete nectar only when their cells are distended, or turgid. The nectar must exert a strong pressure on the elastic cell-walls. If the cells are flaccid no nectar will be secreted. The advent of dry weather has often cut short the honey flow. If want of water causes the leaves and flowers to droop no nectar will be produced. Kenoyer found that plants which had been allowed to dry to the withering-point several times in the course of their growth yielded a little less than one-third as much nectar as normal plants. The nectar of open wheel-shaped flowers, like those of buckwheat, may be largely washed away by rain, as has been proven by the use of an artificial spray.

HUMIDITY

Humidity is the quantity of water vapor in the air. On a rainy day the air is saturated with moisture and the humidity is 100 per cent.; on a dry day it is about 50 per cent. In the semi-arid region of Nevada, where the average rainfall is about 10.6 inches, the average annual humidity is 51 per cent. In Iowa, where the average annual rainfall is 30.9 inches, the average annual humidity is 76 per cent. Honey stored in a humid atmosphere contains more water than that gathered in a dry air. Analyses of honeys from Wisconsin showed an average moisture content 3.79 per cent. greater than was found in honeys from Nevada. Less water evaporates in humid air than in dry air. Nectaries secrete a larger quantity of water in humid air, because the evaporation of water from the leaves is checked and accumulates in the plant cells under greater pressure than when there is little moisture in the air. In a saturated air, water may escape also in drops from the ends and teeth of the leaves. But while the amount of water passing through the nectaries increases, the quantity of sugar secreted remains about the same.

Scholl states that cotton in Texas yields nectar most abundantly in early morning, but that the quantity decreases as the air becomes dryer at mid-day. Toward evening it again increases unless the air is very hot and dry. During cloudy days the yield is good throughout the day. Moist air, says Sladen, with slow evaporation, seems to be necessary for a good gathering from heather in England. Buckwheat flowers secrete much more liquid when placed under a bell jar than in the open air.

TEMPERATURE AND NECTAR SECRETION

Temperature exerts a greater influence on the functions of flowers, including nectar secretion, than does light, humidity, or rainfall. According to recent investigations in France, the movements of the parts of flowers, their opening and closing, are caused by the daily rise and fall of temperature. During a partial eclipse of the sun there was a fall of ten degrees, and flowers ceased to open their corollas. The temperature at which nectar secretion begins, and at which the largest amount of nectar is secreted, varies greatly with different honey plants. In general, high temperature favors nectar secretion, since the membranes of the nectary are rendered more permeable, the solvent power of water is increased, and chemical changes in the plant take place more readily. Records of honey production for 29 years at Clarinda, Iowa, showed that the largest yield from white clover was on days with a temperature from 80 to 90 degrees F. According to the reports of 19 large honey producers in Iowa, Missouri, and Illinois, the largest yields of nectar are obtained when the temperature ranges from 80 to 100 degrees. Alfalfa yields nectar most freely when there is ample moisture in the soil of irrigated land, during a succession of hot days. On the other hand, very high temperature, exceeding 95 to 100 degrees, may be injurious.

Several very important honey plants yield nectar abundantly at moderate, or in some instances relatively low, temperatures. Willow-herb will yield nectar freely during weather that would stop all storing from basswood and clover. "I have seen," says Hutchinson, "bees working on raspberry when the weather was so cool that white clover would not yield a drop of nectar." Sainfoin in England will yield between 60 and 65 degrees, even with very little sunshine. Sladen records that on a September day, in England, bees brought in from heather most honey between 8 a. m., when the temperature was only 48 degrees, and 10 a. m., when it was 52 degrees. I have also a report from British Columbia of bees working on flowers when the temperature did not rise above 52 degrees all day.

Cold or cool nights followed by warm days are better for nectar secretion than a uniform temperature. Sugar is formed most rapidly at a high temperature, but its accumulation in the plant is favored by a lower temperature. On a warm, clear day sugar is made in the cells of leaves more rapidly than it can be carried away, and in the form of starch it is stored in the granules (chloroplasts) containing the leaf-green pigment. During the night, with its lower temperature, it is moved out of the leaves in the form of glucose or grape sugar (dextrose), and is carried to different parts of the plant where it is used or stored. The best, or optimum, temperature for the manufacture of sugar by leaves, in many plants, is between 68 degrees F. and 72 degrees F.; but it may be made slowly in some cases at much higher and lower temperatures. Plants can not make sugar in the night time, because sunlight, the force which drives the machinery of the leaf, is turned off when the sun drops below the horizon.

But growth may take place in darkness as well as in the light. The best temperature for growth with most plants is between 77 degrees F. and 90 degrees F., for there is an optimum temperature for growth as well as for the making of sugar. But it will be noted that a higher temperature is required for growth than for the manufacture of sugar. In the Temperate Zone the temperature at night is usually much lower than in the daytime, and in spring often falls nearly to or even below the freezing-point. During a cold night growth is checked, or stopped, and the sugar formed during the day remains unused, until with the return of light the temperature again rises. There will thus be more sugar available after a cold than after a warm night, during which much of it has been used for growth. Woody shrubs and trees contain the largest amount of starch in the fall, which during the winter is changed into sugar found in the bark and winter leaves. A rise of temperature in the spring causes the sugar to change back into starch. The conversion of insoluble starch into soluble sugar and the reverse is effected by the agency of ferments known as enzymes, the action of which is influenced by temperature.

LIGHT AND THE MAKING OF SUGAR

The secretion of nectar is closely related to the amount of light the plant receives. The force, or energy, required for the manufacture of food material is furnished by sunlight. It is the power which drives the plant, just as the electric current is the power which drives the motor. Deprive the plant of light, and the sugar reserves will soon be consumed and nectar secretion will cease. Kenoyer placed buckwheat flowers under both light and dark jars, and after two days the amount of sugar in the nectar of the flowers under the dark jar began to decrease. After four days approximately only one-fourth as much sugar was secreted per flower. He obtained, also, the same results by covering the leaves of the plant with black paper. The flowers of cucumber, snapdragon, and sage all secreted less sugar in darkness. If, however, the flowers were placed in the dark, and the leaves remained in the light, the amount of sugar in the nectar did not decrease. The secretion of nectar is thus dependent on the food reserves made by the leaves. Darwin observed that the extra-floral nectaries of the common vetch (*Vicia sativa*) ceased to secrete nectar when the sun was hidden by clouds, and the hive-bees left the field, but as soon as the sun broke out again they returned to the feast.

Experiments in illuminating plants with electric lights show that the rate of growth increases with the intensity of the illumination. Clover blooms much earlier in constant light than in shorter daily illumination, though this is not true of all plants. The long days of northern regions promote the manufacture of more sugar, and the same result is produced by the more intense light of alpine heights. The long warm days and cool nights of the north are particularly favorable to bee culture; for example, at Roberval, on Lake St. John, Quebec, an average annual colony yield of 200 pounds is obtained from alsike and white clover. Light is also a direct stimulus to the growth of flowers. In the shade, the colors of many flowers are paler than in the light. The flowers of the chickweed will not open except in light, and some other flowers develop imperfectly in darkness. But too intense light is injurious. Leaves commonly arrange themselves in such a way that they will receive as much light as possible, but if the light is too intense they change their position, so that a smaller amount strikes their surface.

THE INFLUENCE OF ALTITUDE

Flowers at high altitudes have brighter colors and tend to secrete more nectar than in the lowlands. Mueller states that in the lowlands the spur of an orchid (*Platanthera solstitialis*) was only about a third filled with nectar, but that in the Alps it was over half full. According to Bonnier and Flahault, *Silene inflata,* one of the pink family, was much richer in nectar at an elevation of 5895 feet than at an elevation of 1300 feet. They also cite statistics gathered by the French Department of the Pyrenees, from which it would appear that the average honey production per colony was, from sea-level to 1000 feet, 6 pounds 10 ounces, and that with every rise of 1000 feet it increased from 2 to 4 pounds per colony, until at 4000 to 5000 feet it was 19 pounds, 13 ounces. According to Richter, "There is a great deal of evidence pointing to a more profuse nectar secretion of alfalfa at higher elevations."

The increase in nectar secretion is apparently due to the more intense light and the wide range of temperature between day and night. At an altitude of 8510 feet, the chemical activity of the sun's rays is 11 per cent. greater than at sea level. The greater light intensity drives the machinery of the leaf more rapidly, and more sugar is made. It has already been shown that the alternation of warm days and cold nights promotes nectar secretion. In the Alps and in all mountainous regions the nights are much colder than the days. As we ascend a mountain the air grows thinner, or less dense, and at a height of 3.5 miles the pressure per square inch is only one-half as much as on the earth's surface. It also contains much less water vapor. Water vapor acts as a great blanket to the earth's surface and vegetation. If the amount of vapor is small, the temperature will fall rapidly. Hence the cold nights at alpine heights. Dry air is not sensibly warmed by the sunlight passing through it, but if it contains much water vapor its temperature rapidly rises, while the soil receives less heat. In the daytime the earth at high altitudes receives a large amount of sunlight, which the soil absorbs in a much greater degree than does the air. Thus on a clear September day, in the Pyrenees, at a height of 9400 feet when the air was at 50 degrees F. the soil was at 91 degrees F. The warm soil raises the temperature of the layer of air in contact with it so that the dwarfed plants of alpine heights find both soil and air very much warmer than at night time.

EXTRA-FLORAL NECTARIES

Nectaries are commonly associated with flowers, but they may occur on many extra-floral plant organs, and probably first appeared on leaves. They may be found on involucral leaves, as in cotton and poinsettia; on the flower-stalk, as in cowpeas; on the blade of the leaf, as in the turban squash; on the midrib of the leaf, as in cotton; on the leafstalk, as in the cherry, almond, and peach; on the stipules, as in jewelweed; but in the majority of cases they are useless and wholly or nearly functionless. But from the extra-floral nectaries of cotton, partridge-pea, cowpea, vetch, and a number of other plants, a large amount of honey is obtained. The leaf-nectaries of partridge-pea secrete nectar freely, but the flowers are nectarless — a paradoxical phenomenon. Active leaf-nectaries would appear to be a distinct disadvantage to the partridge-pea, since they tend to draw away insects from the flowers; but they are less injurious than would at first appear; for, no matter how abundant the supply of nectar, bees are compelled to gather pollen for brood-rearing. Thus, despite the absence of nectar, the flowers of partridge-pea are well pollinated by bumblebees and very productive of seed.

CHAPTER IV

PLANTS WHICH ARE OF LITTLE VALUE TO BEE CULTURE

Many Flowers Yield Little Nectar — Rare Flowers — Bumblebee-Flowers — Butterfly-Flowers — Hawk-Moth-Flowers — Fly-Flowers — Bird-Flowers.

There is a widespread belief that nearly all plants yield nectar and are of more or less value to bee culture. Accordingly from time to time bumblebee-flowers, butterfly-flowers, moth-flowers, pollen-flowers, and wind-pollinated flowers are reported as excellent sources of nectar. It is not unusual to find bird-flowers, like the wild columbine, the torch-lily, the cardinal flower, and the trumpet honeysuckle in a list of honey plants. A great company of flowers, known as pollen-flowers, are wholly nectarless, and offer their insect visitors no other reward than an ample store of pollen. Several very common plants, as the yellow buttercup and the white daisy, yield floral food in such scanty amount that a bee is seldom seen on the bloom. The potato, the tomato, and the common garden pea, of which there are thousands of acres under cultivation every year, are practically valueless to the beekeeper. A more accurate knowledge of the flora of the United States is necessary.

Thousands of species are so rare as to be of no practical value to the beekeeper. Heather, or ling, which covers vast moors in western Europe, is found in America only locally in the coastal region of New England; while two other species of heaths have become established only in patches on Nantucket Island, Massachusetts. Many subtropical trees and shrubs have migrated into the southern portion of the United States from the warmer regions southward; for example, on the Rio Grande Plain of Texas there are nearly 80 small trees and shrubs, not one of which appears in the Atlantic forests of east Texas. There are scores of flowers which are not found outside of California, as many lupines, and species of *Gilia* and *Sidalcea.* Alpine flowers are restricted to mountainous elevations, and many beautiful orchids live only in cranberry bogs. There is a very general impression that a large flower-garden is an excellent bee pasture; but many of the gaudy exotics of cultivation have either ceased to secrete nectar, or produce it in long tubular nectaries beyond the reach of honeybees.

There is a great company of bumblebee-flowers many of which are useless to the honeybee, as the larkspur, monkshood, snapdragon, turtlehead, bush honeysuckle, scarlet sage, the wood betony, and many handsome orchids. Red clover is also a bumblebee-flower, and under normal conditions honeybees do not resort to red-clover fields, but in exceptionally dry seasons the floral tubes become so short that large yields of nectar are obtained. At Borodino, N. Y., this has happened twice in thirty years. In the fertile limestone soil of western Ohio and eastern Indiana nectar is secreted so abundantly by red clover and the bumblebee horsemints that a portion of it is accessible to honeybees. Not infrequently, in their haste to obtain the nectar, bumblebees bite holes in the spurs of flowers. The tubular nectaries of the fly-honeysuckle, the garden columbine, the skullcap, and the jewelweed are often thus punctured; and when this happens, honeybees take advantage of such opportunities to obtain the nectar. With a little experience honeybees readily recognize flowers which are of no benefit to them, and thereafter pass them by unheeded; yet every season at times they vainly seek to reach the nectar of the larkspur, the nasturtium, and the sweet William.

Many pinks, primroses, milkweeds, orchids, and species of phlox are adapted to butterflies. In the Alps in Switzerland Mueller lists 19 butterfly-flowers. But butterflies visit a great variety of other flowers, and show a preference for flower-clusters which are level-topped, or flat, like boneset.

Hawk-moth flowers are usually yellow or white, since red and blue colors are not easily seen in darkness, and open in the evening when they become sweet-scented, as the evening primrose, various species of catchfly, bouncing bet, the garden honeysuckle, night-blooming tobacco, the Jamestown weed, several white lilies and gentians, the sweet-scented Gardenia, and a number of white or greenish-white orchids. There is in Madagascar an orchid with a slender green nectary 11 inches in length, which is pollinated by a moth with a tongue of equal length. Clearly such a unique flower can never be of value to honeybees. In some instances, however, the tubes fill up with nectar so that a part of it becomes accessible.

The fly-flowers are a peculiar group, often possessing nauseous or unpleasant odors and lurid colors. For the unwary and less specialized families of flies there are pitfall-flowers, prison-flowers, pinch-trap-flowers, and flowers with deceptive odors and colors. Such are Jack-in-the-pulpit, various aroids, Dutchman's pipe, and some milkweeds. The carrion-flower, a graceful vine with smilax-like leaves, common in damp thickets, and the skunk cabbage of the swamps, are examples of flowers with indoloid or nauseous odors. But the honeybee, although it is supposed to avoid repulsive odors, occasionally visits the skunk cabbage to obtain pollen for brood-rearing in early spring. Many flies, however, as the hover or Syrphid flies, visit the same flowers as do bees, so far as they are accessible to them.

Bird-flowers are few in number in this country; but in South America they are very abundant and are pollinated by humming-birds, of which there are more than 400 species described. In Brazil they are on the wing throughout the entire year and visit most flowers. The many bird-flowers of Africa and India are visited by sun-birds, as there are no humming-birds in these regions, while those of Australia depend on honey-suckers. In Europe there are no birds which visit flowers for nectar, and consequently there are no native bird-flowers. The floral tubes of bird-flowers are frequently two or more inches long, so that the nectar is far beyond the reach of honeybees. Occasionally a too adventurous bee attempts to push its way through one of these slender tubes, and, becoming wedged within its narrow walls, perishes in sight of the sweet spoil. Bird-flowers are usually scarlet or crimson in color.

An extended description of these different groups of flowers will be found in the author's book "The Flower and the Bee: Plant Life and Pollination."

CHAPTER V

POLLEN-FLOWERS

Pollen Famines — Common Plants Valuable for Pollen — Many Irregular Flowers of Little Value — Two Groups of Pollen-Producing Plants — Wind-Pollinated Flowers — Deciduous-Leaved Trees, Conifers, Grasses, and Sedges — Pollen-Flowers.

In the absence of honey, bees may be able to live for a time on honey-dew, which in the tropics is produced in enormous quantities; but there is no satisfactory substitute for an ample store of pollen. Brood-rearing occurs from early spring until late fall, and there must be a continuous supply of pollen, or the queen ceases to lay eggs, and the larvae die in their cells. Except occasionally in early spring, bees seldom run short of pollen in the northern states; but in a few localities in the South, of which the best known is the tupelo region along the Apalachicola River in Florida, serious pollen famines occur. Most of the pollen in April and May in northwest Florida is obtained from the willows, maples, elms, and oaks; but, as the tupelo furnishes very little, there is a scarcity from June to September, or for about 75 days. This section well illustrates the importance of a beekeeper's knowing the flora of his location, and how an absence of pollen during the breeding season in spring may nearly destroy a whole apiary. A similar scarcity of pollen occurs in southeast Texas. In Australia pollen famines recur regularly every season. There is a "critical period" in mid-summer when the pollen fails, the queen ceases to lay eggs, and the brood dies of starvation. This shortage is due to the small amount of pollen produced by the gum-trees, or Eucalypti; for example, a colony of bees working on yellow gum will dwindle down to a mere handful, although there is a fine crop of honey. Thus the prosperity of the colony depends upon an ample supply of pollen, and its sources are of vital importance to bee culture.

All flowers which yield nectar, except those which contain only pistils, also furnish more or less pollen. Many supply it in great abundance, as the willows, dandelion, acacia, banana, goldenrod, and sunflower; but others, like the tupelo, basswood, and some species of *Eucalyptus,* produce a very small amount. A large supply of pollen in early spring may be provided by planting staminate willow bushes along brooks or small streams and in low waste land. (Fig. 3.) As they are insect-pollinated, the pollen is so adhesive that none of it is blown away by the wind, as in the case of wind-pollinated plants, like the alders, oaks, and elms. Sunflowers are very easy of cultivation and produce pollen in great abundance for a long time. The acacias furnish a wealth of pollen which bees gather every moment they are able to fly.

Irregular-shaped flowers like the clovers, sages, alfalfa, locust, and catnip are not good sources of pollen, because the anthers are partially or wholly enclosed by the petals in such a manner that bees gather it with difficulty, and usually carry away to the hive only the small amount placed on their bodies by the mechanisms of the flowers. Contrary to the general belief, comparatively little pollen is obtained from white clover, and a bee never visits the bloom for this purpose alone.

Fig. 3.—Staminate catkins of Pussy Willow (Salix discolor). The bushes bearing staminate catkins yield a large amount of pollen in early spring, and are also valuable for nectar.—Photographed by Lovell.

They often return to the hive with only small masses of the pollen, or none at all, in the pollen-baskets. The milkweed has the pollen grains bound up in little packets, attached to clips, which are more of a hindrance than a help to bees. The orchids also have the pollen in little masses, or pollinia, which bees can not gather. This great family is indeed of no value to beekeepers, either for pollen or nectar.

There are not far from 3000 species of plants in North America which are nectarless, but which of necessity produce pollen. They may be divided into two groups, one of which is pollinated by the wind, the other by insects. The first group is very much the larger, and contains the alders, birches, poplars, elms, oaks, hickories, walnuts, bornbeams, sweet gales, and hazelnut; the cat-tails (Fig. 4), bur-reeds, grasses, sedges, and rushes; and many homely weeds like the ragweed, Roman wormwood, pigweed, nettle, pondweed, sorrel, dock, hemp, meadow rue, and mugwort. The flowers are small, dull or green-colored, very numerous, and in the case of the trees are often crowded into cylindrical clusters called catkins. Usually the stamens and pistils are in separate flowers, which are borne on the same plant (monœcious), or on different plants (diœcious). They are all wind-pollinated, and produce immense quantities of pollen, which render many of them of great value to the apiarist.

Wherever there are forests of beeches, oaks, elms, or thickets of alder and hazelnut, or groves of nut-trees, there will be literally acres of bloom, and a dearth of pollen in spring need not be feared. Often the bees may obtain sufficient pollen from the willows and maples, and there may be no occasion for them to visit the wind-pollinated trees; but if the necessity arises they will resort to them by thousands. They have been known to seek a swamp of elms in such numbers that the roar in the beeyard was similar to that which occurs during basswood bloom. Bees also gather pollen from the staminate flowers of the date palm, which is wind-pollinated. In some parts of Europe the wind-pollinated hazelnut furnishes the first pollen in spring. It is composed of 5 per cent. of water, 30 per cent. of albuminous substances, and 60 per cent. of carbohydrates (sugar, starch). The chemical composition of the pollen of the insect-pollinated wild rose is almost identical with that of the hazelnut.

The vast forests of wind-pollinated evergreen or coniferous trees, consisting of pine, fir, spruce, cedar, and hemlock, likewise produce enormous quantities of pollen; but they yield no nectar. The air in pine forests is filled with grains of pollen, which, slowly settling downward, powder the foliage of the trees, the grass, and the ground. The falling of such an immense number of pollen grains has given rise to the reports of sulphur showers. As the pollen of the common pine contains only 16 per cent. of albumen, and the indigestible sac or shell represents 21 per cent., it is much less valuable than the pollen of the flowering plants. The pollen of the conifers is very seldom gathered by honeybees. Large quantities of honey-dew at times cover the foliage of the evergreen trees.

The presence of honey-dew on the leaves of many wind-pollinated trees and shrubs has given rise to frequent reports that they yield nectar. The oaks have been repeatedly listed as fair honey plants, valuable for early brood-rearing. Indeed, the hickory, elm, walnut, and poplar have all been stated in various bulletins to furnish "some honey," while the winged elm is credited with "an amber honey with a strong characteristic aroma." In most cases these mistakes may be explained by the presence of honey-dew.

None of the grasses, sedges, or rushes, of which there are some 7000 species in the world, ever secrete nectar. To the grasses belong the edible cereals — corn, wheat, rye, barley, oats, rice, and millet. The planting of corn by the million acres renders it more important as a source of pollen than any other cultivated plant. It blooms, moreover, at a time when pollen-flowers are apt to be scarce, and honey-

Fig. 4 —Cat-Tail (Typha latifolia). Photographed by Lovell.

bees often visit the tassels in great numbers. Sorghum is another grass which is attractive to bees. Both of these grasses have been again and again erroneously reported to yield nectar. Large areas are also planted with castor-oil beans, the honey yield of which has been reported as "good;" but the flowers are nectarless, although there are on the leafstalks numerous nectaries which are practically functionless. Except the fall-blooming Roman wormwood, few other wind-pollinated herbaceous plants are of much importance to beekeepers.

The second group of nectarless plants is composed of conspicuous flowers, all of which are pollinated by insects. Many of them are of large size, as the rose, poppy, hepatica, anemone, purple clematis, bloodroot, nightshade, elder, rock-rose, St. John's-wort, and species of mullein and Spiraea. They are called pollen-flowers. They contain no nectar, and bees cease their visits as soon as the supply of pollen has been exhausted. Pollen-flowers are of every color, and occur in a great variety of families, as the lily, orchid, buttercup, poppy, nightshade, and honeysuckle families, and it is probable that at some time in their past history they secreted nectar, but have since lost this function.

As in the case of the wind-pollinated flowers, the pollen-flowers are occasionally reported, as the result of careless observation, to yield nectar. The rose has been a veritable thorn in the flesh to both poets and artists, and bees are constantly represented as seeking its sweet secretions. Beekeepers also in the northern and southern states and in southern Europe have reported it as yielding nectar; but, unfortunately, there is no such thing as rose honey. Possibly the sweet-brier rose may produce a thin layer of nectar, but this report needs to be confirmed. There is a rich store of pollen in the flowers, and honeybees, bumblebees, leaf-cutting bees, mason bees, and ground bees hasten to gather it; but as soon as it has all been removed their visits immediately cease. Although an acre of poppies would not produce an ounce of honey, there are occasional reports of bees being stupefied by gathering nectar from poppy blossoms and lying about on the ground unable to fly; but if this statement has any basis of truth, its cause is not poppy nectar. Honeybees devote their attention exclusively to gathering its rich store of pollen, visiting the anthers even before they open. Cultivated flowers which have become double, usually lose the power of secreting nectar. The California poppy in California, and the blue lupine in Texas, carpet large areas with their showy blossoms; and many other pollen-flowers, as the elders and anemones, are very abundant. It is not difficult to understand the origin of a pollen-flower; for plants which in one region, as alfalfa, white clover, and buckwheat, secrete nectar copiously, in another locality may be wholly nectarless.

CHAPTER VI

HONEY-DEW

Origin of Honey-Dew — White Flies — Jumping Plant-Lice — Sugar-Cane Leaf-Hopper — Bark-Lice or Coccidae — Plant-Lice or Aphididae — Life Cycle of Plant-Lice — Excretion of Honey-Dew — Mouth-Parts of Hemiptera — Chemical Analysis of Honey-Dew — Exudation of Sugars — Water Excretion.

Honey-dew is a sweet glutinous liquid excreted in large quantities on the foliage of plants by Hemipterous* insects, chiefly by plant-lice and scale-insects. It is often so abundant on the leaves of trees and bushes that it drops upon the grass and sidewalks, covering them with a glistening coating resembling varnish. At times it falls in minute globules like fine rain. Although readily gathered by honeybees, it has an inferior flavor, and is detrimental to beekeeping. The ancient Roman naturalist, Pliny, supposed that honey-dew fell from the stars, and this belief was generally accepted for centuries. The larger part of the honey-dew gathered by bees is produced by five families of insects belonging to the suborder Homoptera of the order Hemiptera or bugs: plant-lice (Aphididae), bark-lice (Coccidae), lantern-flies (Fulgoridae), jumping plant-lice (Psyllidae), and white flies (Aleyrodidae). A small amount of honeydew is also excreted by a few species of tree-hoppers (Membracidae), but they are seldom sufficiently numerous to do much harm.

The white flies (Aleyrodidae), small winged insects covered with a whitish powder, were formerly classed with the scale-insects, as in their immature state they are scale-like in form. They are widely distributed in both temperate and tropical regions, and infest many common cultivated plants. Honey-dew is deposited in large quantities by the larvae, pupae, and adults, and forms a medium on which grow several fungi. Only the adult females of plant-lice and scale-insects produce honey-dew. In Florida the larvae, pupae, and adults of the woolly white fly (*Aleurothrixus howardi*) excrete such large quantities of honey-dew on the under side of orange leaves that the smaller branches droop slightly under its weight. When an infected branch is jarred, the drops fall in a shower, and the branches and leaves become very sticky. The accumulation of honey-dew causes much annoyance to orange-pickers and to the men and mules engaged in cultivating the groves. The drops are often one-eighth of an inch in diameter, and become overgrown with a dark-colored fungus.

The jumping plant-lice (Psyllidae) are small winged insects about one-sixth of an inch in length. Many of the species form galls. The pear-tree *Psylla* often destroys pear trees by sucking the sap from the twigs. According to Slingerland it excretes honey-dew copiously: "It literally rained from the trees upon the vegeta-

* The mouth-parts of the order **Hemiptera** form a beak, which is used for piercing and sucking. It is divided into three suborders: 1. The **Parasitica** (lice) are wingless insects parasitic on man and other mammals. 2. The **Heteroptera** have two pairs of wings, but the first pair are thickened at the base, while the tips are thin and transparent. They include water boatmen, water-bugs, water-striders, bed-bugs, lace-bugs, and chinch-bugs. 3. The **Homoptera** have the wings of the same thickness throughout their entire length, and include cicadas, spittle-insects, leaf-hoppers, plant-lice, and bark-lice.

Fig. 5.—Bark-lice (Kermes sp., family Coccidae) on California live oak (Quercus agrifolia). Adult females on stem resembling galls; immature males on leaves.—(After Comstock.)

tion beneath; in cultivating the orchard the backs of the horses and the harness often became covered with the sticky substance dropping from the trees. It attracts thousands of ants, bees, and wasps, which feed on it."

More than 400 tons of honey-dew are shipped from the Hawaiian Islands annually, most of which is an excretion of the sugar-cane leaf-hopper (*Perkinsiella saccharidica*), a species belonging to the family Fulgoridae. In 1903 this insect became so abundant as to prove a serious menace to the cultivation of sugar cane. For several years it caused a loss of about $3,000,000 annually to the planters; but it has been brought under control, and to-day the plantations are again producing heavy crops of sugar. The honey-dew from the sugar-cane leaf-hopper is very dark amber in color and slightly ropy. In flavor it strongly resembles molasses. While most honey-dews granulate very rapidly, this type does not granulate at all. Samples several years old are as clear as when first extracted. A small amount of this honey-dew mixed with the light-colored algaroba honey imparts its color and flavor to the entire amount. Bees prefer floral nectar to this excretion; but when the floral nectar is not abundant they gather both, and the honey is a mixture.

The chemical composition of Hawaiian honey-dew honey differs so widely from floral honey that many buyers on the mainland have charged that it was adulterated; but after careful investigation Phillips was convinced that it was a natural sweet product collected by the bees, and shipped without the addition of other sugars. It is not placed on the market in competition with the honeys of the mainland derived from flowers, but is sold to bakers as honey-dew.

Analysis (see table) of the honey-dew honey of the sugar-cane leaf-hopper shows that the ash content is very high, ranging from three to six times the amount found in normal floral honeys. The percentage of dextrine is also very high, and its acidity is three times that of algaroba honey. The percentage of sucrose or cane sugar is a little higher than that of the average of floral honeys. A ray of polarized light is turned to the right by the honey-dew, while pure floral honey turns the ray to the left.

The Coccidae are commonly known as scale-bugs, scale-insects, bark-lice (Fig. 5), mealy-bugs, and coccids. The species are very numerous and infest the bark and foliage of a great variety of plants, and also nearly every kind of fruit. They excrete great quantities of honey-dew both in temperate and tropical regions. Only the adult females exude honey-dew. Not all of the species produce honey-dew, as many excrete wax or resinous substances. In early autumn a large quantity of honey-dew is frequently gathered from oak trees, the limbs of which are covered with a great number of small coccids, gall-like in form, about a quarter of an inch in length, from the ends of which there flows continuously a clear sweet liquid. So profusely is the honey-dew exuded that the trees appear as though they had been sprayed with hundreds of gallons of it. When it dries it solidifies and hangs in small stalactites. This honey-dew is produced, not by galls, as is often reported, but by the adult females of *Kermes galliformis*, which are remarkable for their gall-like form. "So striking is the resemblance," says Comstock, "that they have been mistaken for galls by many entomologists."

Species of *Lecanium*, a genus of coccids found everywhere on plants, attack basswood, tulip tree, maples, and many other trees, covering the leaves with a sweet liquid similar to that yielded by plant-lice. In California a scale-insect (*Lecanium oleae*) covers the foliage of citrus fruit trees with great quantities of shining dew. A fungus often grows luxuriantly on such leaves, forming a dense felt over their surface. At Amherst, Mass., and Guelph, Canada, thousands of bees have been observed gathering from spruce trees the sweet excretions of a scale-insect (*Physokermes piceae*). They are found at the base of new growth, and have the appearance of little buds. Pine trees are likewise at times prolific sources of honey-dew gathered from scale-insects living at the base of the leaves (Fig. 6).

Probably more honey-dew is produced by plant-lice or aphids (Aphididae) than by any other family of insects. They occur on a great variety of herbaceous plants, trees, and shrubs, more than 450 species being listed. A part of the species live on the leaves, a part on the limbs, and others on the roots. Among the decid-

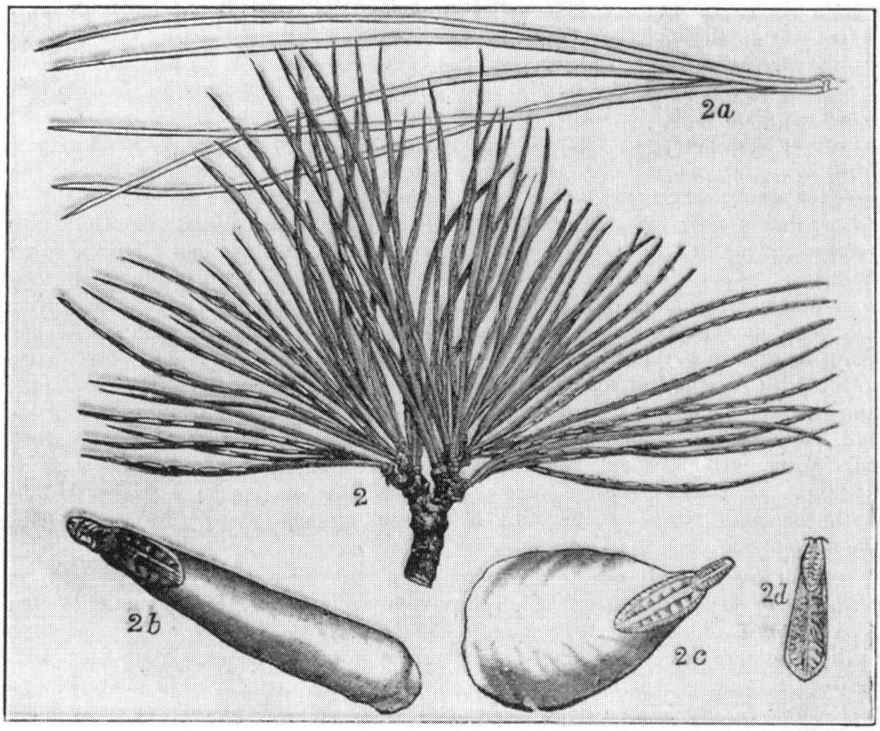

Fig. 6.—Pine-leaf scale (Chionaspis pinifoliae) on leaves of white pine (Pinus Strobus), natural size, leaves stunted; 2a, leaves of white pine not stunted by scale-insects; 2b, scale of female usual form, enlarged; 2c, scale of female, wide form, enlarged; 2d, scale of male enlarged. (After Comstock.)

uous-leaved trees and shrubs on which honey-dew is very frequently found are oak, beech, poplar, ash, hickory, chestnut, maple, willow, basswood, gum trees, fruit trees, grapevine, currant, blackberry, and hazel. A surplus of honey-dew honey is occasionally stored from corn, which at times is attacked by overwhelming numbers of the corn-leaf aphis (*Aphis maidis*). This aphis is found in all parts of the United States where corn is grown, and covers the leaves and silk with honey-dew, which attracts bees, ants, and flies. A small leaf-hopper (*Perigrinus maidis*) also spreads over the plant a saccharine liquid. The mealy plum aphis heavily infests plum trees and covers the leaves, twigs, and fruit with honey-dew. The life of this aphis is passed on two host-plants. In winter it infests plum trees, the eggs hatching in March. In early summer it migrates to the cat-tail rush (*Typha*), returning to the fruit trees in late fall.

The aphids are so well known that they require only a brief description. They are small, thick, usually greenish insects with pear-shaped bodies and long legs. On the back of the sixth segment of *Aphis* and *Lachnus* there is a pair of tubes called cornicles, siphons, or nectaries. They are commonly believed to excrete honey-dew; but this is denied by Forel and other entomologists, who assert that they exude only a gluey substance, which is not sought by ants. The tubes do not connect with the digestive tract, and the liquid which issues from them is produced by glandular cells at their base. In a part of the aphids the tubes are wanting or are greatly reduced in size.

The life cycle of the aphids varies with different species, but the principal points are as follows: In the fall both males and egg-laying females develop. After mating the males soon die; and the females, after depositing one or more winter eggs, likewise perish. The eggs may often be found on the terminal buds of trees, for example, on the terminal buds of the apple-tree three or four minute black eggs are laid. In the spring the eggs hatch and produce wingless females known as stem-mothers. Instead of laying eggs the stem-mothers by budding give birth to living young. The second and many succeeding generations consist, like the first, wholly of females, which may be wingless, but winged females may appear at any time and fly away to new plants. This is the spring migration. The tender upper growing shoots of trees are likely to be infested first, and later the lower branches. Many plant-lice have the singular habit of abruptly changing their food-plants. During the spring they feed on one kind of plant; after the spring migration during the summer on another kind; and in the fall, winged females again reappearing, they migrate back to the first kind of host-plant. This return in the fall to the spring food plant is known as the fall migration. The rosy-apple aphis passes the summer on the leaves of plantains, returning to the apple in the fall. Reproduction without pairing may continue for eight or more generations; but shortly after the fall migration both males and egg-laying females again develop. In warm climates plant-lice are believed to reproduce exclusively by parthenogenesis, years of collecting in tropical regions having failed to disclose the male form.

Plant-lice multiply with extreme rapidity, and it has been estimated that the offspring of each plant-louse, if all survived, would in 100 days amount to over 3,200,000 individuals, and at the end of 200 days to ten trillions. Fortunately they are held in check by a vast number of parasitic insects, such as syrphid flies, lady-beetles, and plant-lice lions, or they would threaten the destruction of all vegetation and drown the world in a flood of honey-dew. Their development is greatly influenced by climatic conditions. Occasionally there comes a year when plant-lice and scale-insects appear in hosts, and there is, consequently, a great abundance of honey-dew, as in 1884 and 1909 in this country, and 1898 and 1907 in Great Britain. In 1909 there was in eastern North America an unprecedented amount of honey-dew, while the crop of white clover and basswood was almost a complete failure. Most of the honey-dew came from the leaves of hickory and oak. While gathering it the bees were exceptionally cross. Since it became alternately partially liquid in the forenoon, and gumlike in the afternoon, they were able to collect it only in the morning hours; the moisture in the air softened it at night, but by noon the sun had again dried it to a viscous state. Honey-dew honey is often stored by the ton, and in certain localities, as in the Sacramento Valley, California, a crop is gathered almost every year.

The dew is forcibly ejected or flipped from the end of the abdomen; and when there are many aphids it falls in a spray of minute globules. If the dew were not thrown a little distance from their bodies they would soon be glued together. As they usually feed on the under side of leaves, the sweet liquid naturally drops on the foliage beneath them. As it is gumlike it may dry and remain on the leaves for a long time, so that the absence of plant-lice is no proof that it is of vegetable origin. If it is very abundant it may drip from the leaves to the ground. In 1891, Busgen observed that a single plant-louse on a maple leaf produced 48 drops in 24 hours (the drops were 1-25 of an inch in diameter); on a basswood leaf 19 drops, and on a rose leaf only 6 drops. The production of honey-dew has been found to be most active in the middle of the day when the temperature is highest.

Hemipterous insects of the families described live wholly on plant sap. The mouth-parts form a jointed beak consisting of four slender bristles enclosed in a jointed sheath, which is a prolongation of the upper lip. With this pointed beak the insect easily pierces the bark or leaf, and sucks out the sap of the plant tissues. The jointed sheath permits of a change of position without the removal of the beak. A part of the sap is digested, and is used for growth and the production of young, while the residue is expelled as a waste substance known as honey-dew. It is thus undoubtedly an excretion which escapes by way of the anal opening. It may not, however, consist entirely of the waste products of digestion. MacGillivray states that in plant-lice, which produce honey-dew abundantly, the posterior portion of rectum is greatly enlarged and is lined with large active cells which may excrete the honey-dew. The objection to honey-dew on the ground that it is an excretion rather than a secretion is largely imaginary, as secretion is the more general term including excretion.

The quality of honey-dew varies greatly according to the plant on which it occurs and the insects producing it. When freshly gathered it may be clear, sweet, and agreeable in flavor, or at least not unpalatable. The better grades find a ready sale to bakers, who prefer it for baking purposes to floral honey. But usually it is very inferior in quality; for when it remains for days on the foliage it gathers many impurities. A black smut often covers the leaves so that the extracted honey-dew is inky black, resembling tar. This type might, perhaps, be used by manufacturers of blacking or of lubricants. It is not a safe food on which to winter bees. If they are left on the summer stands, and can obtain frequent flights, they may winter in fair condition; but if they are placed in a cellar they will all probably perish from dysentery. For brood-rearing in the spring it is unobjectionable, and it is, therefore, advised that it be removed from the hives in the fall, and sugar syrup fed in its stead.

The composition of honey-dew honeys as compared with floral honeys is shown in the chemical analyses given in the following table:

	Water.	Invert sugar.	Sucrose.	Ash.	Dextrine.	Undetermined.	Free acid as formic.
Floral honeys—							
Sweet clover.......	17.49	76.20	2.24	0.12	0.45	3.50	0.12
White clover.......	17.64	74.92	1.77	0.07	0.82	4.78	0.06
Alfalfa	16.56	76.90	4.42	0.07	0.34	1.71	0.08
Honey-dew honeys—							
Hickory	16.05	65.89	2.76	0.78	12.95	1.57	0.12
White oak.........	13.56	55.87	4.31	0.79	10.49	4.98	0.08
Hawaiian sugar cane	15.46	64.84	5.27	1.29	10.01	3.13	0.15

From the above table it is apparent that honey-dew honey contains less invert sugar, but more sucrose or cane sugar, dextrine or gums, and ash, than floral honey. It is because of the larger percentage of gums and ash that it is unsuitable for winter feeding. Honey-dew honey may also be distinguished from floral honey by means of the polariscope. A ray of light passed through a solution of floral honey is turned or rotated to the left; but passed through a solution of honey-dew honey it is turned to the right. If floral honey turns the ray to the right it has been adulterated with glucose. No floral honey is obtained from the wind-pollinated flowers of hickory and white oak.

Besides bees, honey-dew is attractive to wasps, ants, flies, and other insects. Bees pay no attention to plant-lice; but ants care for them and stroke them gently with their antennae in order to induce them to yield honey-dew more freely. This behavior led the botanist Linnaeus to call *Aphis* the cow of the ants (*Aphis formicarum vacca*). Ants defend plant-lice from their enemies, move them to new pastures, care for their eggs, and build over them covers of earth, or cow-sheds, to keep them warm. Ants also extend their protection to scale-insects. It has been shown by observation and experiment that plant-lice and coccids which receive the care and protection of ants increase much more rapidly in numbers than those which are not attended by them.

The term honey-dew should be rigidly restricted to the sweet excretions of insects feeding on plants. Nectar is the secretion of nectaries, whether floral or extrafloral. Many plants have extra-floral nectaries on the flower stalks, leafstalks, and stipules, as cotton, vetch, passion flower, almond, peach, cherry, jewelweed, cowpea, field bean, and partridge-pea, which secrete nectar in large or small quantities. This nectar does not differ from floral nectar, as is attested by the chemical analyses of the honey of cotton gathered from both kinds of nectaries. It should never be called plant honey-dew.

It has been asserted by many beekeepers and not a few botanists in the past, that there is a third sweet liquid which, under favorable weather conditions, is exuded directly by the leaves of certain deciduous-leaved trees and shrubs, as basswood, maple, and oak. Gaston Bonnier has been frequently quoted to the effect that he had often seen trees, on which there was not a single plant-louse, covered with a sweet liquid which exuded from the leaves. Many similar views might be given. But in the majority of cases it has been conclusively shown that the sweet liquid found on the foliage of trees is of insect origin, and that the assertions to the contrary were based on insufficient observation and superficial investigation.

But a variety of sugars or mannas are yielded by a number of trees both in natural and artificial ways. Recent investigations by Davidson and Teit show that from tips of leaves of the Douglas fir in British Columbia, and Washington State, west of the Cascades, there is exuded a sweet liquid in large quantities. By the evaporation of the water it crystallizes quickly into white masses ¼ inch to 2 inches in diameter. This solid again may be dissolved by rain and recrystallized in patches at the base of the tree. "Fir sugar" was known to the Indians of British Columbia long before the discovery of America, and in recent years its presence has repeatedly been reported by beekeepers. But it does not occur every year. The sugar-yielding firs (*Pseudotsuga Douglasii*) are confined chiefly to the very dry belt of British Columbia between the parallels of latitude 50 and 51 degrees and the meridians of longitude 121 and 122 degrees. The sugar is not found on trees in the dense forests, but only on those in the comparatively open areas, on gentle slopes facing east and north, during hot summer droughts. In leaves of the Douglas fir exposed to continuous sunlight a larger quantity of carbohydrates is formed during the day than can be stored or carried away to the growing tissues. In the hot dry atmosphere evaporation ceases and the leaves become gorged with water, which is forced out through their tips. A beekeeper at Victoria states that many of the firs, particularly the isolated trees, are well spattered with the exudation, and the needlelike leaves studded with pale-amber diamonds. A large number of bees gather the liquid, and in some years two or three supers of sections are filled with it. The honey is fair in quality, pale amber in color, with rather dark cappings.

A beekeeper living in the Olympic National Forest, Oregon, 21 miles from Port Angeles, reports that his bees stored 150 pounds of fir sugar during a very dry season. The following winter many bees died from dysentery, which was attributed to the effects of the sugar. This seems very probable, as the composition of this excretion is very different from that of floral honey. It contains, among other constituents, nearly 50 per cent. of the rare sugar melezitose.

It is certainly not improbable that other species of conifers may, under special climatic conditions, exude a sweet liquid. In Switzerland about 40 per cent. of the honey crop is gathered from the weisstanne (*Picea excelsa*), a fir tree. From an excretion found on the leaves of this fir tree the beekeepers in the Vosges Mountains, the Black Forest, and in parts of Switzerland, harvest large crops of *waldhonig* (forest honey). J. A. Heberle believes that this honey is of plant origin, since meteorological conditions seem to determine its production. Unfortunately the observations are too superficial to be conclusive.

On the French Alps a sweet exudation is found on the small branches of young larches (*Larix europaea*), in June and July, in small, whitish, irregular, granular masses, which have the chemical composition of the sap of the trees, both containing the sugar mannite. It is known as Brianson manna, and is used for medicinal purposes. In Sicily the manna ash (*Fraxinus Ornus*) is extensively cultivated for manna or mannite sugar. In July and August incisions 2 inches long are made in the bark. On exposure to the air the sap hardens into flakes, which are secured by scraping. Its chief constituent is manna sugar. Maple sugar, the evaporated sap of the rock maple, is obtained in New England by a somewhat similar process. There are a number of other plants which are reported to yield sugars, but little definite information in regard to them is available. But not all of the sweet white substances sold as manna are plant exudations. Australian manna, obtained from the manna gum tree (*Eucalyptus viminalis*), is excreted by plant-lice. It has been observed falling in minute drops, until the twigs on the ground were covered with a snow-white incrustation a quarter of an inch in thickness.

The following distinctions should be carefully observed: Nectar is the sweet secretion of nectaries, either floral or extra-floral. Honey-dew is the sweet excretion of insects living on plants. The sweet exudations of plants, whether natural or artificial, should not be confused either with nectar or honey-dew. Their composition is the same or nearly the same as the plant sap. They should be known as sugars or mannas. The term *plant honey-dew* should never be used.

The exudation of nearly pure water by the leaves of many plants is clearly very closely allied to the exudation of sweet liquids. The Fuchsia, Indian corn, jewelweed, cabbage, primrose, grapevine, potato, elm, plane-tree, the aroids, and other species of plants often excrete drops of water from the tips and marginal teeth of the leaves. These drops may be observed on lawn grass, the ends of corn leaves, and the margins of jewelweed leaves in the morning, when they are likely to be mistaken for dew. The exudation of drops of water may easily be shown experimentally by placing a young cabbage plant grown in a flower-pot under a bell jar. In a few hours drops of water will appear on the apices or margins of the leaves, gradually increase in size, finally fall off, and new drops form. A surplus of water in the plant thus escapes through the leaf pores. A great amount of water may thus be exuded, and in a single night a leaf may excrete half of its weight in water. A vigorous leaf of Calocasia has been observed to eject water at the rate of 195 minute drops per minute, so that there seemed to be an almost continuous jet of water. The liquid is pure water except for a trace of salts (one-tenth of one per cent.). Certain tropical flowers bear glandular hairs which secrete great quantities of water containing a little sugar, thus exhibiting an intermediate stage between the exudation of pure water and saccharine liquids.

PART II

FLOWERS VALUABLE ONLY FOR POLLEN

I shall feed no more pollen-substitutes. It is a waste of bee energy and strength.—Latham, U. S. A.

We have never been able to rear a strong force of field bees by feeding pollen-substitutes.—Beuhne, Australia.

PART II

PLANTS VALUABLE ONLY FOR POLLEN

There are thousands of species of the higher plants which never secrete nectar, but they all produce pollen. As has been shown in the chapter describing plants which are valuable only for pollen, they may be divided into two groups, one of which is pollinated by insects and the other by the wind. Nectarless flowers pollinated by insects are called pollen-flowers, and are often large and bright-colored, as the poppy and the rose. Insects visit them to gather or feed on the pollen. The nectarless wind-pollinated flowers also produce pollen abundantly. Since they rely upon the wind rather than upon insects nectar would be of no advantage to them. The number of flower-visiting insects is wholly insufficient to pollinate the myriads of little blossoms put forth by the grasses and sedges and vast forests of birches, poplars, oaks, hickories, and beeches. The cereals — wheat, oats, barley, rye, rice, and corn — cover 225,000,000 acres, and manifestly must depend on wind-pollination. Beekeepers should be able to recognize the plants which are valuable only as sources of pollen, lest they mistake them for honey plants. Moreover, from the foliage of many hardwood trees great quantities of honey-dew are gathered, which is often mistaken for floral honey. Thus Brown in his *Chemical Analysis of American Honeys* gives an analysis of the "floral honey" of the poplar (*Populus*) and the oak (*Quercus*), both wind-pollinated trees; but the chemical constituents of these honeys, as well as their polarization, show that they are honey-dew honeys.

While all species of the higher plants produce pollen, nevertheless there is a great number of flowers from which it is wholly absent. For example, there are many plants which have two forms of flowers, one containing the stamens and the other the pistils (alder, willow, poplar, cucumber, squash, gallberry, and tupelo); no pollen can, of course, be found in the pistillate flowers. This division of the essential members prevails very largely among the sedges, many hardwood trees, and many homely weeds and herbs. In the immense family of the Compositae, which includes the thistles, asters, and goldenrods, the marginal flowers of the heads in many genera have lost their stamens, and often also the pistils as well. They have been converted into long bright-hued straps, which frequently differ in color from the disc florets, and whose sole function is to render the flower-cluster conspicuous for the purpose of attracting insects. For the same reason the marginal flowers of the flat-topped flower-cluster of the hobblebush (*Viburnum alnifolium*) have become large and white, and have lost both stamens and pistils, while the inner, nectar-bearing flowers still remain small and green. Under cultivation scores of flowers have become double by the conversion of the stamens into petals, as the double buttercups, roses, and peonies. There are thus myriads of flowers which are wholly devoid of pollen.

The following list of flowers will be found of two-fold value: it will enable beekeepers to recognize plants, often exceedingly abundant, which are nectarless, and thus avoid building false hopes of a honey flow on their bloom; and it will afford them information in regard to many valuable sources of pollen. It should also prevent them from mistaking an inflow of honey-dew from hardwood trees

Fig. 7.—Common Alder (Alnus incana). Photographed by Lovell.

for floral honey. No attempt has been made to list every nectarless plant, as rare species, or cultivated exotics; but only those have been included which it is important that, from a commercial point of view, the apiarist should recognize, or in regard to which frequent inquiries have been received. Many plants which yield nectar freely are also extremely valuable for pollen, as the willows, dandelions, acacias, bananas, goldenrods, and sunflowers, which are listed under Honey Plants. But most nectariferous flowers produce a relatively small amount of pollen — much less than is found in the nectarless flowers. Very little pollen is gathered from basswood, tupelo, and many species of *Eucalyptus*.

PLANTS VALUABLE TO BEE CULTURE FOR POLLEN

ALDER (*Alnus incana*).—Common or hoary alder. A very widely distributed shrub, blooming in early spring before the ice and snow have disappeared. The stamens and pistils are in different catkins on the same plant, and pollination is brought about by the wind. On warm days honeybees may be often seen gathering the pollen. A Pennsylvania beekeeper reports many acres of alder in his locality, and that bees resort to the catkins in large numbers. Newfoundland to Pennsylvania and Nebraska. (Fig. 7.)

ANEMONE (*Anemone quinquefolia*).—Wind-flower. Large white pollen-flowers, blooming in early spring. The pollen is gathered by bees, which have also been observed probing the flowers in an attempt to find nectar. Common west of the Rocky Mountains. None of the numerous species of *Anemone* yield nectar.

BEECH (*Fagus grandifolia*).—A large tree common in woods from New Brunswick to Virginia. The small greenish flowers are wind-pollinated. The stamens and pistils are in separate flowers on the same tree. Honey-dew is found on the foliage.

BLOODROOT (*Sanguinaria canadensis*).—Low perennial herbs common in open woodlands. The large, handsome, white pollen-flowers bloom in April or May, and are visited by honeybees which remove nearly all the pollen.

CALIFORNIA POPPY (*Eschscholtzia californica*).—Large orange-yellow pollen-flowers forming "fields of gold" in the valleys and on the foothills of California. "One of the most common, striking, and widely diffused plants of the California flora, abundant in the spring, but in many portions of the state found in flower in other or in all seasons. On account of its gorgeous beauty it has been favored with an exceptional number of poetic names." (Jepson.) It is extremely variable. Nectarless, but valuable for pollen. (Fig. 8.)

CASTOR-OIL PLANT (*Ricinus communis*).—This plant is a native of India; but it is widely cultivated for ornament, and a large area is planted in Texas for the oil obtained from the seeds. The small flowers are partly staminate and partly pistillate, and the wind is the agency in pollination. Pollen is produced in abundance, and is gathered by a large number of bees. This species has been erroneously reported to be "an excellent honey-producer." Extra-floral nectaries occur on the stems, leafstalks, base of the leaf-blades, and on the teeth of the leaves, but they are practically functionless.

CHESTNUT (*Castanea dentata*).—A large tree bearing small, pleasantly scented, pale yellow flowers; the staminate in interrupted catkins, the pistillate or fertile usually in clusters of three. It has been reported in Rhode Island to yield an bundance of a dark strong honey; but careful examination with the microscope has failed to reveal either nectaries or nectar. The European ecologists all agree that the chestnut produces pollen-flowers. The pollen attracts many honeybees; also many flies and beetles. Honey-dew may be found at times on the foliage, the gathering of which has probably given rise to the belief that the bloom is nectariferous. The structure of the inflorescence would indicate that the genus was formerly wind-pollinated. Extends from Maine to Michigan and Ohio, southward to Delaware and along the mountains to Alabama.

Fig. 8.—California Poppy (Eschscholtzia californica). Photographed by Lovell.

The chinquapins (*C. pumila* and *C. nana*) are reported to yield nectar in the South, but the flowers are similar to those of the chestnut. See Chinquapin under Honey Plants.

CLEMATIS.—This large genus, which comprises about 100 species, growing chiefly in the temperate zone, includes many beautiful cultivated varieties. The varieties of the purple clematis (*C. Jackmanni*) are pollen-flowers, but bees visit them often enough to remove all the pollen. Several species of this genus in Europe are reported to be without nectar. But the common virgin's bower (*C. virginiana*) yields nectar. See Clematis under Honey Plants.

COCKLEBUR (*Xanthium canadense*).—A coarse herb with the staminate and pistillate flowers in different heads on the same plant. Bees obtain pollen from it in summer and fall, when it blooms. Nova Scotia to North Carolina, also Texas and Nevada. It belongs to the Compositae, and, like Roman wormwood, is wind-pollinated.

CORN (*Zea Mays*).—Three-fourths of the corn produced in the United States is grown in the Mississippi Valley, the "corn belt" comprising the states of Kansas, Nebraska, Missouri, Iowa, Illinois, Indiana, and western Ohio. Corn is wind-pollinated, the "spindle" or "tassel" being composed of staminate flowers, and the "ear," of pistillate flowers. It produces no nectar, but is frequently visited by multitudes of bees for pollen. Under special conditions bees may store great quantities of corn pollen, filling whole frames with it. But bees have also been reported as storing honey from corn. Colonies in good condition, according to a Louisiana beekeeper, averaged 100 pounds one season from this source alone, the flow continuing for months. Other reports of a surplus of "pure corn honey" of 10 to 20 pounds per hive have been published. It should be noticed that "corn honey" is obtained only occasionally, sometimes only once in a lifetime; that while it may be light amber and possess a fair flavor, it is often dark-colored, with a poor or peculiar taste. If the weather is favorable, plant-lice (*Aphis maidis*) may infest the plants in vast numbers and cover the foliage and flowers with honey-dew, which bees gather eagerly. A small leaf hopper (*Perigrinus maidis*) may also cover the foliage with a sweet excretion. It is honey-dew, not nectar, which the bees gather from corn, and the sweet substance stored has all the qualities of honey-dew honey. If it were nectar, corn would yield more regularly. Corn is a grass, and none of the 3000 grasses secrete nectar. They are wind-pollinated, or self-pollinated, and offer no allurements to insects.

Like sorghum and sugar cane, corn has a sweet sap; and by grinding the stalks, boiling and refining the liquid, a molasses may be obtained. From an acre of sweet or sugar corn the U. S. Department of Agriculture obtained 900 pounds of sugar. When the stalks are cut, bees may gather the sweet juice which exudes from the cells; for example, two to six bees have been seen sucking the juice from the ends of cobs which had been broken from the stalks. Bees also gather sap from the cut stalks of sugar cane.

CONE TREES (*Coniferae*).—Vast forests of coniferous trees, covering millions of acres, are found throughout the north temperate zone of both the Old and the New World. Well-known species are spruce, fir, juniper, hemlock, cypress, and many species of pine. There are about 350 kinds of conifers in the world, all of which are wind-pollinated. The cones are always unisexual, either staminate or ovulate (producing seed). Both kinds usually occur on the same tree, but in the juniper and yew they are mostly on different trees. The quantity of pollen produced is enormous, and, rising in clouds above the trees, has been mistaken for smoke, or, falling downward and turning the ground yellow, has given rise to reports of sulphur showers. The pollen is often eaten by beetles, and honeybees have been reported as occasionally gathering the pollen of one or two species.

Great quantities of honey-dew are at times gathered from the foliage of the conifers, especially from the spruce and pine. In 1908, Gates observed thousands of living bark-lice (*Physokermes piceae*) on spruce trees at Amherst, Mass. Honeybees by hundreds were humming through the trees and gathering a liberal store of honey-dew. Sweet liquids exuded by the tips of the leaves of cone-trees, which crystallize by evaporation into white sugars, are also gathered. See Chapter on Honey-Dew.

DATE PALM (*Phoenix dactylifera*).—Large areas of Arizona and California are adapted to date culture, and there are extensive plantations of great age in Lower California and Mexico. The staminate and pistillate flowers are on different trees, or the trees are partly "male" and partly "female." The flowers are pollinated by the wind, and under cultivation, in order to economize the pollen, it is the practice to bind a sprig of staminate flowers among the pistillate. The white flowers are small, in large clusters, and bloom in early spring. The date palm has often been declared to be a "splendid honey plant," but the bloom is nectarless. Honeybees, however, visit the staminate blossoms in large numbers for pollen. (See Palm under Honey Plants.)

ELDER (*Sambucus canadensis*).—A shrub 3 to 10 feet tall, bearing numerous broad, flat-topped clusters of small white flowers. The flowers are occasionally reported to yield nectar, but they contain neither nectar nor nectaries. All the flower biologists of Europe class them as pollen-flowers. Bees gather the pollen, which is moderately abundant. It blooms in July, and grows in moist soil from New Brunswick to Texas. Elderberry. Elder-blow. American elder. Sweet elder.

ELM (*Ulmus americana*).—A large and well-known tree, widely planted for shade and ornament. The purplish flowers are wind-pollinated, and appear in spring before the leaves; they are partly perfect and partly unisexual. In a time of scarcity bees have been known to resort to the elms for pollen by thousands. Honey-dew is also gathered from the leaves. Canada to Florida and Texas. (Fig. 9.)

Fig. 9.—American Elm (Ulmus americana). Photographed by Lovell.

FURZE (*Ulex europaeus*).—Gorse. Whin. Introduced from Europe, it is found both in the eastern states and in California. The large, explosive yellow flowers are nectarless, and appear to be adapted to bumblebees. Honeybees collect the pollen, and at times search for nectar. In Australia large areas are "golden" with its bloom in early spring. It belongs to the pulse family.

GRAPE (*Vitis*).—While the flowers of the grape possess five nectaries they do not secrete nectar in Central Europe; and in this country, even where grapevines cover hundreds of acres in western New York, along the southern shore of Lake Erie, a surplus of grape honey has never been obtained. The bloom has been reported to yield nectar in Louisiana, Texas, and California, but on evidence far from satisfactory. In Italy, Delpino states that nectar is secreted abundantly. It has been reported that the flowers are wind-pollinated, but this is certainly not the fact. They are small, green, and inconspicuous, but very fragrant, and are at times visited by many honeybees, bumblebees, solitary bees, and beetles. Further investigation is desirable.

GRASS (*Gramineae*).—This is an immense family comprising some 3000 species, and including the edible cereals, corn, wheat, rye, barley, oats, rice, and millet. The grasses form the garment of the earth and clothe vast areas of its surface. The small greenish flowers are produced in countless numbers, and cross-pollination is effected only through the agency of the wind. An immense quantity of pollen is set free; but, except in the case of Indian corn, it is seldom gathered by bees. There are about 429 species in North America.

HAZELNUT (*Corylus americana*).—A shrub common in thickets and along fences from Maine to Kansas. It is wind-pollinated and blooms in April and May, before the leaves appear, and is of some value for pollen. The sterile flowers are yellowish and in catkins. The fertile flowers are crimson and in clusters. Honey-dew is found on the foliage.

HEPATICA (*Hepatica triloba*).—A small herb common in open woodlands, and blooming in early spring. The white or blue flowers are nectarless, but bees visit them frequently for pollen.

HICKORY (*Carya*).—The hickories are confined to eastern North America, with the exception of one species found in Mexico. The staminate flowers are in catkins, and the fertile are solitary or in small clusters; but both are on the same tree. They are wind-pollinated. Honey-dew is often gathered by bees from the foliage.

LUPINE (*Lupinus*).—The flowers are nectarless, but are frequently visited by honeybees and bumblebees for pollen. On the prairies of Texas in spring the blue lupine, or blue bonnet (*L. subcarnosus*), covers the ground for miles with a solid blue carpet. It yields a large quantity of bright yellow pollen, which is a great help in building up colonies rapidly.

In the pulse family, or Leguminosae, pollen-flowers may be distinguished from those which secrete nectar in the following way: In nectariferous flowers, nine of the ten stamens are united to form a tube, but the tenth stamen remains free and a bee may insert its tongue through the crevice on either side of it. But in pollen-flowers all of the ten stamens are united into a tube or cylinder, and no opening is left for the tongue of the bee.

MEADOW RUE (*Thalictrum polygamum*).—A tall perennial with white or purplish flowers in compound panicles. Conspicuousness is due to the stamens, which are very numerous. The flowers of all the species of *Thalictrum* are nectarless, and are pollinated partly by the wind and partly by insects. Honeybees have been observed gathering the pollen.

MULLEIN (*Verbascum*).—Some of the species have nectarless flowers, and a part yield a little nectar. They are adapted to pollen-collecting bees. Tall perennial herbs with bright yellow flowers in spikes, and large densely woolly leaves. Throughout eastern North America.

Fig. 10.—Red Oak (Quercus rubra). Photographed by Lovell.

OAK (*Quercus*).—A large genus containing about 250 trees and shrubs found chiefly in North America. The red oak (*Q. rubra*), the white oak (*Q. alba*), and the live oak (*Q. virginiana*), are three common and well-known trees. All the oaks are wind-pollinated. The green or reddish flowers are monœcious, or the stamens and pistils are in different flowers on the same tree. The staminate are in drooping catkins, and the pistillate are solitary, or few in a cluster. The trees bloom in spring with the appearance of the leaves. (Fig. 10.)

The bloom of the oaks is entirely nectarless, since this genus relies wholly on the wind for pollination. There is not a trace of a nectary in the flowers, yet there is a very general impression among beekeepers that the oaks are a source of honey. Thus a Florida beekeeper writes: "Last spring the bees stored 150 pounds of oak honey. I have known for several years that the oaks yield nectar." Many other beekeepers have made similar statements. Honey-dew is often very abundant on the foliage of the oaks, and, as has been described in the chapter on honey-dew, the bark-lice, which secrete this sweet liquid, are not infrequently mistaken for buds or galls. When the honey-dew is light-colored and fairly palatable, it may easily be mistaken for honey.

OLIVE (*Olea europea*).—The olive may be grown in most of the foothill sections of the interior valleys of California as far north as Redding. The small white

Fig. 11.—Common Plantain (Plantago major). Photographed by Lovell.

flowers are wind-pollinated and nectarless, but are visited by bees for pollen. Richter states that one of his apiaries was located near 5000 olive trees, but that he never saw more than two or three bees on the bloom of any one tree, and they were gathering pollen.

POPPY (*Papaver*).—The poppies with their brilliantly colored scarlet or purple petals are conspicuous both in gardens and fields. All the species are pollen-flowers, and the abundant pollen is very attractive to bees.

Opium poppy (*P. somniferum*) is cultivated in Europe and Asia for its seed-vessels, from which opium is extracted. There have been reports of bees, while gathering nectar from poppies, becoming stupefied and lying for some time helpless on the ground. An acre of poppies would not yield a drop of nectar; and repeated observations show that bees visit the bloom without injury. A honeybee, which had alighted on a newly opened blossom in which the anthers were still closed, was seen to search unsuccessfully at the base of the petals for nectar.

PLANTAIN (*Plantago*).—Indian wheat (*Plantago ignota*) is abundant over vast areas of sandy and gravelly mesas in Arizona. It is one of the principal forms of vegetation, and affords valuable pasturage to cattle. On the deserts it seldom grows more than 5 or 6 inches tall. The silvery-hairy leaves have a grayish appearance, and it is difficult to determine at first glance whether the plants are alive or dead. The flowers are entirely nectarless and are visited by bees for pollen only. On the slopes another plantain (*P. aristata*) is also very common. None of the plantains, of which there are about 200 species, yield nectar, although several of them have sweet-scented flowers. Honeybees gather pollen from a part of the species. Hovering in the air the bee moistens the pollen with nectar gathered elsewhere, and then brushes it off the anther with the tarsal brushes of the forefeet. Flies are very frequent visitors. The plantains are pollinated both by insects and the wind. (Fig. 11.)

POPLAR (*Populus*).—Aspen. Cottonwood. The poplars bloom in early spring, long before the leaves appear. The staminate and pistillate flowers are on different trees, and pollination is effected by the wind. The anthers are purple, and expel the pollen forcibly, which on mild days is gathered by bees. There are nectar-glands at the base of the leaf-stalks. Honey-dew is gathered from the foliage. Along the rivers of the western plains the narrow-leaved cottonwood and the broad-leaved cottonwood are the common trees instead of the maple, elm, and oak. The narrow-leaved cottonwood (*P. angustifolia*) yields a bright reddish gum, which the bees collect in large quantities and use as propolis. Much propolis is also collected from the balm of Gilead (*P. balsamifera*). Well-known species are the American aspen, the balsam poplar, the white poplar and the narrow-leaved cottonwood. (Fig. 12.) The poplar should not be confused with the tulip tree, often called tulip poplar.

RAGWEED (*Ambrosia artemisiifolia*).—Homely weeds, very common in old fields, blooming in autumn and pollinated by the wind. Ragweed belongs to the Compositae. The stamens and pistils are in separate flowers and in separate heads; but both kinds of heads occur on the same plant, the staminate in long spikes, and the pistillate in clusters of two or three in the axils of the leaves. The flowers are green. As a source of pollen in autumn, ragweed is sometimes of value.

ROCK-ROSE (*Helianthemum canadense*).—Woody herbs with large, solitary, yellow pollen-flowers, and also flowers without petals in clusters. The species are widely distributed.

ROSE (*Rosa*).—The rose as a pollen-flower is described in "*The Flower and the Bee*" as follows: "Even beekeepers generally believe that bees gather nectar from wild roses. There has been some discussion of late, writes one of them, as to whether bees get any honey from roses. 'I believe that I have seen them working freely on wild roses, and I see no good reason why roses should not yield honey, as they belong to the same family as the apple, pear, plum, cherry, and raspberry.

Fig. 12.—Large-Toothed Poplar (Populus grandidentata). Photographed by Lovell.

Fig. 13.—Common St. John's-wort (Hypericum perforatum). Photographed by Lovell.

If one species of a family yields nectar we may expect that they will all do so.' This may seem probable, but it is not the fact. In the buttercup family the buttercups, columbines, and larkspurs all secrete nectar, but the *Anemone* and *Hepatica* do not. Most species of the figwort family (Scrophulariaceae) yield nectar, but some mulleins do not. In the honeysuckle family (Caprifoliaceae) the honeysuckles and

Viburnums are nectariferous; but the elderberries are pollen-flowers. Some orchids secrete nectar, others do not. In the nightshade family (Solanaceae) the nightshade is nectarless, but the ground cherry (*Physalis*) yields nectar."

Although the handsome flowers of most species of rose are devoid of nectar they furnish such an abundance of pollen that they attract a great many visitors, as honeybees, bumblebees, leaf-cutting bees, mason-bees, and ground-bees, as well as flies and beetles. But in the sweet-brier rose (*R. rubiginosa*) nectar, according to Mueller, is secreted in a thin layer on the fleshy margin of the receptacle. The Cherokee rose also has been reported to yield nectar in the southern states. But these so-called exceptions appear doubtful.

RUSH (*Juncaceae*).—The rushes are very common in swamps. They are grass-like herbs with small, greenish, wind-pollinated, nectarless flowers.

ST. JOHN'S-WORT (*Hypericum*).—There are about 31 species of this genus in the United States, all of which have yellow nectarless flowers. The stamens are numerous and are united into clusters. The leaves are opposite and are punctate or black-dotted. The shrubby St. John's-wort (*H. formosum*) thickly covers the ground and presents a solid mass of bloom. Bees gather the pollen mostly in the morning. It grows 2 to 3 feet tall, and spreads by means of running rootstocks. Many of the smaller species of this genus are seldom visited by insects. (Fig. 13.)

SEDGE (*Cyperaceae*).—A great family comprising some 3000 species with small greenish flowers, all of which are nectarless and wind-pollinated. The sedges are abundant along the margins of rivers and lakes, and in swamps, where they form the principal part of the vegetation. Tule (*Scirpus lacustris*, variety *occidentalis*), one of the bulrushes, with dark-green stems 4 to 10 feet tall, covers 500,000 acres of brackish and fresh-water marshes in California. In the delta region of the San Joaquin and Sacramento rivers alone there are 50,000 acres of tule. A so-called "tule honey" has been placed on the market, but tule bloom does not secrete nectar.

SKUNK CABBAGE (*Symplocarpus foetidus*).—Flowers small, crowded on a spadix, and inclosed in a hood-shaped leaf, striped with purple, called the spathe. Common in bogs from Nova Scotia south to North Carolina and west to Iowa. The spathe or hood barely rises above the ground, and the flowers bloom before the leaves appear. The skunk cabbage is one of the first flowers from which bees gather pollen, and as many as seven have been seen at one time in one of the hoods. As the hoods are close to the ground and protect the bees from the wind, they are able to work on cool or cloudy days, when the temperature in the shade is only 42 degrees F. This early pollen at once starts brood-rearing, and assures prosperity to the colony. For this reason Doolittle, one of America's pioneer beekeepers, valued the skunk cabbage more highly than any other pollen-producing plant. If any nectar is secreted the quantity is so small that it is unimportant. The strong odor is attractive to flies, which have been found in the hoods in large numbers.

WALNUT (*Juglans*).—The stamens and pistils are in separate flowers; the staminate in long drooping catkins; the pistillate solitary, or two or three together. There are five species of walnut in the United States. The black walnut (*Juglans nigra*) and the butternut (*J. cinerea*), both common in eastern woodlands and blooming in April and May, furnish pollen for early brood-rearing. Honey-dew is found on the foliage of the black walnut.

WATER CHINQUAPIN (*Nelumbo lutea*).—Large aquatic plants with pale yellow flowers, five to ten inches broad. Found in rivers and ponds from New England to Texas. At Elkhart Lake, Texas, in the spring of 1919 bees were reported to have secured thirty pounds of surplus from this species. But according to both American and European flower ecologists the flowers are nectarless and bees gather pollen from them exclusively. Honeybees are very common visitors, as are also solitary bees, flies, and beetles.

WATER LILY (*Castalia tuberosa* and *C. odorata*).—The white water lilies have been reported in Texas to yield nectar; but all flower ecologists describe them as pollen-flowers. They are visited by bees exclusively for pollen, which is very abundant. On top of the stigma, when the flower first opens, there is a large drop of water. This drop of water is not sweet and is not gathered by bees, although small bees are sometimes drowned in it. By the second day it has dried up. The large flowers are often very abundant in small ponds and are visited by honeybees in great numbers for pollen. In the yellow cow-lily (*Nymphaea advena*) nectar is secreted on the outer side, near the top of the short thick petals, the nectariferous portion being orange-colored. No honeybees, however, have been observed to visit the flowers in Maine, Illinois, Wisconsin, Florida, and Germany.

PART III

HONEY PLANTS OF AMERICA, NORTH OF MEXICO

There is no subject of more importance to the beekeeper, nor is there one that gives him more pleasure, than the study of the honey-producing flowers.—Doolittle.

PART III

THE HONEY PLANTS OF AMERICA, NORTH OF MEXICO

In the following list of descriptions, all North American honey plants north of Mexico, of known importance, have been included. Owing to the great extension of the peninsula of Florida, descriptions of a few tropical species of general interest have also been given, as the royal palm, campanilla, and logwood. Plants valuable for pollen only have been segregated, and should be looked for in Part II. A brief account of nectariferous plants of little or no value to bee culture will be found in Chapter IV. A few common species erroneously reported to be good honey plants have been added, lest their omission should lead to continued misapprehension as to their value. There is a large number of genera, especially in the Compositae, which are more or less visited by honeybees for pollen or nectar, and in the aggregate are of moderate importance, but which, in the absence of any definite information as to their exact value to the apiarist, it has not seemed worth while to list. Very likely a few of these may deserve more particular mention at some future time, but the majority will always remain of secondary importance. Great care, however, has been taken to prepare accurate descriptions of the more important honey plants, and to this end numerous beekeepers in every state in the Union have contributed. Thousands of letters have been sent out by the author requesting information in regard to the plants yielding a surplus of honey, their reliability, the length of the honey flow and the conditions affecting it, and the qualities of the honeys; and it is due to the cordial co-operation of the beekeepers of the United States that a great amount of important data has been obtained.

In describing the different honeys an effort has been made to give the opinion of the people generally, rather than a local viewpoint or the estimate of an individual. Doubtless in some instances basswood honey would be regarded as having too strong a flavor, but by the majority of the people it is held in very high esteem. In like manner buckwheat honey in Schoharie, Albany, and a few other counties in New York, is considered as equal or superior to that from clover or basswood; but the general market outside of the state of New York looks upon buckwheat as an inferior honey, dark in color and unpleasant in flavor. It is the color and flavor preferred by consumers generally which determine the demand and largely the selling price of a honey. The wholesale market is thus an excellent gauge of the preference of the American public in honeys. Local beekeepers are apt to overestimate the honeys produced in their sections of the country, and to compare them favorably with the best grades of honey. Clover and basswood honeys are by most buyers in the East placed at the head of the list. Alfalfa, sweet clover, sage, and orange in the West, and willow-herb and raspberry honeys in Michigan, are very fine flavored and are considered equally good. Great care has been taken to avoid exaggeration in the discussion of the qualities of the various honeys.

ACACIA.—Two species of *Acacia* in Texas, catsclaw (*Acacia Greggii*) and huajilla (*A. Berlandieri*), yield a large surplus of heavy white honey of fine quality. Huajilla honey is considered the best honey in the state.

There are 300 species of *Acacia* in Australia, and about 150 in other parts of the world. Some 60 species have been introduced into California, and after fifty years have become fully acclimated. Half a million trees have been planted in Golden Gate Park, San Francisco. The species consist of herbs, shrubs, vines, and stately trees, which are well adapted to semi-arid regions, or the borders of deserts. They are valuable for tanbark, gums, dyes, medicines, and fiber.

Leaves two-pinnate; flowers generally yellow, in globose heads or spikes. They are not usually rich in nectar, but are wonderful pollen-producers. Wherever the species will thrive, they will furnish a wealth of pollen in spring, which is exceedingly rich in proteids. Many of the acacias are of no practical value to the apiarist.

Silver wattle (*A. dealbata*) and black wattle (*A. decurrens*) are extensively planted for ornament in California. When in full bloom during the winter months the golden-yellow clusters of flowers present a most beautiful appearance and are very fragrant. On bright days in January and February bees are always at work on the feathery bloom. If planted extensively, black wattle will assure a valuable supply of pollen for early brood-rearing. See Catsclaw and Huajilla.

AGARITA (*Berberis trifoliolata*).—A much branched shrub, 6 to 8 feet tall, with evergreen, leathery leaves of three leaflets, the teeth spine-tipped. The small, fragrant yellow flowers are in clusters, and bloom from January to April. The bright-red berries have a mild acid flavor. It is abundant in southern Texas, where it yields a large amount of pollen and nectar. A super of light-amber honey of good flavor is reported to have been stored in a favorable season, but usually it is chiefly valuable for building up the colonies. There is only one blooming period, but that is of long duration. Bees also gather the pink juice of ripe berries, which have fallen to the ground. See Barberry.

AGAVE.—This genus comprises over 100 species, all native to America, but most abundant in Mexico. In this country it is limited to the southern and southwestern regions. The most familiar species is the century plant or maguey (*Agave americana*). The plant consists of a rosette of fleshy leaves from the center of which, after a number of years, there grows up a stalk which may reach a height of 40 feet and produce as many as 4000 densely clustered flowers. The flowers are yellowish, and secrete nectar copiously, attracting great numbers of bees and other insects. Many gallons of sweet sap may be collected by cutting out the flower stalk and collecting the liquid, which flows into the cavity thus formed amid the leaves. Also called American aloe. Several species are valuable for fiber.

ALFALFA (*Medicago sativa*).—Alfalfa honey is white to amber-colored, with a pleasant minty flavor which has long made it a favorite with the public. The body is heavy, running 12 to 13 pounds to the gallon, while the weight of other honeys seldom exceeds 12 pounds. It granulates soon after extracting, forming a creamy solid, which is often retailed in cartons and tin pails. The color of alfalfa honey varies from water-white to amber, according to the character of the soil, latitude, altitude, and the season of the year. In the intermountain country of Colorado, Utah, Nevada, Wyoming, and Idaho, the honey is white; while in Arizona, New Mexico, and Imperial County, California, it is amber-colored. An analysis of alfalfa honey differs from that of white clover chiefly in the higher percentage of sucrose, or cane sugar, a slightly higher percentage of dextrose, or grape sugar, and in the lower percentage of dextrine, or gums.

While alfalfa is grown in every state in the Union, it requires a fertile lime soil and a compact seed-bed. The largest surplus of honey is obtained in arid and semi-arid regions, where irrigation is practiced. With ample moisture in the soil, the largest surplus is obtained when there is a hot dry atmosphere, and the temperature ranges from 80 to 100 degrees F. In Colorado, honey is seldom stored from the third crop of alfalfa, owing to cool nights. At a high altitude the temperature may be so low that it checks the flow of nectar; but a moderate elevation may favor its secretion. A beekeeper at Grand Junction, Colorado, states that three apiaries, in 1919, were in that locality barely self-supporting; while three others, 40 miles away and 1800 feet higher, produced a good crop of honey.

Alfalfa, however, secretes nectar freely in the Imperial Valley 200 feet below the level of the sea. In the eastern United States and in England it does not produce nectar to a very great extent except in a few localities. According to the census of 1920 there are 8,600,000 acres of alfalfa in the United States, of which 775,000 acres are east and 7,825,000 west of the Mississippi River. As the acreage in the eastern states is practically nectarless, the distribution of alfalfa in this section is relatively of little importance to the beekeeper. But near East Syracuse, N. Y., during very hot seasons, from the second crop in midsummer, from 10 to 30 pounds of surplus have been secured. Also at Peru, Indiana, honeybees have been observed gathering nectar from the bloom. It was estimated that there were about ten bees to the square rod. But alfalfa is seldom a good honey plant east of the Missouri River.

East of the Mississippi River, where there is an abundant rainfall, the cultivation of alfalfa is largely determined by the character of the soil. On fertile limestone soils the acreage is large, but on the acid soils of the swamplands, ricelands, and sandy pine barrens of the southeastern states it can not be grown successfully. New England has 9000 acres of alfalfa, and it is noteworthy that of this acreage 5000 are in Maine — over 4000 in Kennebec County. New York, with 120,000 acres, has a larger acreage than any other eastern state, while Florida, with less than 25 acres, has the smallest acreage. The summer temperature is not an important factor, since on the prairie lands of Mississippi there are 30,000 acres, and on the calcareous soils of Kentucky, 56,000 acres. North Carolina, South Carolina, Georgia, Florida, and Alabama report only a few thousand acres. Northward on the fertile prairie and glacial soils of Ohio, Michigan, Indiana, Illinois, and Wisconsin, there are 390,000 acres of alfalfa.

Bordering the Mississippi River on the west are the five states of Minnesota, Iowa, Missouri, Arkansas, and Louisiana, with a combined acreage of 440,000 acres. According to Pammel, in Iowa, where 172,00 acres are grown, alfalfa is of no importance as a honey plant. The beekeepers of Missouri also report the bloom as nectarless in that state.

In the semi-arid tier of states west of the Missouri River, comprising the Dakotas, Nebraska, and Oklahoma, the acreage of alfalfa shows an enormous increase; but, as it is largely grown by dry-farming methods, the surplus of honey stored is very variable. In South Dakota, with 462,000 acres in the Black Hills, where irrigation is practiced, a crop of 100 to 200 pounds of honey per colony is obtained nearly every year. The densest area of alfalfa in the United States is in Nebraska and Kansas, the former reporting 1,215,000 and the latter 1,315,000 acres. In the valley of the Platte River, Nebraska, where irrigation is possible, alfalfa is the main reliance of the beekeeper. But in localities where it is grown without irrigation the yield varies greatly in different years. The surplus comes chiefly from the second blooming in July. On the Republican River, in the southern portion of the state, alfalfa is the most important honey plant. In the eastern rainbelt of Kansas alfalfa is practically nectarless. It is most dependable in the valleys of the rivers and smaller streams west of Topeka. On the bottom-lands it will yield the entire season if water can be reached at a depth not exceeding 10 feet. On the uplands alfalfa secretes nectar only after showers. In Oklahoma alfalfa secretes nectar under the same conditions as in Kansas.

In Texas, alfalfa, of which there are 58,000 acres, is valuable as a honey plant only in the irrigated areas of the Trans-Pecos region. In the vicinity of El Paso there has not been a failure in the crop during 10 years; but at Barstow, Ward County, it is not always dependable.

In the eleven remaining western states 4,000,000 acres of alfalfa are under cultivation, Colorado and California each reporting over 700,000 acres. The 375,000 acres of irrigated alfalfa in Montana are found chiefly along the Yellowstone River and its southern tributaries, and in Ravalli County in the mountains. In no state are larger crops of alfalfa honey obtained than in Wyoming. The larger portion of the 330,000 acres of irrigated alfalfa is in Big Horn County and on the north-central border, and around Laramie in the southeast section. Beekeeping in Colo-

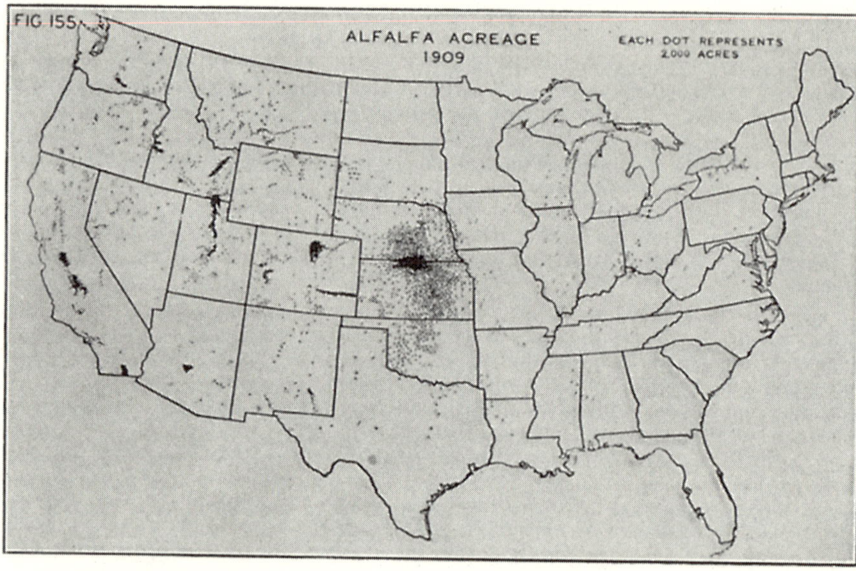

Fig. 14.—Acreage of alfalfa in the United States. Ninety-four per cent. of the crop is grown west of the Missouri River, and nearly one-half of it under irrigation.—(After Baker, Geog. World's Agr.)

rado is restricted chiefly to the valleys of the South Platte and Arkansas rivers and the Western Slope. There are few bees in the mountains. Most of the honey is stored from the first and second crops of alfalfa. On the Western Slope beekeeping is confined to the alfalfa areas, as sweet clover grows only along the ditches. In New Mexico alfalfa is grown chiefly in the valleys of the Rio Grande, San Juan, and Pecos rivers; in Utah, in the Uintah Basin, south of the Uintah Mountains, and in the central mountainous belt, where irrigation is extensively practiced; in Nevada, in the western portion of the state; and in Arizona, chiefly in Maricopa County.

Eight-tenths of the irrigated land in Idaho lie in the valley of the Snake River, where there are 650,000 acres of alfalfa. The largest surplus of honey is obtained in the Boise Valley and in the vicinity of Twin Falls and Idaho Falls. Washington and Oregon each report over 200,000 acres of alfalfa, grown almost entirely in the irrigated sections in the eastern portions of the two states. The Yakima Valley in Washington and Umatilla County in Oregon are two well-known centers. Of the 718,000 acres of alfalfa in California, the largest areas are grown in the Great Central and Imperial Valleys. Along the coast it is not important. In the San Joaquin Valley the second and third crops yield most of the honey, while the first and last crops are of little value. In dry seasons alfalfa is the chief dependence of the California beekeeper. (Fig. 14.)

Alfalfa is a perennial, herbaceous plant with trifoliate leaves; violet purple, irregular flowers in short racemes; and spirally twisted pods, each containing several kidney-shaped seeds. A most important character of alfalfa is the taproot, which may extend downward to a depth of 15 feet, enabling the plant to obtain food materials and water accessible to few other field crops. The genus *Medicago* contains more than 100 species and varieties, natives of the Caucasus and West Siberia, the Mediterranean region, and Northern Africa.

The flowers of alfalfa are known as explosive flowers. The anthers and stigma are held in the keel under elastic tension; and when a bee alights on a flower they are suddenly released, and fly forcibly upward, the anthers discharging their pollen against the under side of the bee. The stigma, which stands a little in front of the anthers, strikes the bee first; and, if the bee is covered with pollen from another

Fig. 15.—Alfalfa (Medicago sativa). Photographed by Lovell.

flower previously visited, cross-pollination takes place. A single normal visit is sufficient to effect pollination, and all subsequent visits are useless. After the flowers have been exploded, or "tripped," they continue to secrete nectar and receive insect visits. This is clearly an imperfection, since the visitors are no longer beneficial. Honeybees usually thrust their tongues sidewise between the petals, and gather the nectar without touching the anthers or stigma. The flowers are very frequently visited normally by leaf-cutting bees and bumblebees. For climatic reasons seed-growing is restricted chiefly to Utah, Kansas, Nebraska, California, Arizona, Colorado, Oklahoma, Idaho, Montana, and Wyoming, named in the order of yields. According to Alter, "The climate is the limiting factor in seed production generally, and current weather is the major control affecting any yield." A humid climate is unfavorable to the production of seed. (Fig. 15.)

The common alfalfa is probably of Asiatic origin, as it has been found growing wild in Afghanistan, Persia, and the region south of the Caucasus. In China it has been under cultivation from a very early date. The plant was brought into Greece at the time of the Persian war, 470 B. C., from Media, whence the generic name *Medicago*. In Italy it has been cultivated from about the first century, and is well described by Virgil and Pliny. During the Middle Ages it received the vernacular name of lucerne, from the valley of Luzerne in Piedmont in Northern Italy. It was long popularly known under this name in Europe, outside of Spain, and in eastern North America.

It was also very early introduced into northern Africa, where it was called alfalfa, a word of Arabic origin signifying "the best fodder." During the Moorish invasion it was carried into Spain, and later was brought by the Spaniards to Mexico and South America, and finally, in 1854, was carried from Chile to California. It was first heard of in England about 1650. Under the name of lucerne the early colonists introduced it into eastern North America, where it still grows spontaneously in fields and waste places; but the first attempts to cultivate it proved unsuccessful. Other common names are Spainsh trefoil, Burgundy, Brazilian, and Chilian clover. It is also known as purple Medick, from the color of the flowers and the Latin word Medica (Media), and snail clover from its twisted pods.

ALFALFA, WILD.—See Wild Alfalfa.

ALFILERILLA (*Erodium cicutarium*).—Also called pin-clover, stork's-bill, heron's-bill, and pin grass. Alfilerilla is derived from the Spainsh word for pin. The fruit slightly resembles a heron's bill, whence the generic name *Erodium,* the Greek for heron. This valuable forage plant is cultivated in several western states, and as a weed is very abundant from Oregon to Texas. The pink flowers are a valuable source of honey, and also furnish much pollen. Nectar is secreted on the outer side on the base of each of the five outer stamens, and collects in a little hollow at the base of each sepal. Honeybees have been often seen gathering both pollen and nectar. Alfilerilla is readily eaten by stock, and compares well with red clover in nutritious value.

Also called filaria and filaree, contractions of the Spanish alfilerilla, a pin, so-called from the form of the carpels or seed pods.

The same English names are also applied to another species of *Erodium* (*E. moschatum*), which very closely resembles the preceding species, but differs from it in the broader segments of the leaves and the whiter stems. In California it is very abundant in the fertile lands of the valley orchards and vineyards, and in the northern Coast Ranges, where it forms extensive growths. It begins blooming in March and April. Musk clover.

ALGAROBA.—See Mesquite.

ALKALI-HEATH (*Frankenia grandiflora*).—A herbaceous or shrubby plant, common on the alkaline plains of the interior of California, which yields nectar in late summer and fall. The Yerba Reuma of the Spaniards.

ALMOND (*Prunus Amygdalus*).—The conspicuous sessile flowers appear before the leaves. Large groves are valuable in spring both for nectar and pollen. The

Fig. 15A.—Filaree (Erodium moschatum). Photographed by Richter.

Fig. 16.—Small White Aster (Aster vimineus). Common in fields. Photographed by Lovell.

almond-growers have learned that, by placing colonies of bees in their orchards, much larger crops can be obtained, as the result of the more perfect pollination of the flowers.

ALSIKE CLOVER.—See Clover.

ANAQUA (*Ehretia elliptica*).—A honey plant in southeastern Texas, especially along the Rio Grande, and in northern Mexico. A small tree, or often a shrub, on sterile ridges. The small white flowers in panicles are produced in profusion from autumn until spring. It is often planted as an ornamental tree. Knockaway.

ANGLE-POD.—See Blue-vine.

AMERICAN LINDEN.—See Basswood.

ANDROMEDA.—See Fetterbush.

APPLE (*Pyrus Malus*).—The honey from apple bloom is pale yellow, with an agreeable flavor, and granulates quickly. It is of great value in stimulating early brood-rearing, and there is a proverb in New York, "As goes apple bloom, so goes the season." More than half a century ago a noted beekeeper, Moses Quinby, of St. Johnsville, N. Y., wrote: "In good weather a gain of 20 pounds is sometimes added to the hive during the period of apple bloom." But if, instead of continuous fair weather, it is cold and rainy, the stores may show a loss rather than a gain. In most of the apple-growing sections of the country bees get a little more than a living from apple bloom four years out of five. Occasionally a large surplus is reported.

APRICOT (*Prunus armeniaca*).—The pinkish-white flowers appear very early in spring, and yield both nectar and pollen. Cultivated in southern California.

ARTICHOKE (*Cynara*).—Globe artichoke (*C. Scolymus*) is freely visited by bees, but the honey is unknown. Cultivated; blooms June-July. Cardoon (*C. cardunculus*) is very common in Argentina. Jerusalem artichoke (*Helianthus tuberosus*), cultivated in New York for its tubers, is also a good honey plant.

ASPARAGUS (*Asparagus officinalis*).—Yields an amber-colored honey in the Sacramento Valley, California. The pendulous yellowish-green flowers secrete nectar freely, and are very attractive to bees. On the Russian steppes it grows wild in such abundance that the cattle feed on it like grass.

ASTER (*Aster*).—Pure aster honey is light — as light, according to beekeepers familiar with it, as white clover honey; but it is seldom obtained pure. Usually it is colored amber or yellow by honey from goldenrod or other late-blooming autumn flowers. The asters remain in bloom later than the goldenrods. When newly gathered it has a rank odor, but this disappears when it has ripened. The flavor is aromatic, and is quite strong when mixed with honey from other fall flowers. It is so thick that at times it is extracted with difficulty, and it granulates quickly with a finer grain than goldenrod. It has been stated to be unfit for table use; but many beekeepers describe the flavor as agreeable.

Aster honey is gathered chiefly from the very common species, *Aster multiflorus, A. vimineus, A. lateriflorus, A. Tradescanti,* and *A. paniculatus,* all of which produce dense clusters of small white or nearly white-rayed heads, except *A. multiflorus,* which has the rays white or purplish. Over large areas in Kentucky, Indiana, and other central states, the bloom is so abundant that the fields in autumn look as though covered with snow. The plants are often very bushy, growing from six inches to three feet tall. When the weather is favorable colonies will pack their combs with aster honey, or, if the combs have already been filled from an earlier source, a surplus is often stored. (Figs. 16, 17, 18.)

Many beekeepers have complained that their colonies suffered more or less loss when wintered on aster honey. So strong has been the opposition to it for this purpose that its removal, with the replacing of the stores by feeding sugar syrup, has been repeatedly advised. It is not improbable that aster honey, gathered so late that it only partially ripens and remains unsealed, is liable to deteriorate and become injurious before spring; but any other honey under similar conditions would be open to the same objection. Its tendency to granulate quickly and solidly has also added to its poor reputation as a winter food. But if this honey possessed

Fig. 17.—Panicled Aster (Aster paniculatus). Common in waste places. Photographed by Lovell.

Fig. 18.—Dense-Flowered Aster (Aster multiflorus).

properties which were actually injurious to bees, they would appear uniformly wherever it is largely stored, but this is not the case. The experience of scores of beekeepers, continued through many years, proves that aster honey well ripened and sealed is not a bad winter food for bees. The asters are bland, innocuous herbs which are readily eaten by domestic animals, either dried as hay or green in the pastures. Many beekeepers have testified that they have wintered bees successfully year after year on aster honey with very little loss. It has been suggested that perhaps different species of aster yield different kinds of honey, but there is no ground for such a supposition; on the contrary, the nectar of the various species, as in the case of the goldenrods, is very similar.

In Georgia several species of aster (the most common are *A. adnatus* and *A. squarrosus*) grow all over the state, and in many places are the main reliance for winter stores. In a few localities a surplus is obtained. The honey is medium in

Fig. 19.—Frost-Weed Aster (Aster ericoides). Common from Maine to Florida and west to Kentucky.

quality, of fair color, but granulates quickly in the comb if not sealed. The blooming time is from September to November. Over 100 species of aster occur east of the Mississippi River, but they are less important westward, although still abundant. (Fig. 19.)

BACCHARIS.—See Willow Baccharis.

BACHELOR'S BUTTON (*Centaurea Cyanus*).—The blue flowers are very frequently visited by honeybees both for nectar and pollen. In waste places, escaped from gardens. Bluebottle. Cornflower. In Oregon, where it is called French pink, it yields a dark-amber honey with a greenish reflection. The flavor is moderately strong with a bitter after-twang.

BALL SAGE.—See Sage.

BANANA (*Musa Sapientum and Musa Ensete*).—The flowers yield large quantities of nectar and pollen. The pollen may be dipped up by the spoonful. Cultivated in Florida. In Queensland, Australia, the banana yields a light-colored honey of fair quality.

BARBERRY (*Berberis vulgaris*).—Common in thickets in New England. The flowers are yellowish, in drooping clusters; each flower contains two orange-nectar glands from which the nectar flows into the angles between the stamens and ovary. May-June. In Texas the three-leaved barberry (*B. trifoliolata*), or agarita, is valuable in January and February for both pollen and nectar. The California barberry (*B. pinnata*), in March and April, in Monterey County, yields a surplus of amber-colored honey. See Agarita.

BASIL (*Pycnanthemum virginianum*).—Mountain mint. Minnesota to Georgia and Alabama. Bees visit it freely.

BASSWOOD (*Tilia americana*).—Basswood honey is white and has a slightly aromatic flavor. It is easy to tell when the blossoms are out by the odor about the hives. The taste of the honey also indicates to the beekeeper the very day the bees begin to work on the flowers. Honey extracted before it is sealed over has a rather strong flavor; but when sealed and fully ripened in the hive it is considered one of the best table honeys, especially in localities where it is known. Pure extracted basswood honey can often be blended with advantage with a milder-flavord honey, as mountain sage. (Fig. 20.)

Excepting white clover and alfalfa, basswood at one time furnished more white honey than any other honey plant in this country; but the trees have been so largely cut for timber that to-day very little basswood honey reaches the wholesale market, although a large amount is still gathered in many localities. It is a variable source of nectar, and is not reliable every year. Rarely in eastern New York, early in spring, a drop in the temperature sufficient to freeze ice has been known to kill all the flower-buds on low ground, and greatly injure those on the hills. Even when the trees are laden with flowers, if the weather is cold, cloudy, or windy, no surplus will be obtained. Hot, clear weather and a humid atmosphere are most favorable for the secretion of nectar. Small drops may then be seen sparkling in the bloom, and the bee may obtain a load from a single blossom. During a favorable season nine tons of basswood honey have been obtained as surplus at Delanson, N. Y. According to E. R. Root basswood yields nectar more rapidly than any other northern honey plant. There are other honey plants which are the source of more honey in a season, but none which yield so large a quantity in so short a time. Doolittle has recorded a gain of a hive on scales of 66 pounds in a day. The length of the honey flow from basswood may vary from 5 to 25 days, while the date of blooming is influenced by locality, altitude, and temperature. The flowers open ten to fifteen days earlier in a hot season than they do in a cold one. In localities where basswood grows both in the valleys and on the high hills, the bees will have a much longer time to gather the nectar, since the trees in the lowlands will bloom earlier than those at a greater height.

The most common species is *Tilia americana*, a tall tree growing in forests

Fig. 20.—Basswood (Tilia americana). Photographed by Lovell.

from New Brunswick to Nebraska and Texas, and especially abundant in the Alleghanies. *T. Michauxii* extends from Connecticut to Florida and Texas; and *T. heterophylla* from Pennsylvania to southern Illinois and Alabama, but is most common in limestone regions. The three species may be separated by the following leaf characters: *T. americana* has the leaves smooth on both sides; in *T. Michauxii* the leaves are smooth above and pubescent or woolly beneath; and in *T. heterophylla*

Fig. 21.—Basswood (Tilia americana). Photographed by Lovell

the leaves are smooth above and silvery white beneath. The European linden (*T. europea*) is widely cultivated in this country as an ornamental tree. Other vernacular names are linden, lime tree, bee tree, whitewood, and whistlewood. (Fig. 21.)

The clusters of from six to fifteen flowers are drooping, and are protected from the rain by the broad leaves. The stem of the cluster is adnate to an oblong membranous bract. The nectar is secreted and held in the fleshy sepals; and it is often so abundant that it appears like dewdrops in the sunlight. The blossoms are small, light greenish yellow, and exhale a honeylike fragrance. The stamens are numerous, and the anthers contain a small amount of pollen; but honeybees seldom gather it when the nectar is abundant. If, however, the nectar supply is scanty, then both honeybees and bumblebees may be seen with balls of pollen on their thighs. In England basswood seldom sets seed.

BEAN (*Phaseolus vulgaris*).—Both the pole and the bush varieties of the common garden bean are bumblebee-flowers; but, as they yield very little nectar, they are only occasionally visited by bumblebees. They are thus largely dependent on self-fertilization for the production of seed. Honeybees are not able to depress the keel. Bumblebees sometimes make holes in the flowers of the scarlet runner (*P. multiflorus*), and honeybees then in large numbers rob them of nectar. Lima bean

Fig. 22.—Beggar-ticks (Bidens frondosa). Photographed by Lovell.

belongs to another genus. It is an important honey plant in Ventura County, California, where it yields a honey of good quality, but not equal to that of sage, orange, or alfalfa. See Lima Bean, also Castor-oil Bean.

BEARBERRY.—See Manzanita.

BEARD-TONGUE (*Pentstemon laevigatus*).—A smooth perennial herb belonging to the figwort family, with slender stems, 2 to 3 feet tall, opposite oblong leaves, and two-lipped white or purple-tinged flowers. At Brandon, Florida, this species is very abundant and has spread over many acres of land. Colonies of bees are reported to have stored 200 pounds of surplus comb honey from this source alone, on which account it is locally called a "wonder honey plant." After the middle of the day the flow of nectar ceases, and a bee is rarely seen on the bloom until the next morning. The honey is mild in flavor, white in color, and very slow to granulate. Beard-tongue in this locality blooms from June to August. It is found in open woods from Pennsylvania to Florida and Louisiana.

BEE-BALM (*Melissa officinalis*).—A culinary herb, widely cultivated in gardens. While the flowers abound in nectar they are better adapted to bumblebees than to honeybees. The tongue of the honeybee is not long enough to reach the bottom of the tubular corolla.

BEEWEED.—See Rocky Mountain Bee Plant.

BEGGAR-TICKS (*Bidens frondosa*).—This species is common in damp land throughout the northeastern states. Many species of *Bidens* are abundant in wet

Fig. 23.—Bitterweed (Helenium tenuifolium). Flowers natural size. Photographed by Lovell.

land; but they are not of great value as honey plants, except in the case of Spanish needles (*Bidens aristosa*). (Fig. 22.) See Spanish Needles.

BITTERWEED (*Helenium tenuifolium*).—This plant yields a canary-yellow honey of heavy body and attractive appearance, but as bitter as gall. It is so extremely bitter that it can not be eaten; and, if it is mixed with any other honey, it renders it unmarketable. It granulates quickly after extracting. In central Alabama and Mississippi a large amount of honey is stored for winter from this source, and it is not necessary to feed sugar. Beekeepers engaged in the shipping of pound packages of bees northward regard it as invaluable for brood-rearing and find their colonies very strong in the spring if there is an abundance of bitterweed honey in the fall.

Bitterweed is a herbaceous plant 1 to 2 feet tall, with threadlike leaves and yellow flowers in heads. It belongs to the Compositae and to the same genus as sneezeweed, extending from Virginia to Kansas, and southward to Florida and Texas. It is abundant in waste land and along the roadsides, blooming from mid-July to October, and yielding nectar freely and also much pollen. When cows eat bitterweed the milk and butter are so bitter that they can not be eaten. See Sneezeweed. (Fig. 23.)

BLACKBERRY (*Rubus allegheniensis*).—The genus *Rubus,* of the rose family, comprises the raspberries and blackberries. There are many species widely distributed in the North Temperate Zone. The value of blackberries as honey plants varies widely in different parts of North America. In New England the wild species are seldom sufficiently common to be of much importance, nor do the cultivated varieties yield nectar freely. As visitors the solitary bees far outnumber the honeybees, which show a preference for collecting pollen rather than the scanty supply of nectar. In New York, New Jersey, and Pennsylvania little mention is made of these bushes as nectar-producers. In Tennessee, although plentiful, they are almost entirely neglected by honeybees. In Michigan, after the hardwood forests of beech and maple have been lumbered, there speedily springs up a luxuriant growth of brambles, many acres being covered with raspberries and blackberries; but, while the former is an excellent honey plant, the latter is seldom noticed by bees.

The failure of the blackberry to obtain a place among the honey plants in so many states might seem to warrant its omission; but, contrary to expectation, in several states it stands well up in the honey flora. At Bogart, in northern Georgia, the larger part of the surplus comes from wild blackberries, which are abundant in the woods. The flow is at its height about the middle of April, and 25 pounds per colony is reported to have been stored from this source. In the Coastal Plain of North Carolina, South Carolina, and Georgia blackberry is very abundant, and yields an ash or smoky-colored honey of inferior quality. In California the common blackberry is *Rubus vitifolius,* which grows chiefly along streams in southern California, and is of great value. The honey is light amber, and has a good flavor. Both Stockton and Chiles, of this state, report surplus crops. In Oregon, during the latter part of June, the wild blackberry furnishes a dark inferior honey that would not be salable were it not mixed with honey from white clover. In Australia a blackberry (*R. fruticosus*) has been introduced which has completely overrun some districts. It yields a thin white honey. (Fig. 24.)

BLACK GUM.—See Tupelo.

BLACKHEART.—See Smartweed.

BLACK HAW (*Viburnum prunifolium*).—There are in North America some 19 species of arrow-wood, or *Viburnum,* which bloom in spring or early summer. They are widely distributed throughout both the northern and southern states. The honey is not certainly known; but that from *B. dentatum* is supposed to be light-colored. This species is very abundant in southern Rhode Island. (Fig. 25.)

BLACK WATTLE.—See Acacia.

BLUEBERRY (*Vaccinium corymbosum*).—Blueberries and huckleberries yield an amber-colored honey of good quality. In New England the honey flow comes late in May or early in June. It lasts for about ten days, and strong colonies store from 60 to 100 pounds. There are large areas of both blueberries and huckleberries in southeastern Massachusetts, Rhode Island, and Connecticut, and beekeeping is in localities largely dependent upon these shrubs. In Rhode Island the blueberries are all of the bush type, the low blueberry (*V. pennsylvanicum*) being seldom found. Blueberries and huckleberries are also valuable honey plants in North

Fig 24.—Blackberry (Rubus allegheniensis). Photographed by Lovell.

Carolina, Georgia, northern Florida, Oregon, Washington, and many other states. (Figs. 26, 27.)

BLUEBOTTLE.—See Bachelor's Button.

BLUE CURLS (*Trichostema lanceolatum*).—Turpentine-weed. Vinegar-weed. Camphor-weed. The honey is milk-white, and after it has granulated is very attractive for table use. It is remarkable for the quickness with which it granulates, often becoming solid before the cells have been sealed. It is called turpentine-weed and camphor-weed because of the strong scent of the foliage. The flowers are blue, and the stamens are spirally coiled in the bud, whence the name blue curls. It is common on the foothills of the Coast Range and Sierra Nevada and in dry fields and plains, to which it gives a soft purplish tinge. Beginning with September it blooms throughout the fall, and adds not a little to the winter stores. Abundant in Fresno and Ventura counties and in the Sacramento Valley, California.

BLUE GUM.—See Eucalyptus.

BLUE THISTLE (*Eryngium articulatum*).—Honey mild-flavored, but dark-colored. Along the lower Sacramento River and in Sacramento County, California, bees store many pounds of honey from blue thistle from August to September. A perennial strong-smelling herb, 2 to 3 feet tall, with small bluish flowers in umbels, belonging to the carrot family, or Umbelliferae.

BLUE THISTLE.—See Blue-weed.

BLUE VINE (*Gonolobus laevis*).—Also called sand-vine; angle-pod from the angled fruit; blue vine from the bluish color of the flower; dry-weather vine, since it secretes nectar most freely in dry weather, and shoestring vine. A twining herbaceous vine with a tough stem which may attain a length of 40 feet. The oval heart-shaped leaves resemble those of the morning-glory or sweet potato. The small bluish white flowers are in numerous axillary clusters, and resemble the flowers of the milkweed. The pollen grains are in waxy masses. The pods are very large, thick, tapering to a point; the seeds bear a tuft of long silky hairs, and are carried for miles by the wind.

Climbing milkweed, or blue vine, is widely distributed over the central and Gulf states from Iowa to eastern Texas, eastward to the Appalachian Ranges. As a honey plant it is important chiefly in southwestern Indiana and in central Missouri. It grows luxuriantly on the rich alluvial soils of the river-bottom lands, but does not thrive equally well on upland or thin clay soils. Commercially it is most valuable to the beekeeper in extreme southwest Indiana, along the lower portions of the Ohio, Wabash, and White rivers. In this region it is the main source of surplus. At Bloomfield, on the west fork of the White River, blue vine is reported to be spreading each year. As it is a perennial it dies down in the fall, but comes up again in the spring. It yields well only in dry seasons, and in the wet season of 1915 bees neglected the bloom entirely and no honey was secured. In Daviess County, also on the White River, there are thousands of acres of river-bottom cornfields, which give an unlimited pasturage of blue vine. Early in the season it is held in check by the cultivator; but as soon as cultivation stops blue vine climbs the cornstalks, twining around the spindles, and reaching across from one row to another. It begins to bloom during the latter part of July, and by August 15 the honey flow is at its best. The plant is killed by the first frost. Sixty to eighty pounds of honey per colony are not infrequently obtained. A hive on scales recorded a daily gain of 4 pounds for 15 consecutive days. The honey is nearly white, or has a slight pinkish tinge, and an aromatic flavor. The flowers have a pleasing fragrance, which is very noticeable in the evening; and when the sections of honey are removed this delightful fragrance is at once apparent. It does not granulate readily, even in cold weather.

Blue vine should prove a good honey plant along the Ohio River in southern Illinois. At Brunswick, Missouri, at the junction of the Grand River with the Missouri River, blue vine or climbing milkweed is very abundant, and in a cornfield of 1200 acres there was not a stalk on which there was not a vine. It blooms from July to about September 10, and in dry seasons yields well. The honey is described as having the color of Colorado alfalfa honey, and a fine flavor.

BLUE-WEED (*Echium vulgare*).—Viper's bugloss. Blue thistle. In fields and by the roadside from Maine to Virginia and Nebraska; a common weed in some sections. Naturalized from Europe. This species was formerly very abun-

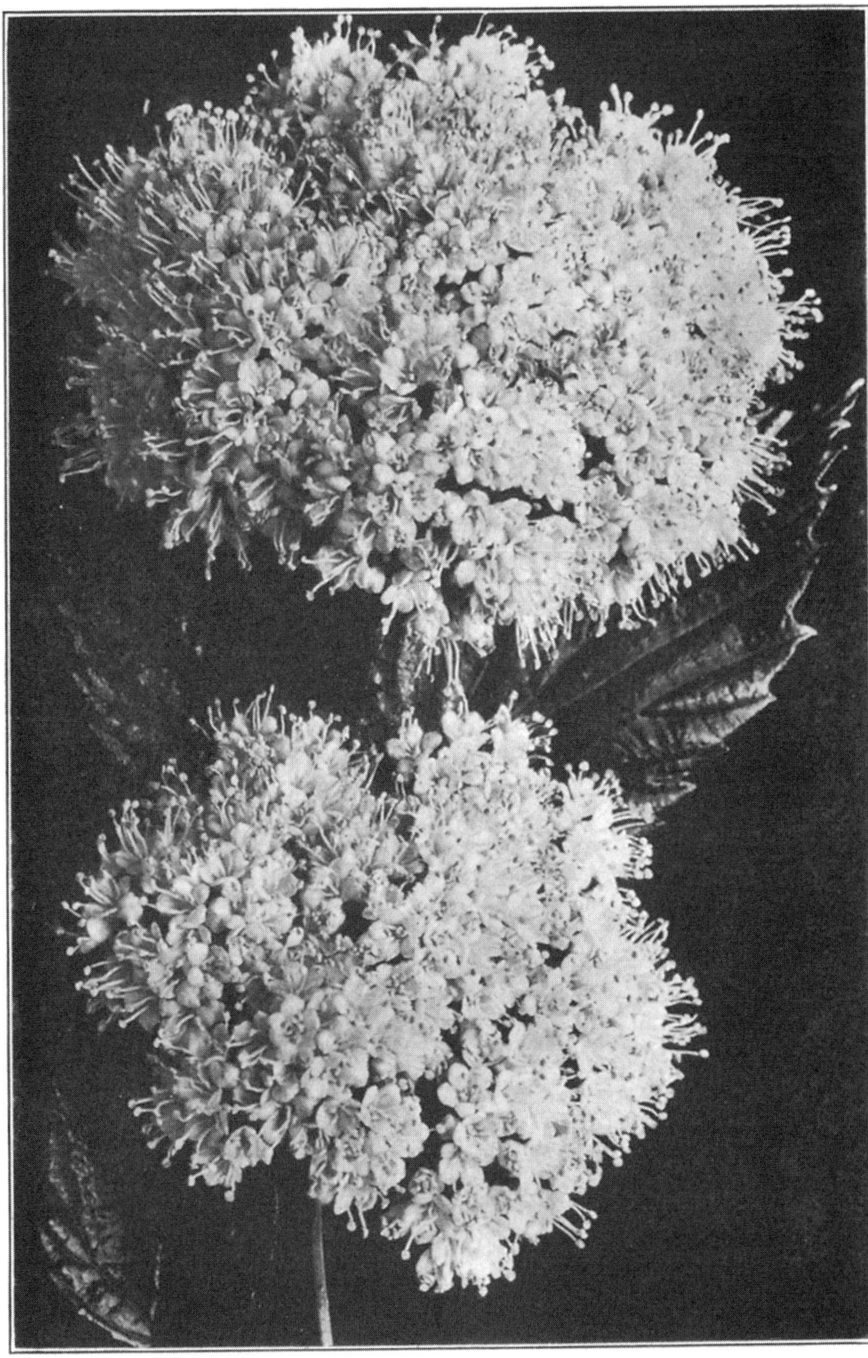

Fig. 25.—Arrow-Wood (Viburnum dentatum). Photographed by Lovell.

Fig. 26.—Low-Bush Blueberry (Vaccinium pennsylvanicum). Photographed by Lovell.

dant in northern Virginia, especially in the Shenandoah Valley, but is to-day rapidly decreasing, owing to the planting of large orchards. The honey is a light amber, and has a good flavor and body. The plant is about two feet tall, and bears blue flowers which bloom in August and September.

BOKHARA.—See Sweet Clover.

Fig. 27.—Huckleberry (Gaylussacia baccata). Photographed by Lovell.

BONESET (*Eupatorium perfoliatum*).—Thoroughwort. The color of boneset or thoroughwort honey is dark reddish amber. It is very thick and heavy, and has almost the consistency of molasses. The odor and flavor are at first very rank and strong, and after several months it still retains an herby taste and odor. A decoction of the plant is an old-fashioned remedy, reputed to be beneficial as a tonic and a diaphoretic; and it is not improbable that the honey possesses medicinal qualities. (Fig. 28.)

In 1917 at Mt. Pleasant, Alabama, more than 2000 pounds of boneset honey was stored from Sept. 14 to Oct. 3 from *E. urticaefolium*, rainy weather bringing the flow to a sudden close on the latter date. F. L. Pollock, a well-known author of fiction relating to beekeeping, after tasting a sample wrote, "I don't think the honey is bad. It has the queerest flavor I ever tasted. It is good and thick, but too dark; but I am sure it will sell as baker's stock."

The thoroughworts are tall, coarse plants with large, resinous-dotted, aromatic leaves, and white or purple flowers grouped in large flat-topped clusters. The two most widely known species are common thoroughwort (*Eupatorium perfoliatum*), also called boneset, Indian sage, and ague weed; and purple boneset (*E. purpureum*), other English names of which are Joe-Pye weed, gravel root, queen of the meadow, and kidney root. These two species will usually be found listed in the honey floras of most of the eastern states; but, owing to the fact that many other flowers bloom at the same time, pure thoroughwort honey is seldom obtained. Thoroughwort is an important component of the dense masses of weeds which cover the great saw-grass flats of Florida; but wild sunflower, goldenrods, asters, Spanish needles, and smartweed bloom contemporaneously with it, so that the honey is a blend from many flowers. In most of the North, goldenrod and aster greatly exceed thoroughwort in abundance.

In Tennessee several species of *Eupatorium* are common along the northern edge of the state and yield heavily, especially white thoroughwort (*E. album*) and white snakeroot or white sanicle (*E. urticaefolium*). The former species is confined largely to the southern states, but the latter is a woodland plant from New Brunswick to Louisiana. Farther south, although these plants are abundant, they produce very little honey. Throughout Kentucky thoroughwort is common on damp ground; but the genus is best represented in the southwestern part of the state. In Todd County at least 75 per cent. of the field flow comes from boneset (*E. serotinum*), which yields a light-amber honey.

BORAGE (*Borago officinalis*).—A great favorite with honeybees, which visit the flowers constantly. The honey has apparently an excellent flavor. The plants can be easily grown from seed, and produce a profusion of flowers from midsummer until frost. It would not, however, be profitable to cultivate borage for honey alone. The leaves are sometimes used as a salad and in medicine. (Fig. 29.)

BOSTON IVY (*Psedera tricuspidata*).—This hardy climber is very extensively trained over the walls of stone and brick buildings; and in the neighborhood of cities there may be acres of it accessible to bees. In Massachusetts it sometimes yields a small surplus. The honey has an offensive odor, but an agreeable flavor. It blooms between the 15th and 20th of July, when the clover flow is about over, and continues in bloom from 4 to 6 weeks. The vines on the north side of a building bloom a week later than those on the south side. As soon as the blossoms begin to appear the clover honey should be removed from the supers in order that it may not be scented or mixed with that from the ivy. Flowers small, greenish; leaves ovate, cordate, or 3-lobed.

BOX ELDER (*Acer Negundo*).—A small tree with small drooping clusters of flowers which appear before the leaves. Yields nectar, also honey-dew in the fall. New England to Manitoba and westward.

BRAZILWOOD (*Condalia obovata*).—A small tree, about 30 feet tall, but often only a spiny shrub, forming dense thickets, and growing in dry land in central and southern Texas. It blooms in spring, and the greenish-white flowers yield fairly

Fig. 28.—Boneset or Thoroughwort (Eupatorium perfoliatum). Photographed by Lovell.

well. It remains in bloom about 15 days or longer. The honey is dark amber, and does not granulate readily.

BROOMWEED. (*Gutierrezia texana*).—Coarse herbs with yellow flowers from which in Texas a dark-amber, strong-flavored honey is obtained in September and

October. It is excellent for stimulating brood-rearing and for winter stores. Dry plains of Texas and Arkansas; it is less common in dry seasons. Honey granulates quickly. It is a slender weed with clean stiff branches, a handful of which makes a good broom or brush.

BRUNNICHIA (*Brunnichia cirrhosa*).—A perennial vine, 10 to 20 feet long, climbing by tendrils. The leaves are ovate, and the greenish bell-shaped flowers are in clusters. In the Yazoo Delta in Mississippi it is abundant, and in June produces great sheets of bloom which attract many bees. It is classed among the surplus-honey plants of the state. In the lowlands of Texas it is also reported to yield a small surplus. This species belongs to the buckwheat family (Polygonaceæ), and is closely allied to smartweed.

BUCKBRUSH (*Symphoricarpos occidentalis*).—Wolfberry. A branching shrub, three to five feet tall, common in Washington and Idaho. In Idaho it is one of the most important honey plants, yielding in some sections a surplus, on an average, of 25 pounds per colony. Near Fraser, in the northern section of this state, the honey is secured in large quantities. The extracted honey is water-white with a very pleasant flavor, and is slow to granulate. After three years a bottle of it had not crystallized. Buckbrush blooms from June 15 to July 20. The flowers are white tinged with pink, bell-shaped, woolly within, and occur in small clusters in the axils of the leaves. They secrete nectar freely and are very attractive to wasps, whence the flowers of this genus have been called wasp-flowers. The fruit is a white berry which is eaten by pheasants and cattle. It is also abundant in western Iowa, where it yields well in dry weather. The honey is very similar to that of white clover. In the Missouri River basin, especially on the loess bluffs, this is a very common species.

Coralberry (*S. orbiculatus*). Indian currant. This species extends southward from Iowa to Texas, and is abundant along the Missouri River. It produces small red berries resembling red currants. The flowers, which are smaller than those of the snowberry, appear for two or three weeks, and secrete a large quantity of nectar.

Snowberry (*S. racemosus*). A northern species found from Alaska to Nova Scotia, and southward on the east coast to Pennsylvania, and on the west coast to California. In Iowa in summer a large amount of excellent honey is obtained from it. It has large white berries, and is frequently cultivated for ornament.

BUCKEYE (*Aesculus glabra*).—The Ohio buckeye is a large tree with small nectariferous flowers, found from Pennsylvania to Kansas. Considerable honey is obtained from the California buckeye (*A. californica*). The horse-chestnut (*A. Hippocastanum*) also belongs to this genus. It is adapted to bumblebees, but honeybees obtain both pollen and nectar. (Fig. 30.)

BUCKTHORN (*Rhamnus catharticus*).—The common buckthorn is often planted for hedges. Cascara sagrada (*R. purshiana*), the bark of which is so extensively used in the manufacture of drugs, occurs in Mendocino and Humboldt Counties, California, northward to Washington. In Sonora, California, it is the principal honey plant. It blossoms about three weeks after fruit bloom is over, and on an average lasts for about 25 days, although there are stray bushes near ditches or cultivated grounds which send out new shoots of bloom until September or October. The comb honey is so dark that it does not sell readily, but is well liked by those accustomed to it. It is not purgative, but is an excellent remedy for chronic constipation. It does not granulate.

Coffeeberry (*R. californica*) is a smaller shrub, 4 to 5 feet tall. Common in the foothills of the Sierra Nevada and Coast Range. Honey amber-colored, flavor good, slightly cathartic; also called pigeonberry and Herba del Oso.

Redberry (*R. crocea*) is only 1 to 3 feet tall, blooms from February to May, and is valuable for early brood-rearing. Southern California.

BUCKWHEAT (*Fagopyrum esculentum*).—Buckwheat honey has a dark purplish color, and looks much like old New Orleans or sorghum molasses. The body is usually heavy. The flavor, to one who is accustomed to clover and basswood honey, is more or less nauseous; but those who have always used buckwheat honey, or at least many of them, prefer it to either of these standard honeys. In New York and Albany it brings almost as high a price as the fancy grades of white honey; but in the western market it sells as an off-grade honey. (Figs. 31 and 32.)

Buckwheat honey occasionally contains 33 per cent. of water, and is, therefore, too thin to meet the requirements of the National Pure-food Law, passed June 30, 1906, which limits the amount of water in honey to 25 per cent. In such cases thin honey should be evaporated to make it conform with the law. This may be done by means of a honey-evaporator, or by storing it in open tanks for a while in a hot dry room.

Buckwheat can be cultivated throughout the north temperate zone. It is extensively grown in Asia, especially in Japan, and is also widely cultivated in Europe. An immense quantity of buckwheat honey is gathered in Russia. In North America, while it is grown to some extent in Canada, it is chiefly valuable for grain in the United States. It is best adapted to New York, Pennsylvania, Ohio, Michigan, Wisconsin, and New England, and to the mountainous sections of Maryland, West Virginia, Virginia, Kentucky, North Carolina, and Tennessee. About two-thirds of the crop is now raised in New York and Pennsylvania. In the South it succeeds better in the uplands and in mountainous sections than in the lower lands. In 1919 there were 742,000 acres of buckwheat in the United States, of which 467,000 acres were in the states of New York and Pennsylvania. (Fig. 33.)

In New York and Pennsylvania there are thousands of acres within a radius of three miles. On one hilltop in Schoharie County, N. Y., bees were reported a few years ago to have access within a radius of three miles to 5000 acres of buckwheat, all of which was within the range of the eye. So great is the acreage in New York that from 2000 to 3000 colonies can be kept in some counties. There are hundreds of farmers who keep a few colonies in order that they may get the honey as well as the grain.

The flowers of buckwheat secrete nectar usually only in the morning; toward noon the flow lessens, and ceases entirely during the afternoon, but begins again vigorously the next morning. After all of the nectar has been gathered, the bees often make a few vain flights to the fields, and then remain idle until the following day. Thus in the afternoon, in spite of the great expanse of bloom and the strong fragrance, only a few bees can be found in the fields. But in the province of Quebec, about Sept. 1, the flowers bloom later in the day, and much of the honey may be stored in the afternoon. But the bloom yields only a few hours during a day. The flowers are blasted by high temperature, especially by hot sunshine after showers.

In New York, buckwheat can be depended upon almost every year to yield a crop of honey; but in the West it is more uncertain, some years yielding no honey, and in others doing fairly well. Since in the East it is almost always reliable, and yields well when even clover and basswood fail, as they do sometimes in any locality, the beekeeper is usually able to make his expenses and at least a small profit. In New York it is seldom that he is not able to make a fair living from buckwheat alone.

Among cultivated crops there are few which will afford a better artificial honey-pasture than buckwheat. The beekeeper who raises this cereal largely for honey should plant at three different times in order to prolong as much as possible the flow of nectar. On an average a single crop will occupy the land about sixty days. It will commence to yield nectar in fifteen or twenty days from the time it is planted, and take about ten days to mature after the honey flow ceases. If the first crop is sown on the 20th of June, the second crop on the 4th of July, and the third about the 18th of July, the beekeeper will be assured of a good bee pasturage from the middle of July, when basswood and clover are past, up to the middle of September, when the fall bloom of wild flowers commences.

The plant is a smooth annual, growing from 1 to 3 feet tall. The small flowers are clustered, and possess a strong fragrance; the petals are wanting, and the sepals are white or tinged with rose. The nectar is secreted by eight round yellow glands placed between the same number of stamens. This species has two forms

Fig. 29.—Common borage (Borago officinalis). Photographed by Lovell.

of flowers — one with long stamens and short styles, and the other with short stamens and long styles — an arrangement for promoting cross-pollination. The flowers, according to Darwin, possess the power of self-fertilization; but when covered with nets they were, early in the season, almost wholly self-sterile, and produced hardly any seed. Flowers cross-pollinated artificially at the same time produced seed in abundance. Later in the season, in September, both forms of flowers became highly self-fertile. They did not, however, produce as many seeds as some neighboring uncovered plants which were visited by insects. Thus the crop of seed is largely dependent on honeybees, which are estimated to make 90 per cent. of the visits of insects to the flowers.

BUCKWHEAT, WILD (*Eriogonum fasciculatum*).—See Wild Buckwheat.

BUR CLOVER (*Medicago hispida*).—Occasionally yields a surplus, but of more value in stimulating early brood-rearing, as it begins blooming in March. In moist places it blossoms throughout the summer. The spirally twisted pods are

Fig. 30.—Horse-chestnut (Aesculus Hippocastanum). Photographed by Lovell.

Fig. 31.—Buckwheat (Fagopyrum esculentum). Illustration from A B C and X Y Z of Bee Culture.

beset with hooded prickles, whence the name, bur-clover. Common on the hills and plains of California. Belongs to the same genus as alfalfa. (Fig. 34.)

BUTTONBUSH (*Cephalanthus occidentalis*).—Also called buttonball, button tree, honey-balls, globe-flower, pond dogwood, and buttonwood shrub. Produces a mild light-colored honey of fine flavor. A very widely distributed shrub, found from New Brunswick to California, and southward to Florida, Texas, and Arizona. There is only one species of buttonbush in North America, but five others occur in Asia and elsewhere. An important honey plant on the overflowed lands of the Mississippi River, and along streams and in swamps in many eastern states. For example, at Mayfield, Massachusetts, there is a large swamp ten miles long, in which buttonbush, or buttonball, blooms profusely in August. At about 11 o'clock in the morning the bees usually leave buckwheat, also in bloom at this time, and start for the great swamp, where they work buttonbush until night. The honey greatly improves the flavor of that from buckwheat, with which it is mixed.

The corolla-tube of the flower of buttonbush is 9 mm. long (a millimeter is about 1-25 of an inch), while the tongue of the honeybee is only 6 mm. in length; but, as the tube flares a little at the mouth, the bee is probably able to obtain all, or nearly all, of the nectar, which is abundant. Bumblebees, solitary bees, and many butterflies, which easily reach the nectar with their long tongues, also visit the flowers. (Fig. 35.)

CACHANILLA (*Berthelotia sericea*).—A shrub about three feet tall with silky entire leaves and pink flowers, which are reported to yield a small amount of light-colored honey. It is found along streams in the Rio Grande Valley, New Mexico.

CACTUS.—There are over 1000 species of cactus, which are found almost exclusively in the hot, dry lands of the southwestern states, in the arid regions of Mexico, and in tropical South America. They are fleshy plants, globose or columnar

Fig. 32.—A field of buckwheat in full bloom. From A B C and X Y Z of Bee Culture.

in form, without leaves, but producing usually large brilliantly colored flowers. They may be divided into two groups according to the forms of the flowers — one with long floral tubes, and the other with short floral tubes. The long tubular flowers are adapted to birds, but they secrete nectar so copiously that it falls in drops to the ground from the pendulous blossoms. Well-known forms are the Turk's cap, the hedgehog cactus, and the torch thistles which belong to the genus *Cereus*. The flowers are red, white, yellow, and purple in color.

The prickly-pear cactuses, or opuntias, are the most important to bee culture. There are between fifty and sixty species in the southwestern states. The flowers are wheel-shaped, without a floral tube, and are usually yellow, or yellow inside and red outside. Throughout southern and western Texas the prickly pear (*Opuntia Engelmanii*) is common, and yields both nectar and pollen, but it is chiefly important for pollen. The prickly pear in Texas yields great quantities of honey, according to E. G. LeStourgeon, only under certain conditions. About one year in four a surplus is obtained from this source. Extremely hot and humid weather is required during the blooming period. The honey is very heavy, and is almost viscous in consistency, having a ropy appearance as it is extracted. The color is light amber, and the flavor very good. The prickly-pear bloom is a great source of pollen, and one of the most dependable of Texas plants for this purpose.

CAMPANILLA (*Ipomoea sidaefolia*).—White campanilla is an important honey plant in Cuba. It is a perennial vine, growing over shrubs and trees or along fences and stone walls. The white bell-shaped flowers open about Christmas time, and at this season of the year it is a common sight to see almost every tree, shrub,

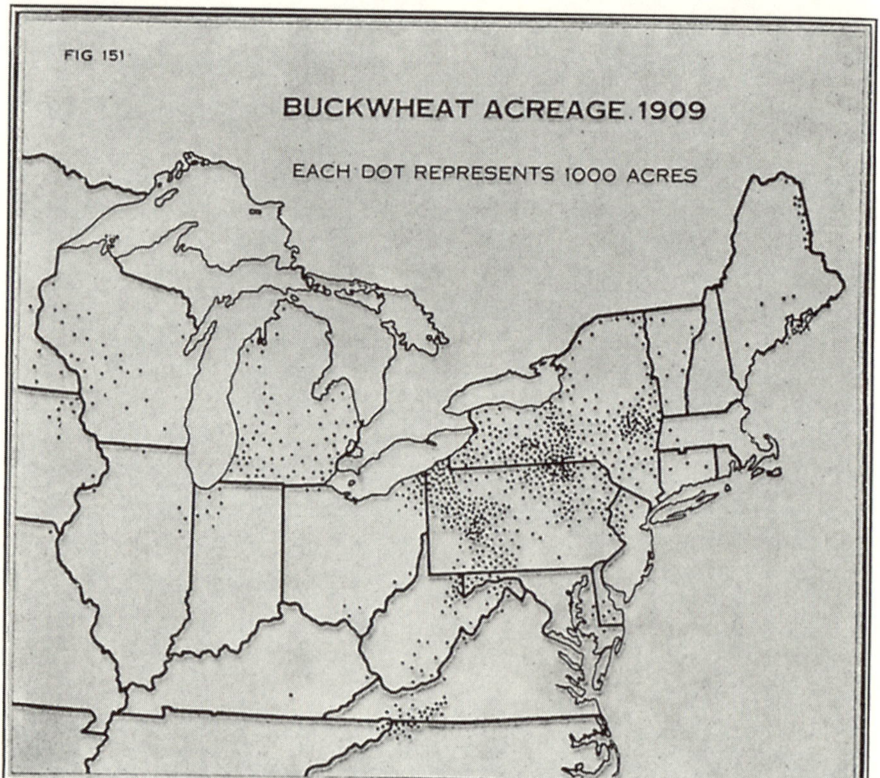

Fig. 33.—Acreage of buckwheat in the United States; production centers chiefly in New York and Pennsylvania —(After Baker.)

Fig. 34.—Bur Clover (Medicago hispida).

and fence along the road one mass of campanilla bloom. The period of blooming lasts for about a month or six weeks. The flowers open very irregularly. Sometimes every vine is in full bloom every other day; at other times, for several days in succession. Then there are days when not a single vine can be seen in bloom during miles of travel. The honey is white, or nearly white, and has a fine flavor. The comb is pearly white, and yields a white wax. Also called campanilla blanca, aguinaldo de pascuas, and Christmas pop.

No. 35.—Buttonbush (Cephalanthus occidentalis). Photographed by Lovell

Fig. 36.—Canada thistle (Cirsium arvense). Photographed by Lovell.

Pink campanilla (*I. triloba*) blooms in October and November. It is found principally in western Cuba, in the region known as Vuelta Abaja, the great tobacco district, and its abundance is a result of the method of growing tobacco. Tobacco seed is, as a rule, sown on virgin soil, and large tracts of land are cleared every year on both the mountains and coast for this purpose. As soon as the plants are large enough to be transplanted, they are pulled up and shipped to the fields where the tobacco is grown. These tobacco-seed beds are the next year and for years afterward covered with the pink campanilla, which in western Cuba springs up everywhere in cultivated soil. Also called campanilla morada and aguinaldo rosada.

CAMPHOR WEED.—See Blue Curls.

CANADA THISTLE (*Cirsium arvense*).—Honey very white, of a very fine quality, and a most pleasant flavor. Like many other pernicious weeds, the Canada thistle belongs to the Compositae. It is the source of a small quantity of honey in parts of Canada, and is increasing in some of the central states. The heads are small but very numerous. The nectar is so abundant that it rises in the corolla tubes to a point where it can be reached by nearly all insects. It yields nectar well in dry, warm weather. (Fig. 36.)

In Australia the Canada thistle has spread over large areas and become a "proclaimed weed." The honey, says Tarlton-Rayment, is water-white, candies quickly with a fine grain, and is usually associated with clover honey. It has a very mild flavor.

CARPET GRASS (*Lippia nodiflora*).—The honey is white or sometimes light amber, heavy, and granulates with a fine grain. The quality is fully equal to that of alfalfa. The small white flowers are very attractive to bees, and the flow of nectar is steady and abundant.

Carpet grass is a prostrate, creeping, herbaceous perennial, only a few inches high, which forms dense mats in damp soil and on river banks. It spreads rapidly by means of "runners," and is of great value in preventing the erosion of sandy land, and is, in consequence, known as a "sand-binder." It belongs to the verbena family, or Verbeniaceae, which also includes the garden verbena. Common carpet grass is widely distributed in the warmer regions of North America, extending from Central America and the West Indies to Florida, Georgia, and Texas. It is very abundant in Sutter County, and in the Sacramento River Valley, where it produces a large amount of honey, and blooms from May until September. In Texas carpet grass grows along rivers and small streams, but is of little importance as a honey plant.

The carpet grasses, of which there are about 100 species, belong chiefly to the warmer regions of the Old and the New World, but are most abundant in tropical and sub-tropical America. They yield nectar in Central America, and are also valuable honey plants in the West Indies and the Bermudas. There are about 9 species in the United States, distributed over an area extending from New Jersey to Nebraska and Kansas, southward to Georgia and Texas, and westward to Arizona and California.

In 1900 *Lippia repens* was introduced from Italy into California, where it now covers thousands of acres. Because of its thickly matted growth it is widely used for covering lawns and tennis courts. Only one or two cuttings are required during the summer. It thrives in the poorest soils, smothers weeds, requires but little water, and looks as well as any grass; but during the winter it turns brown and ceases to grow, a new growth appearing in early spring. The small flowers are visited by many honeybees, and probably the honey does not differ from that of the common carpet grass. Also called fog-fruit and mate grass. (Fig. 36A.)

CARROT (*Daucus Carota*).—Introduced from Europe, this weed is often very abundant in worn or neglected fields and by the roadsides throughout the eastern states. It is steadily increasing, and in localities where, ten years ago, it was rare, it is now common, as in pastures along the lower course of the Missouri River. The small white flowers are in umbels often with a single purple flower in the center. In the Sacramento Valley, California, it is an excellent honey plant, yielding a white honey, with the flavor of the foliage, which granulates in a few months. In the eastern states this species, known as "wild carrot," is a noxious weed. It yields nectar about once in ten years, but usually it is of no value.

CASSIA.—Herbs, shrubs, and trees with pinnate leaves and nearly regular, often yellow, flowers, which are nectarless and usually scentless. About 25 species occur in this country, but they are very abundant in tropical America. The flowers are visited chiefly by bees, and furnish only pollen. But there are extra-floral nectaries, as in the partridge-pea, wild senna, and coffee-senna, which are the source of a great quantity of honey. See Partridge-pea.

CATALPA (*Catalpa speciosa*).—A large tree growing in Tennessee, Missouri, and Arkansas, and extensively planted for timber in the North. The clustered bell-shaped flowers are nearly white, and so large that honeybees can easily reach the nectar, which can be seen at the bottom of the flower. Reported to be a good honey plant in localities. (Fig. 37.)

CATNIP (*Nepeta Cataria*).—The veteran beekeeper, Moses Quinby, once said that, if he were to grow any plant extensively for honey, it would be catnip. Several have grown it in small plots, and have reported that, in a state of cultivation, it apparently yields more honey than in its wild state. The flowers are two-lipped, white spotted with purple, and are very attractive to bees. (Fig. 38.)

CATSCLAW (*Acacia Greggii*).—Also called Paradise flower and devil's claws. This species blooms twice a year, the first flow coming in April and the second in July. The dense, thorny vegetation of southern Texas secretes nectar better in a dry than in a wet season, as too much rain causes the bloom to fall from the stems.

Fig. 36A.—Carpet Grass (Lippia nodiflora). Photographed by E. R. Root.

Fig. 37.—Catalpa (Catalpa speciosa). Photographed by Lovell.

There must be sufficient rain, however, during the preceding fall and winter to enable the plants to make the required growth. Catsclaw honey is white to light amber in color, very heavy, with a rich mild flavor, and is regarded by the native Texans as one of the finest honeys in the world. It granulates with a smooth, waxy grain. Commercial quantities of either catsclaw or huajilla honey are seldom unmixed, as they bloom at nearly the same time. At Uvalde, Texas, they furnish the larger part of the surplus, and 200 pounds per colony has not been unusual. Late cold weather sometimes cuts short the flow. This species occurs

Fig. 38.—Catnip (Nepeta Cataria). Photographed by Lovell.

from southern Texas to southern California near the Mexican border, and extends southward into Mexico. (Fig. 39.)

Catsclaw is a bushy shrub, or small tree, with low-spreading branches, attaining a height of 15 to 20 feet, and armed with short, curved, pricklelike spines, whence the name, catsclaw. The semi-desert region of Texas, where these thorny shrubs grow, can not be used for farming without irrigation; and the Acacias and mesquite are, therefore, likely to remain a permanent pasturage for bees, unmolested by the onward march of civilization, which will fail to displace them with more profitable farm crops.

CAT'S-EAR (*Hypochaeris radicata*).—A perennial herb bearing one or several heads of yellow flowers. The toothed, hairy leaves, which resemble the leaves of the common dandelion, are clustered at the base of the naked stems. A weed, naturalized from Europe, and common in California from Marin County to Humboldt County, also abundant east of the Cascade Mountains in Oregon and Washington. It yields an amber-colored honey, which darkens the white honey from fireweed (*Epilobium angustifolium*). Also called California dandelion and gosmore. It blooms from June to August.

Fig. 39.—Catsclaw leaf, twig, and blossoms; life-size. Photographed by E. R. Root.

CEANOTHUS.—While no surplus has ever been reported from *Ceanothus*, the species are so many, so common, so widely distributed, and furnish so much nectar and pollen that they are undoubtedly of great benefit to the apiarist. Early-blooming forms greatly stimulate brood-rearing. There are some 35 species of this genus, all natives of North America. They are more common in the West than in the East; but New Jersey tea (*C. americanus*) is found in open woodlands throughout the eastern United States. It produces a great abundance of small white flowers in clusters, which are very sweet-scented. More species of insects (382) have been collected on them than on the inflorescence of any other American plant. In

Fig. 40.—Chicory (Cichorium Intybus). Photographed by Lovell.

California these shrubs are common on the mountain slopes and on the foothills of the great ranges, where they are constituents of the chaparral. They are locally known as mountain lilacs.

CELERY (*Apium graveolens*).—Flowers white, in umbels. Along the Sacramento River, California, it yields a surplus. Some of the finest garden country in the world is found for many miles along the Sacramento River, California. Truck farming is here an important industry, and celery and parsnips are grown by the acre for seed. Both of these plants yield nectar freely, little drops of which can be readily seen gleaming in the flowers. Large quantities of celery honey are stored, which has at first the well-known flavor of this vegetable.

CENTURY PLANT.—See Agave.

CHAPMAN'S HONEY PLANT.—See Globe Thistle.

CHAMISE (*Adenostoma fasciculatum*).—Also called greasewood. An evergreen spreading bush not more than 10 feet tall, very abundant on the Coast Ranges and Sierra Nevada, where it grows above the foothills, excluding all other vegetation. The small white flowers are produced in dense terminal clusters, and are very attractive to bees. Blooms in June.

CHERRY (*Prunus*).—Cultivated cherries are popularly divided into two groups — the sweet cherries (*P. avium*) and the sour cherries (*P. Cerasus*). In Oregon, where the cherry ranks fourth in importance among the cultivated fruits, and there are orchards of 10 to 100 acres, the cultivated cherry is an important honey plant. Of wild cherries, the choke cherry (*P. virginiana*) extends from New England to Texas and Colorado, and the white bloom is eagerly sought by bees. It is common in thickets. In Florida it grows on both high pine lands and the hammocks, and seldom fails to yield bountifully. A surplus of this honey is more hurtful than beneficial, as it is dark red and as bitter as wormwood, having nearly the same flavor as the cherry-pit. A very little of it will spoil the flavor and color of the first orange honey. Up to the beginning of the flow from orange, it is largely consumed in feeding the brood.

CHERRY LAUREL (*Laurocerasus caroliniana*).—A beautiful lustrous-leaved evergreen tree, 25 to 40 feet tall, growing in the river valleys of Florida and widely cultivated for ornament. The greenish-white flowers are in short rather dense racemes, and in the region of the orange trees bloom in February and early March. It is an excellent source of nectar for stimulating early brood-rearing and building up the colonies of bees for the orange flow. If the weather is fair during the blooming-time, the flowers are alive with bees, and their humming can be heard for a long distance. The tree is known locally as "wild olive."

CHICORY (*Cichorium Intybus*).—The showy heads of bright blue flowers open early in the morning. Common in fields and by the roadsides. A great favorite of bees. July to October. In Phillips Island, off the Victorian Coast of Australia, it is grown as a commercial crop, and yields nectar well in showery weather. Chicory has been extensively cultivated in Germany for the roots, which are used as a substitute for coffee. (Fig. 40.)

CHINA-TREE.—See Soapberry.

CHINQUAPIN (*Castanea nana*).—The honey is thick and dark, resembling dark molasses in color, and strong and very bitter in flavor. It is ranked by E. R. Root as the poorest of honeys, and he compares its taste to that of a mixture of cayenne pepper and quinine. It grows at its best in Georgia. Through the great sand-ridge section of Florida this shrub grows luxuriantly to a height of 12 to 16 feet. This section is usually burned over in winter, and all sprouts killed; but each spring new shoots spring up from the roots, often in clusters of 50 or more. The leaves are dark green upon the upper side, but dingy white underneath. The staminate catkins, in which the nectar is said to be plainly visible, are long,

Fig. 41.—Chinquapin (Castanea nana). Photographed by Lovell.

ill-scented, and plumose. The pistillate catkins are above the staminate, in the axils of the leaves, and consist of only two or three flowers. The bloom lasts for four or five weeks. This species grows on sandy hills and barrens from Florida to Louisiana. *C. pumila,* also called chinquapin, is a large shrub, often a small tree, and has a wider distribution, extending from New Jersey to Florida and Texas. It blooms later in May and June. (Fig. 41.)

CHITTAM.—See Cascara Sagrada, under Buckthorn.

CHRISTMAS BERRY (*Heteromeles arbutifolia*).—Toyon. Produces a thick amber-colored honey with a pronounced flavor, granulating with a coarse grain within three or four months. The small white flowers open in July. An evergreen shrub growing throughout the Coast Ranges and Sierra Nevada and in southern California. A surplus in Monterey, Colusa, and Nevada counties. "One of the most handsome of California shrubs when covered from November to January with its fine clusters of crimson berries."

CLEMATIS (*Clematis virginiana*).—A climbing vine common in damp thickets, with white diœcious flowers. In both the staminate and pistillate flowers nectar is secreted in small drops on the inner side of the flat filaments, which are yellowish-green at first. If not removed the drops run together and accumulate between the stamens. Beginning with the outer row, the stamens gradually double in length, turn white, and cease to secrete nectar. Nectar is thus found only in young flowers, and the same flower-cluster may contain both nectarless and nectar-yielding blossoms. The pollen is white. The purple clematis of the garden is quite devoid of nectar, as are many other species of this genus.

The Virginia clematis is very abundant in some sections of Rhode Island. According to A. C. Miller, it is a heavy yielder of sparkling golden-colored honey with the fragrance of the flowers. It is not reliable every year. (Fig. 42.)

CLEOMELLA (*Cleomella angustifolia*).—An erect, branching herb, 2 to 4 feet tall, with the leaves divided into three oblong narrow leaflets, and small fragrant yellow flowers. It blooms for two months or longer, during the dry season, on the prairies from Nebraska to Texas and New Mexico. Bees visit it in large numbers. It belongs to the same family as the Rocky Mountain bee plant.

CLETHRA (*Clethra alnifolia*).—White alder. Sweet-pepper bush. A common source of honey from Nova Scotia to Florida in wet land along the coast. Comb honey white; extracted honey tinged with yellow, thick, with a fine slightly peculiar flavor suggestive of the bloom. The honey often fills with bubbles, which force off the cappings soon after it is stored. It is an excellent honey to blend with that of white clover. It is also slow to granulate, honey two years old still remaining liquid. In eastern Massachusetts this shrub is much valued as a honey plant, hundreds of pounds of Clethra honey being gathered annually. As much as 900 pounds has been gathered by three colonies at Westport, Massachusetts. Often there are 7 or 8 supers on one hive. Here it blooms the last of July or the first of August, the time becoming earlier southward. There is about one poor year in every three or four, owing to a late frost, cloudy and rainy weather, or other causes. On the coastal plain of Georgia and northern Florida a surplus is also obtained. A shrub growing 6 to 8 feet tall, with white flowers in racemes. (Fig. 43.)

CLIMBING BONESET (*Mikania scandens*).—Also called climbing hempweed. A twining vine, 5 to 15 feet long, with heart-shaped leaves, growing in swamps and along streams from New England to Texas. The white or purplish flowers are clustered like those of thoroughwort, which they closely resemble. It is abundant in marshy land in Alabama, Mississippi, and Louisiana, climbing over logs, fences, and shrubs. The bloom is eagerly visited by bees but is almost unknown to beekeepers. As a honey plant it is probably only of secondary importance. Like boneset it blooms in late summer and fall. The genus *Mikania* belongs to the Compositae.

CLOVER (*Trifolium*).—White clover is the most important species; *Trifolium,* the most important genus; and the Leguminosae, or pulse family, the most important family among the honey plants of North America. There have been described in the world about 250 species of clover, of which some 64 occur in North America. Many of the species are of little value to bee culture, being rare, or local in distribution, or yielding very little nectar, as the yellow hop clover and rabbit clover. The species which deserve description are white, alsike, crimson, red, and, in a less degree, sour clover. Even these few species are in many regions of no value, either because they will not grow or do not secrete nectar. The great abundance of the clovers is due partly to their great vegetative vigor, which may be partially explained by their strong root system and the presence of nitrogen-fixing bacteria on the roots, and partly to the production of a large quantity of viable seed. All the clovers require lime, since the bacteria which form tubercles on the roots can not live in an acid soil. An acid condition is indicated by the presence of field sorrel, or by the change in color from blue to red of a strip of litmus paper placed in damp

Fig. 42.—Virginia Clematis or Creeper (Clematis virginiana). Photographed by Lovell.

Fig. 43.—Sweet Pepperbush (Clethra alnifolia). Photographed by Lovell.

soil. Lime may be applied in various forms, as air-slacked lime, or ground limestone; but the latter is advised, as it is equally efficient, and cheaper in price. From 2000 to 4000 pounds to the acre may be used to advantage, which should be thoroughly harrowed into the soil. Alsike requires less lime than either white or red clover.

Alsike Clover (*Trifolium hybridum*). The honey from alsike clover is fully equal to that of white clover, and probably can not be distinguished from it. It is generally conceded that alsike clover yields nectar more freely, and is a more re-

Fig. 44.—Alsike clover (Trifolium hybridum). Photographed by Lovell.

liable honey plant, than white clover. It has repeatedly been observed that apiaries in the vicinity of alsike yield more honey per colony than those having access only to white clover even in great abundance. The period of bloom of alsike is also much longer than that of white clover, lasting, when pastured, nearly all summer. But the late bloom is seldom of much value except in stimulating brood-rearing. During the first year it does not often make a heavy growth, not attaining its full luxuriance until the second and third season. (Fig. 44.)

Alsike clover is far more hardy than red clover, and will grow on land too wet for the latter. It is adapted to moist clay soils, and sandy loam soils rich in humus, but it will thrive in dry sandy or gravelly land. It requires lime, but will succeed with less than either white or red clover. In Ontario, Canada, where hundreds of acres were formerly grown exclusively for seed, it was regarded as the foremost honey plant, and in many localities it is the only source of honey in quantities. There is probably no region in this country where it produces larger yields than in the vicinity of the Great Lakes. As the home of alsike clover is in Sweden, this species doubtless had its origin in northern latitudes; and, consequently, it thrives better in a cold than in a warm climate. For this reason it is more or less generally cultivated in the northern half of the United States, while it is almost unknown in the southern half. It is extensively grown from New England to the east border line of the Dakotas and Nebraska, and as far south as the Ohio River. Other dense but small areas occur in central Tennessee, east Kansas, western Colorado, southern Idaho, and the Pacific coast region of Washington and Oregon. To a more limited extent it is grown in Virginia, North Carolina, Kentucky, Missouri, Wyoming, and Montana, and in portions of the adjoining states. In Michigan the crop of clover honey would be small if it were not for alsike clover, as white clover has disappeared to a large extent. It is a great benefit to beekeeping in many other states, and in the future will surpass all other species of *Trifolium* in importance. It is growing in favor with farmers, for it endures well adverse conditions of weather, and is well fitted for grazing. Its planting instead of red clover should be encouraged. Alsike clover is grown only to a very small extent in the hot climate and sandy soils of the Gulf and southwestern states. In localities a heavy rainfall may enable alsike to withstand hot weather.

The substitution of alsike for red clover promises to be of great importance to American beekeepers. Many locations will be literally transformed. Beekeepers should not only preach the gospel of sowing alsike, but should offer to pay a part of the cost of the seed. After alsike has once been introduced it becomes self-sowing, and springs up where other clovers fail to make a satisfactory growth. The quality of the hay is improved and the quantity of the honey increased, and less feeding of sugar syrup in the fall is necessary. The seed is always saved from the first crop of flowers, which should be allowed to stand two weeks longer than when the clover is cut for hay. On an estate in Sweden, where twenty acres were set apart for raising seed, the average annual production for five years was 133 pounds per acre. It should be mowed either early in the morning or late in the evening, when wet with dew, otherwise the riper pods with the best seed fall off and are lost.

Alsike clover was called *hybridum* by Linnæus, who supposed it to be a hybrid between white clover and red clover; but it is now believed to be a distinct species. It was named alsike from the parish of Alsike in Upland, Sweden, where it was first discovered, and where it is very abundant. In 1834 it was introduced into England. It is a very hardy perennial plant adapted to cultivation in a cold climate. The structure and pollination of the flowers are similar to those of white clover.

White Clover (*Trifolium repens*). In the central and eastern states no other honey plant is so universally known as white clover, and white-clover honey is the honey *par excellence* — the honey with which all other honeys are compared. It is a delicious white honey of the finest quality; while not so thick and heavy as goldenrod, nor so pronounced in flavor as buckwheat or basswood honey, yet it possesses the qualities which satisfy the largest number of consumers, and fills most perfectly the demand for a table honey of the highest grade. It is given the preference by most purchasers, and the highest praise which can be bestowed on any honey is to pronounce it equal to that of white clover.

THE DISTRIBUTION OF WHITE CLOVER

Like alsike clover, white clover thrives better in a cold than in a warm climate, and consequently secretes nectar more abundantly in northern than in southern regions. In favorable seasons a surplus of 200 pounds of clover honey per colony, according to Sladen, is common at Lake Temiskaming. At Roberval, on Lake St.

John, Quebec, an average colony yield of 200 pounds is obtained from alsike and white clover. In Minnesota Wisconsin, and Michigan white clover is a very reliable honey plant. A beekeeper in Fillmore County, Minn., writes that a complete failure of white clover has not been known in that locality in 20 years. In Wisconsin, Michigan, and in St. Lawrence and Jefferson counties, New York, it is also usually dependable, yielding from a medium to a large crop of honey. But farther southward in Illinois many beekeepers report that a full crop is obtained only one year in three. In Maury County, Tennessee, in 1913, 500 colonies in one apiary could not have gathered all the nectar available; but since that year not more than one-fourth of a crop has been obtained. Although abundant in portions of Mississippi and Louisiana, it yields much less bountifully than in northern regions. In general, assuming that the soil and rainfall are suitable, in localities where the mean annual temperature does not exceed 44 to 50 degrees F., white clover is reliable nearly every year; but in warmer regions it is less dependable.

White clover is widely distributed throughout the northern portion of the United States, but is much less common in the southern states. In New England it is abundant on the limestone soils of Aroostook County, Maine, in the Champlain Valley, Vermont, and the Berkshire Valley, Massachusetts. There is an excellent white-clover belt on the calcareous glacial soils of St. Lawrence and Jefferson counties in northern New York; and from Buffalo nearly to the Hudson River there is a strip of land 20 to 50 miles wide, on which white clover is usually dependable. In the southeast corner of Pennsylvania the clovers thrive on a productive limestone soil in York, Lancaster, and Chester counties; while within the Appalachian Mountains are many fertile valleys with limestone floors where all the clovers flourish. (Fig. 45.)

White clover is at its maximum in "the white-clover belt," which includes the states of Ohio, Indiana, Illinois, Michigan, Wisconsin, Minnesota, and Iowa. But it is not equally abundant in all parts of these states, nor are the limits of the belt strictly defined by their boundaries. The area covered by white clover in this section has been greatly reduced by the intensive cultivation of the cereals; and in districts where dairy farming is largely practiced, as in southern Wisconsin and

Fig. 45.—A field of white clover in Iowa. From A B C and X Y Z of Bee Culture.

Michigan, it has been largely replaced by alsike clover. Northwestern Ohio and eastern Indiana are both good clover regions. Throughout Iowa white clover is reported to be the principal source of honey, and in many localities no other plant yields a surplus. In the eastern part of the state it is hardly reliable more than two years in three; but at Colo, the state center, there have been only four years in twenty-three in which it was a total failure. In no other part of Illinois does white clover succeed so well as in the northwest corner, especially in Stephenson County. Central Illinois is largely devoted to growing corn and oats.

Northern Missouri is also largely a grain-growing country, and, owing to dry weather, a crop from white clover is obtained only about once in four years. The larger part of the soil-forming rocks of the Ozark Plateau in southern Missouri are

Fig. 46.—White clover blossom, first stage. Photographed by Lovell.

limestone; but as they are the oldest soils in the state, and the land is very hilly, much of the lime has been removed by leaching. White clover is in this section not of much value. In the famous blue-grass region of Kentucky, or the Lexington Plain, and the central valley of Tennessee, or the Nashville Basin, both limestone areas, the temperature is too high for nectar secretion by white clover to be dependable; but in the eastern valley of the Tennessee River, where the climate is colder, a bountiful crop may be obtained.

In the Atlantic and Gulf Coastal states white clover is of little importance. Throughout the Coastal Plain from New Jersey to Florida and Texas the pine barrens and swamps have acid soils, and the climate along the coast is subtropical. The leading forage crops are cowpeas and corn, and only a few thousand acres of alsike clover and red clover are grown — in South Carolina only 375 acres. The largest areas of white clover in the southern states are found in Mississippi, Louisiana, and northeastern Texas. In northwestern Arkansas and the bottom-lands of the rivers in the eastern part of the state white clover is valuable in favorable seasons. In the great Limestone Valley of Virginia west of the Blue Ridge Mountains, and in the smaller limestone valleys of southwestern Virginia, when the summer temperature is not high, it is abundant and a fair honey plant.

Over much of the region west of the Mississippi River white clover will not grow, as the climate is too dry, and over a large area too hot. In the semi-arid regions of Wyoming, Colorado, New Mexico, Utah, Nevada, Arizona, and Texas there are less than 4000 acres of clover alone under cultivation. In the eastern portions of the Dakotas, Nebraska, and Kansas white clover is frequently reported as a source of nectar. Both white and alsike clovers grow sparingly in the river valleys of Montana. White clover is abundant in northern Idaho, and at Moscow furnishes the larger part of the surplus. West of the Cascade Range in both Washington and Oregon it is increasing in abundance; but after July it often dries up. It is reported to yield nectar less freely than in the Mississippi Valley. White clover is rapidly extending over northern California; and, as it is a dependable source of nectar, this section will probably soon offer excellent localities for the production of honey.

THE SECRETION OF NECTAR

In order to obtain a large surplus from white clover, it is not sufficient that the plants produce a profusion of bloom. "As an actual fact," writes an experienced beekeeper, "the amount of clover honey is not measured by the quantity of bloom; for I have seen the fields white with an abundance of it, with only a fair crop. I remember one year when there was a great scarcity of bloom, and yet there was a good crop of clover honey. I have also seen fields white with clover, but no honey." In England white clover is usually a good honey plant, but "in France and Switzerland," according to C. P. Dadant, "one may travel for several kilometers and not see a bee on the flowers. At Rouen, France, during one day of white-clover bloom a hive on scales actually lost 300 grains in weight."

Nectar secretion in white clover is largely influenced by temperature, soil, rainfall, and winter protection. The relation of temperature to the distribution of white clover in the United States has already been briefly discussed. At Clarinda, Page County, Iowa, J. L. Strong carefully recorded from day to day for 29 years, from 1885 to 1914, the weight of one hive on scales and the weather conditions. White clover is the most important honey plant in this locality. According to an analysis of these statistics by L. A. Kenoyer, the largest amount of honey, or 46 per cent., was secured on days when the temperature was between 80 and 90 degrees F., while 17 per cent. was stored on days when the temperature was less than 80 degrees, and 37 per cent. when the temperature was over 90 degrees F. According to the above figures a high temperature is less unfavorable to nectar secretion in white clover than is commonly supposed; and it is not difficult to understand how large crops in certain seasons may be obtained in localities much farther south. Days with a wide range of temperature are best. A very cold spring may cause a failure of the crop, even if there is a normal rainfall. In 1907 in parts of New York the average temperature of May and June was four degrees below the respective means for these months in other years, and there was no white-clover honey.

A second important factor influencing nectar secretion is the character of the soil. As has already been pointed out, white clover as a honey plant is restricted to limestone soils. On the calcareous glacial soils of the north-central states it is more abundant and yields a larger surplus of honey than in any other section. On an acid soil it can not be grown successfully. Vigorous plants secrete more nectar than those which have been stunted by a sterile soil.

A cold rain no doubt checks the secretion of nectar; but clearly the quantity of nectar brought in by the bees can not be regarded as an index of the amount secreted, since the bees will not fly freely in rainy weather. At Clarinda, Iowa, during 29 years the largest amount of honey was brought into the hives on clear days. Of the entire gain in weight, 61 per cent. was made on clear days; 13 per cent. on partly cloudy days; 13 per cent. on cloudy days, and 13 per cent. on rainy days. On a part of the rainy days there was practically no increase.

There must be sufficient rain to insure a vigorous growth of the plants, otherwise there will be a scarcity of nectar; or, if the dry weather is long continued, there may be no surplus, as has frequently happened. In the northern states, if there is a good stand of white clover in early spring, a drought in May or June, if copious rains follow, will only retard the bloom and delay the harvest. Cold rainy weather during the honey flow will both lessen the quantity of honey and prevent the bees from working on the bloom. Great humidity of the atmosphere will increase the quantity of nectar secreted by checking the transpiration of the leaves, but will not increase the sugar content.

On light sandy soils, as in Kentucky, it is well established that there will be a very small honey flow if the preceding season has been dry. If there is no rain after July 1, the drought destroys the old plants of feeble vitality, checks the growth of offshoots, prevents the growth of seedlings, and retards the formation of an extensive root system, with the result that there are few blossoms and little nectar the following season. Although the injury wrought by drought does not become apparent until the next season, it should not be attributed to winter-killing, but to the correct cause — the absence of sufficient moisture in the soil.

A fourth important factor is the protection of the roots of white clover in winter. In wet clay ground in regions where the winters are severe, the roots may be much broken and drawn out upon the surface, or the plants killed outright by repeated "lifting," caused by alternate freezing and thawing of the soil. The destructive work of the frost, however, is much lessened by the natural mulch afforded by the dead vegetation found in waste places and in meadows which have not been cropped too closely. Snow also offers excellent protection; and, where it covers the ground for most of the winter, clover suffers little or no damage. Winter-killing from freezing in well-drained sandy soil, or in warmer climates, is practically unknown.

An ideal season, when one of the largest crops of comb honey per colony on record was secured, occurred at Marengo, Illinois, in 1913. From 72 colonies, spring count, 266.47 pounds per colony, or a total of 19,182 sections of chiefly white-clover honey, were obtained. The three best colonies yielded 390, 385, and 402 sections respectively. The flow began about June 1, and continued until August, the bees then gradually changing to sweet clover and heartsease. The season consisted of a succession of hot humid days; and up to September 1 there were only two rainy days. At other times rain came during the night, the weather becoming clear again before the bees were ready to begin work in the morning.

THE POLLINATION OF WHITE CLOVER

The flowers of white clover are familiar to every one, since the plant finds a congenial habitat in the vicinity of human dwellings. It carpets the lawns, fringes the paths and roads, and is common in the fields and pastures. There have been counted in the heads or flower-clusters from 57 to 89 small florets. At first all the florets stand erect; but as the marginal ones are pollinated they cease to secrete nectar, and bend backward and downward against the stem. By preventing useless visits this change in position is beneficial to both flowers and insects. When they expand the flowers are white, but they often turn reddish after they are reflexed, and finally become brown. It is manifest that the individual florets of a white-clover head are far too small to support a honeybee. The bee clings with its legs to several flowers, and only its head rests on the flower from which it is sucking nectar. As the nectar is not deeply concealed, other insects can obtain it besides bees. The pollen is yellow; but after it has been moistened and packed by the bees in the pollen-baskets it assumes a brownish color. Bees never visit the flowers for the purpose of gathering the pollen alone, since the anthers are inclosed in a sac or "keel" formed by the two lower petals, and emerge only when the bee's head is pushed into the flower. If the point of a lead pencil is thrust into a mature flower, when it is withdrawn a little mass of pollen will be found on the under side. Compared with its value as a honey plant, white clover is of much less importance for pollen, and many of the home-coming bees enter the hives with empty pollen-baskets. (Figs. 46 and 47.)

There is no more important or more interesting subject to the beekeepers of the "white-clover belt" than the life-history of white clover and its problems. It is propagated both by seeds, and by runners, which root at the nodes and finally become independent stocks. In the absence of bees the flowers remain self-sterile and produce no seed. Clover raised from seed is more valuable for nectar the second season than during the first.

Fig. 47.—White Clover (Trifolium repens). Last stage, in which the flowers are all bent downward and have turned brown. Photographed by Lovell.

Crimson Clover (*Trifolium incarnatum*). Other English names are Italian clover and carnation clover. Crimson clover is a "winter annual." It is seeded in late summer and makes its early growth in autumn, passes the winter in a green state, begins growing again very early in the spring, and matures its seed and dies before summer. It grows wild in southern Europe, and is widely cultivated in France, Germany, and Great Britain. It was introduced into this country in 1822, and during the last thirty years has been extensively grown on the sandy soils of the middle and southern states. In the northern states it is usually killed by the severe winters.

Crimson clover has been listed as a good source of honey in North Carolina and several other southern states; and a surplus is reported to have been obtained at Dover, Delaware. At Medina, Ohio, almost as many honeybees have been seen on

Fig. 48.—Crimson Clover (Trifolium incarnatum). From the A B C and X Y Z of Bee Culture.

the flowers as on the bloom of buckwheat. The honey is apparently the same as that of red clover. (Fig. 48.)

Red Clover (*Trifolium pratense*). The flowers of red clover are almost wholly self-sterile, and, in the absence of cross-pollination, produce very little seed.

As the extensive cultivation of red clover is of primary importance to the agricultural prosperity of this country, its manner of pollination becomes of great practical interest. There has been much discussion in the agricultural and bee journals as to the relative value of bumblebees and the honeybee as pollinators of this species, and a brief review of the available data is desirable as an encouragement to further investigation.

In Europe and America red clover is very largely self-sterile. In England Darwin protected 100 heads of red clover from insects during the blooming period and not a single seed was produced; while a similar number of heads, exposed to insects, produced an average of 27 seeds per head. In the United States, Westgate obtained from 757 heads covered with tarlatan an average of only one-tenth of one per cent. seed per head. Many similar results have been obtained by other experimenters. But Waldron and others have found that a small quantity of seed may be produced by natural self-pollination. As in the case of many other plants, climate may exert an influence on seed production.

In Great Britain and continental Europe, where there is an abundant rainfall and the floral tubes attain their normal length, bumblebees, according to Darwin, Mueller, Knuth, and all other flower ecologists, are the chief pollinators of red clover. Mueller gives a list of ten species of bumblebees taken on the flowers. But he also reports seeing hundreds of honeybees collecting pollen on red-clover bloom, and effecting cross-pollination. Many species of solitary bees also gather the pollen and effect cross-pollination.

Previous to the introduction of bumblebees into Australia and New Zealand red-clover seed could not be produced commercially. (See "History of the Humble-Bee in New Zealand," by I. Hopkins, Chief Government Apiarist, N. Z. Dept. Agr. 1914. Also "New Zealand's Experience with the Red Clover and Bumblebees," by S. Graenicher, Bull. Wis. Nat. Hist. Soc., Vol. VIII, 1910.) Hopkins observed at Metamata that honeybees gathered a considerable amount of pollen from red-clover blossoms. In order to make a thorough test of their value as pollinators of red clover, a number of strong two-story colonies were placed in the center of a 700-acre tract of this plant. The second crop of flowers was just opening. The bees gathered both pollen and nectar; but, when the flowers were later examined for seed, a head here and there contained a good many, others very few, but the great majority of heads contained none. This experiment was made before the importation of bumblebees. After several lots of bumblebees had been brought from England and they had become common, many farmers reported that almost every clover head was full of plump seed, and as high as 720 pounds to the acre were obtaind at Waterlea. A large amount of additional evidence is given in the papers mentioned above. It is also well known that in Australia honeybees were not reliable pollinators of red clover, and profitable crops of seed were not produced until after the introduction of bumblebees. A similar experience is reported from the Philippine Islands.

In the United States, according to the reports of many beekeepers and farmers, honeybees gather a large amount of nectar from red clover, when the floral tubes are shortened by drought, or when the nectar is secreted so copiously that it largely fills the tubes and a portion of it is readily accessible to honeybees. In many localities hundreds of pounds of red-clover honey have been secured. On the fertile limestone soils of the white-clover belt, red clover, like white clover, yields nectar very freely, and is visited by thousands of honeybees. Under these conditions many beekeepers believe that they are important pollinators of the bloom. It is the opinion of E. R. Root, after numerous interviews with county extension agents and farmers,

Fig. 49.—Red Clover (Trifolium pratense). Photographed by Lovell.

that honeybees should receive more credit as pollinators of red clover than has previously been accorded to them. "In 1925," he writes, "I found that red-clover seed was produced in Ohio, West Virginia, and Kentucky. In some of the localities visited there were apparently few bumblebees, while in other places they were very abundant. In all cases the number of honeybees working on red clover far exceeded the number of bumblebees."

Also in the Boise Valley, Idaho, according to E. F. Atwater, the production of red-clover seed is so profitable that the acreage of alfalfa has been greatly reduced. Honeybees are believed to be very important pollinators of the bloom. "When the alfalfa has all been cut," he writes, "and honeybees are seen all over the red-clover blossoms, I shall believe until the contrary has been proved that much of the seed crop is due to their work." On the other hand Waldron asserts that bumblebees are responsible for 95 per cent. of the red-clover seed produced in North Dakota. There are no reliable numerical data in existence showing how large a percentage of red clover flowers are pollinated by honeybees when gathering nectar alone. It is, however, well established that short-tongued bumblebees, the honeybee, and short-tongued solitary bees are not reliable pollinators of flowers having the nectar concealed in tubes longer than their tongues when they devote their attention wholly to gathering nectar and do not collect pollen. *Bombus terrestris,* one of the bumblebees introduced into New Zealand, has a short tongue, and has proved an unsatisfactory pollinator of red clover in that country. Thus the fact that thousands of honeybees have been observed gathering nectar does not prove that they were pollinating the flowers.

But honeybees in large numbers at times gather pollen from red-clover bloom, and then almost invariably they effect cross-pollination. In 1911 (Bull. No. 269, U. S. Dept. Agr.) honeybees collected large quantities of red-clover pollen, springing the keels, and pollinating the flowers. Westgate erected in a red-clover field a screened cage 12 feet square, in which he placed a small colony of bees. An average of 37 seeds per head was obtained. The bees collected pollen, as well as nectar, and it is certain that the crop of seed was in part, and it may have been wholly, due to pollination effected while the bees were seeking pollen. H. Mueller, as has been mentioned, saw hundreds of honeybees gathering red-clover pollen. Hopkins also noticed that they brought into the hives more or less pollen from red clover. A beekeeper in Illinois says that at times his bees have brought in red-clover pollen, but that he never could be certain that they gathered nectar. Pammel (Trans. St. Louis Acad. Sci., Vol. V., p. 248) saw honeybees gathering pollen at La Crosse, Wisconsin. In Iowa, in 1915, honeybees were observed collecting pollen from red-clover bloom on two or three days during the summer. Pammel is inclined to the opinion that honeybees do not gather nectar from red clover in Iowa, notwithstanding the opinion of many beekeepers in this state to the contrary. When honeybees gather red-clover pollen they are reliable pollinators of the flowers; but, as the anthers are enclosed in a keel or carina, this occurs apparently at irregular and rather infrequent intervals. (Fig. 49.)

The long-tongued bumblebees are admitted by everyone to be better adapted to the pollination of red clover than honeybees. A bumblebee can pollinate from 30 to 35 red-clover flowers in a minute, and thus in two or three days a comparatively small number of them could pollinate millions of flowers. Taking into consideration all the localities throughout the world where red clover flourishes, the author is of the opinion that this species of clover is dependent chiefly on bumblebees for pollination. But in certain sections, where special conditions prevail, it is possible that honeybees may equal or exceed them in value. Basing his statement on many inquiries and observations, E. R. Root would give a larger amount of credit to the honeybee at least in the United States. "While the evidence at present," he says, "is not conclusive, I believe that further inquiry and research will show beyond question that the honeybee in many localities has taken the place of the bumblebee as a pollinator of red clover. While the former is not so effective as the bumblebee,

Fig. 50.—Yellow Hop Clover (Trifolium procumbens). Photographed by Lovell.

the great number of them present in the clover fields, in my opinion, offsets to a large extent their inability to reach the bottom of the floral tubes of red clover."

More than a score of clovers occur in California, several of which are common and widely distributed. Sour clover (*T. furcatum*), which grows rankly on low alkaline land, is of some value for nectar.

COLIMA (*Xanthoxylum Fagara*).—This species of prickly ash is a thorny shrub or small tree, found in southern Florida and along the Gulf Coast of Texas, and very common in the West Indies. The small yellowish-green flowers in southern Texas occasionally yield a surplus of honey of good body and flavor. Usually it is only moderately important. The staminate and pistillate flowers are on different trees. Also called wild lime.

COMA (*Bumelia lycioides*).—A small tree, or in Texas often a shrub, growing along streams from Illinois to Texas and Florida. It is reported to be a good honey plant in the Rio Grande Valley from Brownsville to Del Rio. It is usually a thorny, rough shrub covering hundreds of acres, and so dense are these thickets that they can not be penetrated by a man on foot or horseback. The small white flowers are in clusters in the axils of the leaves, completely covering the limbs. Coma blooms from December to April, and yields a honey of light-amber color and fair flavor. Although lighter colored, it has a twang suggestive of buckwheat; but, by those accustomed to it, coma honey is preferred to the milder honeys. Also called southern Bumelia, chittinwood, ironwood, and southern buckthorn. There are many other species of *Bumelia* in tropical America.

CORN FLOWER.—See Bachelor's Button.

CORAL BEAN (*Sophora secundiflora*).—A shrub, or small tree, growing on limestone bluffs from New Mexico to Texas and Mexico. The racemes of strongly fragrant, violet-blue flowers yield nectar freely. The red bony seeds are very poisonous. It is known in Mexico by the name of frigolito.

CORALBERRY.—See Buckbush.

COTTON (*Gossypium*).—The number of species of cotton have been placed at from five to fifty-four, but conservative authority admits of seven well-defined species. The number of varieties with English names is very large, but the common names give no assistance; they even tend to lead the botanist astray as to the origin of the species. For example, a cotton called Siamese comes from America. Only four or five species are of interest to the beekeeper. The Asiatic cottons (*Gossypium herbaceum*) are extensively cultivated in India and China, and have been known for more than two thousand years. American upland cotton was long referred to this species by mistake. Tree cotton (*G. arboreum*), a taller species with purple flowers, is a native of Africa, but was held sacred by the Hindus of India and cultivated about their temples. The botanical name of the Egyptian cottons is uncertain, but by many they are considered forms of *G. barbadense*.

Two species of cotton are extensively cultivated in the United States. They are Sea Island cotton (*G. barbadense*) and American upland cotton (*G. hirsutum*). Sea Island cotton yields a very fine long staple (1½ to 2 inches in length), but it is grown only along the coast of South Carolina and inland in southern Georgia and northern Florida. Upland cotton (*G. hirsutum*) forms more than 99 per cent. of the cotton crop of the United States. Two principal commercial types are grown in the United States — short-staple upland cotton (fibers under 1⅛ inches in length), which has by far the largest acreage; and long-staple upland cotton (fibers 1⅛ to 1½ inches long), which is largely confined to the Yazoo Delta, Mississippi, a few counties in South Carolina, and the Imperial Valley of southern California. Egyptian cotton, which has a very long staple (1¼ to 1¾ inches) is grown in the Salt River Valley, Arizona. Cotton was cultivated in Mexico and Peru at the time of their discovery by the Spaniards, and the American species probably originated in tropical America. The Asiatic cottons have white seeds, while the American cottons are black-seeded. There is a valid species of cotton indigenous to the Sandwich Islands and another to the Society Islands.

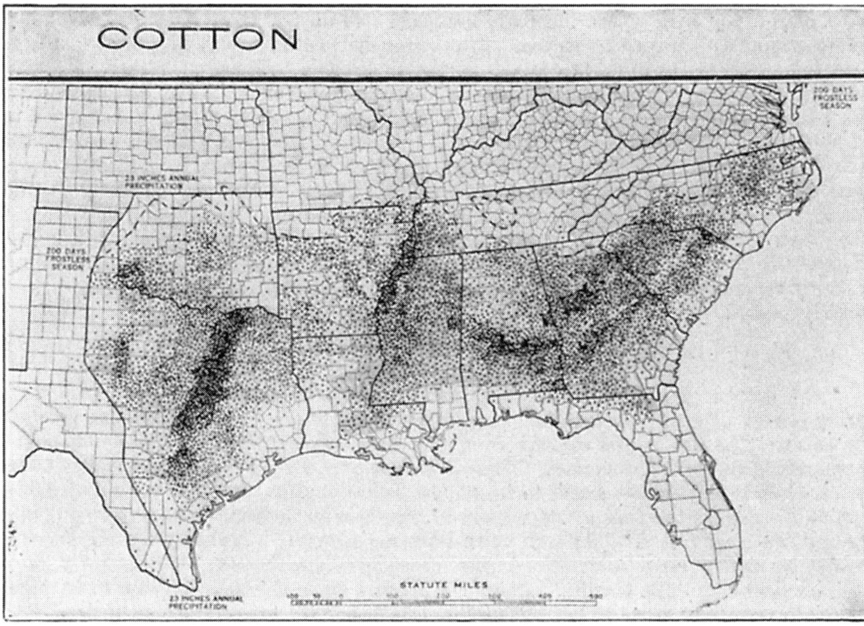

Fig. 51.—Acreage of cotton in the United States. In 1919 there were 33,960,000 acres. (After Baker. Geog. World's Agr.)

Upland cotton (*G. hirsutum*) is a perennial, but is commonly treated as an annual. It requires at least six months free from frost. The plant grows from 3 to 10 feet tall, and bears alternate, palmate leaves with 3 to 5 lobes. The large erect flowers are 3 inches across, and on the first day are a creamy white or pale yellow; but soon after midday they begin to turn reddish and on the second day are a deep reddish purple. The flowers of Sea Island cotton are yellow with a reddish-purple spot at the base of each of the five petals. The 5 sepals are united into a cup or calyx, and below the flower there is an involucre or whorl of 3 green leaflike bracts. The involucre becomes dry and brittle, and is often torn off with the boll by careless pickers.

LONG-STAPLE AND SHORT-STAPLE COTTON

When ripe the boll, or seed vessel, splits into 3 to 5 lobes, locules, or locks, containing the seeds covered with long fibrous hairs or cotton, a provision for their dispersal by the wind. A lock seldom contains more than 9 seeds. The cotton fiber is known as lint, floss, or staple, and varies greatly in length — the greatest difference between short-staple and long-staple varieties being nearly an inch. When the fibers are under 1⅛ inches in length the cotton is short-staple, and when more than this length it is long-staple. In Sea Island cotton the staple, or lint, may be 2 inches long. The lint varies greatly in length, and in color from white to yellowish brown. As the fiber dries it becomes flattened and twisted into a spiral like a corkscrew, when it can be spun into a thread even by the fingers, since the spirals adhere to each other.

THE COTTON BELT

Cotton ranks second in value among the crops of the United States, and in the cotton belt its value exceeds that of all other crops combined. In 1920 the acreage was 35,504,000 acres, and the average annual number of bales produced during five years (1914-1918) was 12,424,000. The northern limit of cotton growing in this country follows closely the mean summer temperature line of 77 degrees, and very little cotton is grown where there are less than 200 days without frost. There must be an annual rainfall of 23 inches. The cotton belt comprises chiefly eastern North Carolina, South Carolina, Georgia, northern Florida, Alabama, Mississippi, the western lowlands of Tennessee, Arkansas, Louisiana, Oklahoma, and eastern Texas; the densest areas are in the Piedmont Plateau and upper Atlantic Coastal Plain of South Carolina and Georgia, the Black Prairie of Alabama and Mississippi, the Yazoo Delta in northwestern Mississippi, the Red River Valley in Arkansas, and, most important of all, the Black Prairie of Texas. American upland cotton (*G. hirsutum*) is almost exclusively planted over this area. More than 600 varieties have been named and described, which are divided into groups according to the size of the boll, the length of the staple, and earliness of fruiting. The big-boll group is the most popular and widely grown, since the cotton can be more easily and quickly picked.

NECTARIES

The cotton plant has both floral and extra-floral nectaries. The floral nectaries consist of a narrow band of papilliform cells at the base of the inner side of the calyx. The five petals overlap except at base, where there are five small openings leading down to the nectar. These gaps are protected by long interlacing hairs which exclude insects too small to be of use as pollinators, but present no obstruction to the slender tongues of long-tongued bees and butterflies. There is also above the nectary a ring of straight stiff hairs pointing upward. Trelease saw the flowers visited by many bees, and Allard saw honeybees, bumblebees, and solitary bees (*Melissodes*) enter the corolla. After the flowers have changed in color from pale yellow to red, they cease to secrete nectar, and bees pay little attention to them.

The statement has been made that honeybees gather the surplus of cotton honey wholly from the leaf nectaries, but this is incorrect. Many beekeepers report that a large quantity of honey is gathered from the bloom. At Waxahachie, so little honey is gathered on the uplands until cotton blooms that it is necessary to feed the colonies. Late in the fall of 1909, at Trenton, Texas, cotton bloomed profusely from the middle of October until mid-November and two supers of honey were secured. The bees were laden with pollen as well as nectar and the queens laid as in the spring. Prominent beekeepers at Cordele, Georgia, and at Sulphur Springs, Arkansas, report that cotton blossoms yield a great amount of excellent honey.

There are two sets of extra-floral nectaries — the involucral nectaries and the leaf nectaries. Below the flower there are the three leaflike bracts called the involucre. At the base of each of these bracts there is a nectary both on the inner and outer side — six in all. The three inner involucral glands are situated between the calyx and the involucre, and are present both in the American and Asiatic species of cotton, but are sometimes absent in individual flowers. In form they are round, shield-shaped, or heart-shaped. The three outer involucral glands are at the base of the bracts on the outside. They are entirely absent in the Asiatic cottons. Greatly magnified, they strikingly resemble a shallow round dish with the bottom covered by a layer of large shot. According to Trelease the involucral nectaries secrete nectar abundantly, which in the daytime attracts bees, ants, and humming-birds, and at night two species of moths.

The leaf nectaries are located on the under side of the main rib of the leaves, and vary in number from one to five. They are absent from individual leaves, and entirely wanting in *Gossypium tomentosum*. They are small pits, oval, pear-shaped, or arrow-shaped with long tails running down toward the base of the leaves. In the tropics they are soon overrun and blackened by a growth of mold. (Tyler, J. T. The Nectaries of Cotton, Bu. Pl. Ind. Bull. 131, Pt. 5, 1908.) The leaf-glands seem to be most active at the time the leaf reaches full maturity. When the conditions are favorable nectar will collect on these glands in such large drops that it can be readily tasted and a bee can obtain its load in a very few visits. Honeybees then neglect the blossoms and honey comes in very rapidly. The honey secured from the foliage of the cotton does not differ in color or flavor from that gathered from the flowers. Samples submitted to the United States Bureau of Chemistry were reported to be normal pure honey.

The surplus obtained depends largely upon locality, the soil, season, and atmospheric conditions. There are many factors which influence the nectar flow, and cause it to vary in different places and at different times. One of the most important factors is the soil. Cotton is grown on a great variety of soils, as sandy loams and clay loams. Rich alluvial soils and black prairie soils are admirably adapted to its culture; but by the use of fertilizers the poor pine lands of the Atlantic slope and in the vicinity of the Gulf can be made to produce a crop. Lime seems to be required, since the Black Prairie of Texas, the most important area in the United States, is underlaid by Cretaceous limestone. Little nectar is secreted by cotton on light sandy soils, and even in the black-land area on the lighter soils the plant is unreliable. The growth of the plant may be as luxuriant as on the heavier soil; but, no matter how promising its appearance, no cotton honey is obtained. A beekeeper at Levita, Texas, states that on the river locations in the timber regions he never obtains any surplus from cotton, but that five miles southward on the black land of the prairies he secures a large amount of honey. On the dryer soils of the uplands the color of the honey is reported to be lighter than on the bottom-lands. (Fig. 51.)

Throughout the larger portion of the Atlantic and Gulf Coastal Plain cotton is a poor honey plant. Opinions differ widely as to its value in different localities, and are often contradictory. In North and South Carolina the cotton belt is the poorest section of the state for beekeeping. At Cordele, Georgia, from one to three supers of cotton honey may be obtained, but in northern Georgia cotton is of minor importance. In Alabama, Mississippi, and Louisiana only a very small quantity of cotton honey is reported. But in the Arkansas River Valley, Arkansas, where there is an immense acreage of cotton, 96 pounds of honey per colony in an apiary of 12 colonies were obtained from this source. In Pulaski County the surplus from cotton per colony in 1918 was 40 pounds, and 20 pounds in 1919. In Oklahoma the humid conditions required to stimulate the secretion of nectar by cotton bloom occur only occasionally, and it is in consequence a very unreliable honey plant. Temperature and rainfall permit its cultivation in every part of the state except along the north border.

It is in Texas that cotton rises to the rank of a great honey plant, where it yields nearly one-fifth of the entire crop of honey produced in this state. Although there are 10,000,000 acres of cotton under cultivation in Texas, it is chiefly in the Black Prairie that cotton secretes nectar abundantly. Either to the east or west of this belt the honey flow shows a marked decrease. In Metagorda County, on the coast, cotton secretes nectar well only occasionally. At Bay City cotton is not dependable, but in some seasons good yields are secured from it. At New Braunfels and northward to Waxahachie, cotton is the main dependence for honey. "In an average season," according to Scholl, "a good yield may be expected from cotton in the black-land districts and the river valleys. Under favorable conditions it is not excelled by any other nectar-yielder in the cotton-growing belt. On poor soil and on sandy land it does not secrete nectar plentifully and in some sections under certain weather conditions not at all." On the bottom-lands of the Brazos River there are cotton plantations which are several thousand acres in extent. Cotton is the only source of nectar and an average of about 75 pounds of bulk comb honey is secured annually; one season the surplus exceeded 100 pounds per colony. In 1919, one of the larger producers of the cotton belt had taken off 20,000 pounds before the beginning of the fall flow, and there still remained in the hives 10,000 pounds to be extracted. For the fall flow 5000 pounds would not be an overestimate. Waco, McLennan County, is near the center of the cotton belt; and in this county and around Waxahachie more than 500,000 acres of cotton are cultivated. The apiaries are numerous and often average per colony 60 to 70 pounds of honey. In Hunt County, northern Texas, cotton is also the main dependence for a honey crop.

HONEY FLOW

The honey flow may last from July until long after the first frosts, yielding in some localities as much surplus as all other sources combined. Even after the first frosts, if there is pleasant weather, the bees may continue for two weeks longer to work upon the plants and make a large increase in the honey crop. The surplus obtained depends largely upon locality, the soil, the season, and atmospheric conditions. There are many factors which influence the nectar flow and cause it to vary in different places and at different times. Cotton yields best when the atmosphere is warm and damp. The yield is most abundant in the early morning, and decreases toward the middle of the day as the atmosphere becomes drier. In the afternoon, unless the season is very dry and hot, the yield begins to increase again. During cloudy days, or when the atmosphere is damp, nectar is secreted abundantly throughout the entire day. The flow has also been observed to increase toward the close of the season.

COTTON HONEY

Cotton honey is very light in color and mild in flavor when thoroughly ripened, and compares favorably with the very best grades of honey. When first gathered cotton honey has a flavor very characteristic of the sap of the cotton plant itself, but this disappears as the honey ripens. During a heavy flow there is a strong odor in the apiary like that produced by bruising cotton leaves. At Trenton, Texas, in 1909, during a very long drought a very fine and pure grade of cotton honey was obtained from cotton growing on rich bottom-land. It was so thick that it was almost impossible to extract it, and entirely out of the question to strain it through even a single thickness of cheese-cloth. It was light in color, mild in flavor, and very heavy. and was considered superior even to the famous huajilla honey. Ordinarily cotton honey granulates quickly, and in the granulated form is almost pure white and very fine-grained.

COTTON IN THE SOUTHWEST

There is evidence that cotton was grown in Arizona by the prehistoric cliff-dwellers before the discovery of America. The Indians and early settlers, likewise, attempted the cultivation on a small scale of short-staple cotton. About 1900 a variety of a long-staple cotton was introduced from Egypt, where in the valley of the Nile it had been grown successfully for many years. At the Government Experiment Station at Sacaton a new variety, known as the Pima, was developed from the Egyptian plant. This cotton has a greater length (1⅝ inches) and a greater degree of fineness than any other cotton in the world. In 1911 about 400 acres of Pima cotton were planted in the Salt River Valley. This was the beginning of the growing of long-staple cotton as a commercial crop in Arizona. The acreage gradually increased until 1917, when the supply of long-staple cotton used in the manufacture of automobile tires became wholly inadequate, and the price increased to one dollar per pound. One of the large American tire companies in the spring of 1918 bought several thousand acres of land in the Salt River Valley and seeded them with the American variety of Egyptian cotton. The alfalfa growers ploughed up their fields and raised cotton instead. In 1920 about 110,000 acres of long-staple cotton were growing in Maricopa County and it was expected that the crop would be 100,000 bales. The average yield is one-half a bale per acre, but on fertile soil one bale per acre is not unusual.

The high price of cotton also greatly stimulated its production in southern California. Imperial Valley, Palo Verde Valley, and Kern County are recognized as cotton-growing centers. In Imperial Valley and Lower California, it is estimated that, in 1920, there were 120,000 acres of short-staple and 33,000 acres of long-staple cotton. The total acreage in California was about 200,000 acres. It was demonstrated that long-staple cotton can be grown satisfactorily in this state where there is a season of 250 frostless days and high temperatures occur while the crop is maturing. But the recent decline in the price of cotton has checked its cultivation throughout the Southwest, and many acres in the Palo Verde Valley were not harvested in the fall of 1920.

Cotton does not yield as much nectar per acre as alfalfa, and in localities where it has largely supplanted alfalfa beekeepers are not securing as large a surplus as formerly. In the vicinity of Chandler, Arizona, 90 per cent. of the alfalfa has been ploughed up and the land planted with cotton. In the Buckeye Valley alfalfa has been largely replaced by cotton, which here yields so little nectar that 2000 colonies have been moved out of the valley. The larger acreage of cotton and the longer blooming season will to some extent compensate for the decrease in the alfalfa acreage.

ASIATIC COTTON

About two-fifths, or 40 per cent., of the world's cotton is grown outside of the United States, chiefly in Egypt, India, and China. India is the most ancient cotton-growing country, and five centuries before the Christian era the clothing of the people consisted chiefly of cotton garments. It produces about 18 per cent. of the total cotton crop of the world. In Egypt, which ranks third in the production of commercial cotton, the crop can be raised only under irrigation. The land suitable for this purpose is restricted to the Delta and a strip along the river about a mile wide. The Asiatic cottons, cultivated as a commercial crop, are varieties of *G. herbaceum*. China produces about 16 per cent. of the world's cotton; and Russia, Peru, and Brazil also yield a small amount.

COW-ITCH (*Cissus incisa*).—A fleshy vine with warty bark, sometimes growing 30 feet long; leaves trifoliate, coarsely toothed; flowers small, greenish, in clusters (cymes). On sandy shores from Florida to Texas. A surplus is reported from Gunnison, Mississippi, and from Texas. It grows on low heavy bottom soil. At Bay City, Texas, it was formerly the main dependence, but a succession of dry seasons have nearly destroyed it. Where abundant, it furnishes a surplus of good light-amber honey. The black berries are sweet, and in the fall honeybees often gather the juice. Cultivated for ornament. Snow-vine, or pepper-vine (*C. arborea*), also a honey plant, belongs to this genus. See Snow-vine.

COWPEA (*Vigna sinensis*).—Nearly 7,000,000 acres are planted with cowpeas in the southern states. Nectar is obtained from nectaries on the flower-stalks, and not from the flowers. At Hollis, N. C., cowpeas are a most valuable source of honey in late summer. The honey is thick, deep yellow in color, and has a very strong flavor. At Fremont, Mo., honeybees have been observed working on the extra-floral nectaries throughout the day. Cowpeas are grown for forage, for food, and soil improvement. A trailing or twining vine with white flowers in small clusters. It belongs to the pulse family. (Fig. 52.)

CRANBERRY (*Vaccinium macrocarpon*).—There are thousands of acres of cranberry bogs on Cape Cod, Mass., and in New Jersey and Wisconsin. In June they are covered with innumerable pinkish-white blossoms, which are adapted to pollination by bees. Pollen falls from pores in the ends of the anthers on the head of a bee seeking nectar, but otherwise the anthers do not open. While bees greatly increase the crop of berries by their visits, they do not obtain a large amount of nectar in return. (Fig. 53.)

CREOSOTE-BUSH (*Larrea tridentata*).—At the close of the flow from mesquite thousands of bright-yellow flowers come out on the creosote-bush, which yield enough nectar to keep up brood-rearing and to furnish a little bluish-yellow honey. Grows in the arid region of the Southwest and in Mexico.

CROCUS (*Crocus vernus*).—A beekeeper in New York states that near his apiary there are some 10,000 bulbs of crocus, which when in bloom present a beautiful sight. Early in April he found in every hive at least three pounds of unsealed honey. As there were no other flowers open in the vicinity and the bees worked freely on the crocus bloom, the honey must have come from this source. Where

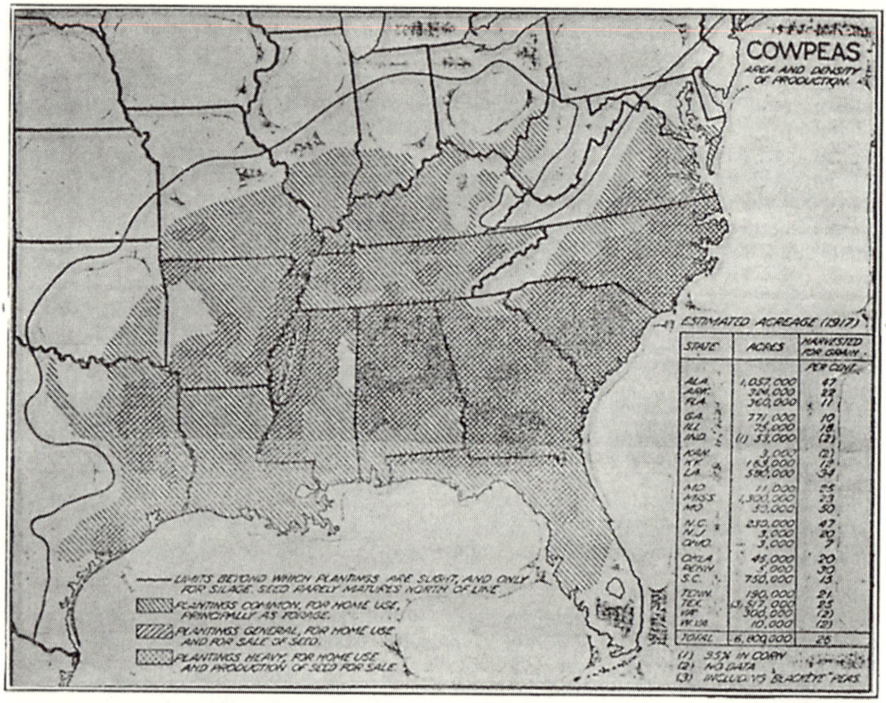

Fig. 52.—Acreage of cowpeas in the United States. In 1917 there were 6,800,000 acres. Prepared by the United States Department of Agriculture.

sufficiently abundant the flowers of crocus should be excellent for stimulating early brood-rearing.

CROCIDIUM MULTICAULE.—At Hermiston, Oregon, this little Composite plant grows on arid or desert land in great abundance. It is one of the earliest sources of pollen; and, in 1919, beginning with the 18th of March, there was for ten days a honey flow which yielded not far from five pounds of honey per colony. This unexpected flow occurred during a period of night frosts followed by bright warm days. The plants are only three or four inches tall, and the stems terminate in rather small heads of bright-yellow flowers. It does not occur east of the Pacific Coast states. For a vernacular name it may be called yellow crocid.

CROWNBEARD (*Verbesina virginica*).—On moist land in Texas crownbeard is very abundant, the white flowers blooming in the fall and yielding a large amount of amber-colored honey which is largely used for winter stores. Two other species of *Verbesina* occur in Texas, Texas crownbeard (*V. texenis*) and sunflower crownbeard (*V. helianthoides*). On the yellow flowers of the latter species there have been observed in the central states many species of bees gathering nectar. The crownbeards are mostly tall perennial herbs with winged stems and opposite ovate leaves and clustered or solitary heads of flowers.

CUCUMBER (*Cucumis sativus*).—Cucumbers, muskmelons, cantaloupes, watermelons, squashes, and pumpkins are all of value to bee culture both for nectar and pollen. In the vicinity of pickle-factories large areas are devoted to growing cucumbers. Two factories at Marengo, Ill., are supplied by 600 acres, which yield from 75,000 to 100,000 bushels of cucumbers annually for picking. The fields vary in size from one-half an acre to three or four acres, and the ground is completely covered by the large heart-shaped leaves. The total number of acres cultivated for cucumbers throughout the country must be very large. As many as

300,000 cucumbers have been produced on a single acre, but this is more than double the average crop.

The honey obtained from cucumber blossoms is pale yellow or amber, and has at first a rather strong flavor, suggestive of the fruit. In localities where there are pickle-factories beekeepers find the cucumber a valuable addition to the honey flora. Sufficient honey for wintering is often secured, besides a small surplus. In the absence of bees, cucumber blossoms, whether in the field or hothouse, remain barren. The stamens and pistils are in different flowers on the same vine, the staminate flowers being more abundant on the main stems, and the pistillate on the lateral branches. The former are sometimes incorrectly called "male" and the latter "female" blossoms. The nectar is secreted at the bottom of a cup formed by the fusion of the floral leaves at base. In the staminate flowers this cup is covered by the fleshy expanded stamens, and access to the nectar is gained through three narrow lateral passages between the stamens. When an insect inserts its tongue in one of these passages both sides of its head are dusted with pollen. In the pistillate flowers, the pistil rises from the center of the cup. The staminate flowers are the larger and open first.

In order that the pistillate flower may be fruitful, pollen from the staminate flowers must be brought to the stigmas; and in the fields this work is chiefly performed by honeybees, other insects than bees being of little importance. A market-gardener in Manitoba states that during three years he was unable, without colonies of the domestic bee, to obtain more than a dozen cucumbers, and in the case of these exceptions the flowers were pollinated by hand. A colony of honeybees was bought, and subsequently the number of hives was increased to eleven, and that year cucumbers to the value of $55.00 were sold. "No bees, no fruit," may be regarded as an axiom in cucumber-growing.

Cucumbers raised under glass must either be pollinated by hand or by hives of bees placed at each end of the hothouse. In Massachusetts cucumbers are grown very extensively in hothouses, and more than 2000 colonies of bees are required annually to pollinate the blossoms. A single grower is reported to have forty

Fig. 53.—American Cranberry (Vaccinium macrocarpon). Photographed by Lovell.

acres under glass. Many colonies are also required for the same purpose in New Jersey, Ohio, and several other states. Pollination was formerly effected by hand; but bees have proved more efficient pollinators, and enormous crops of cucumbers are obtained, which are more uniform in size and shape. Unfortunately the bees often beat against the glass in their efforts to escape into the open, until they fall exhausted to the floor and die by hundreds. Many also perish from lack of sufficient stores, so that, as a matter of course, new colonies are required each year. The cucumber has been in cultivation in India for over 3000 years, and was known to the Greeks and Romans. According to De Candolle, it was one of the fruits of Egypt regretted by the Israelites in the desert.

Muskmelons and cantaloupes are cultivated chiefly in New Jersey, Delaware, southern Illinois, Georgia, Florida, Colorado, and California. In the northern states the watermelon acreage is largest in New Jersey, Delaware, Indiana, southern Illinois, and eastern Iowa. In the southern states the largest acreage is in the Carolinas, Georgia, Florida, Oklahoma, and eastern Texas. There is also a large area in Imperial Valley, California. These species yield nectar freely and for a long time, often until frost.

CURRANT (*Ribes*).—More than fifty species of currants and gooseberries are widely distributed in North America. They bloom early and furnish pollen and nectar. The cultivated varieties occur chiefly in the northeastern states. They are largely dependent on bees for pollination, and in their absence would produce little fruit. The Missouri gooseberry (*R. gracile*) is considered a fair honey plant in the middle states.

DANDELION (*Taraxicum officinale*).—Other English names are lion's tooth, blowball, yellow gowan, and priest's crown. Widely distributed over Europe, Asia, North America, the Arctic regions, and in many other parts of the civilized world. The dandelion is a better honey plant than is commonly supposed. On many farms in Ontario and Quebec it produces more honey in spring than any other plant. At Ottawa, if the weather is warm and fair during the last two weeks of May, a strong colony will store 30 to 40 pounds in the super. In Vermont, at Middlebury, the brood-chambers are at times packed with dandelion honey; and when the supers are put on the hives it is carried above to the detriment of the white-clover surplus. Comb made during dandelion bloom is colored by the pollen a beautiful shade of yellow. At Boulder, Colorado, it is common for the hives to be filled with dandelion honey, and a few beekeepers have extracted and placed it on the market. A surplus of dandelion honey has also been reported from central Illinois and western Iowa. But while always valuable for pollen, in many localities it yields little nectar, as in northern Ohio and southern Maine.

The honey is deep yellow, and sometimes granulates in a week or two after extracting. It has a strong not to say rank flavor, with the aroma of the flowers, and probably would not be liked by persons accustomed to a mild honey. It improves with age. Owing to its tendency to granulate it would not be a desirable winter food for bees; but, as it is gathered so early, none is left by fall.

The dandelion belongs to the Compositae, and is related to the hawkweed and chicory. The head, or capitulum, consists of from 100 to 200 florets. The corolla of each floret is strap-shaped, but at the base unites to form a short tube which holds the nectar. (See Frontispiece.) At night and in damp weather the head closes so that there is little visible except a protecting whorl of green bracts. The pollen and nectar are thus completely sheltered from dew and rain. In fair weather the hour of opening in the morning varies from 6 to 8 or 9 o'clock, and the time of closing from 2 p. m. until sunset, according to the month and latitude. The flowers open much later in September than in mid-summer, and in northern than in southern regions. The dandelions often bloom a second time in the fall, but much less freely than in spring. (Fig. 54.)

As the nectar and pollen are readily accessible, a great variety of insects are attracted, and more than one hundred different species of bees and flies have been observed seeking the flower food of this species. The supply of nectar is usually

Fig. 54.—Dandelion (Taxaricum officinale). The blossom here shown is larger than the average. The usual size is about two inches across. Photographed by E. R. Root.

scanty; but honeybees visit the bloom frequently to gather pollen, which is abundant. The grains are large, many-sided, and spinose, and so firmly do they hold together that bees can carry large packets of them.

DEWBERRY.—See Blackberry.

DEWDROP, GOLDEN (*Durantia Plumieri*).—An introduced plant for hedges; very attractive to bees.

DOGBANE (*Apocynum cannabinum*).—Indian hemp. This species grows 2 to 3 feet tall and bears many clusters of small bell-shaped flowers. An acre of this plant near Cleveland, Ohio, was constantly visited by bees and other insects from

June until frost. According to W. J. Shephard there are in the interior of British Columbia hundreds of acres of spreading dogbane (*A. androsaemifolium*). It is the source of a large quantity of honey in dry seasons and in districts where there is not a large amount of rain. The pinkish white bell-shaped flowers have a pleasant odor, and remain in bloom for a long time. The honey is very white, like that of fireweed, but it is usually more dense and has a better flavor. In 1922 there was a very large yield of honey from this source.

DOGWOOD (*Cornus*).—Shrubs with small white flowers in flat-topped clusters. There are about 18 species in America north of Mexico. They are not usually common enough to be of much value, but at Divide, West Virginia, the hillsides are in spring white with the bloom, which is frequently visited by honeybees. Nectar is secreted in a thin layer by a ring at the base of the style. Also called cornel.

DROUGHTWEED.—See Turkey Mullein.

DUTCH CLOVER.—See White Clover.

ELDERBERRY.—See Pollen Flowers.

EPHEDRA (*Ephedra antisyphilitica*).—Also called Mexican ground pine, "buchu," and switch plant. A low straggling shrub, 1 to 4 feet tall, growing on mesas and dry hills in the semiarid regions of New Mexico, western Texas, and Mexico. The leaves are reduced to two small, dry, thin, nearly transparent scales at the joints of the stems. The older stems are woody and light gray in color; the young stems are hollow, long-jointed, striate, about the size and length of a knitting-needle. They are green in color and perform the function of leaves. The small flowers are in little clusters, and are unisexual. At Robert Lee, Coke County, Texas, it blooms in February and March, and yields a large amount of pollen and nectar, strongly stimulating brood-rearing. A snow-fall of several inches only temporarily interrupted the nectar flow. In this locality it flourishes on rocky hills, where it attains a height of about two feet. Three other species of *Ephedra* occur in New Mexico and the other southwestern states, where they are probably of some importance to beekeeping. The young stems of these plants strongly resemble those of the horsetails. This is a very unique genus of honey plants, since it belongs to the Gymnosperms.

ERYNGO.—See Blue Thistle.

EUCALYPTUS.—A large genus of evergreen trees growing chiefly in the coast regions of Australia and New Guinea. About 150 species have been described, of which not far from 100 have been introduced into California. At the Forestry Station at Santa Monica there have been planted nearly 70 species and varieties, the qualities of which are being tested and compared. In the groves in the Berkeley Hills, Alameda County, there are from 1500 to 2000 acres, and another large center is around Newark. To a much smaller extent they have been planted in Arizona and the Gulf region of Texas. Few Eucalypti will endure a temperature below 20 degrees, or above 120 degrees F. They grow very rapidly, and a few species may become very valuable sources of timber and other commercial products; and all are very effective as avenue and landscape trees. Some of the species are popularly known as gum-trees because a resinous gum flows from incisions in the bark; others are called iron-bark trees, from their very hard bark; and still others, from their fibrous bark, are termed stringy-bark trees. To this genus belongs the tallest tree in the world, *E. amydalina,* which attains the height of 480 feet. (Fig. 55.)

The Eucalypti, popularly called "gum-trees," from which the honey crop of Australia is chiefly obtained, yield a large amount of nectar. The blooming-time of the various species in California varies so widely that there are at least from three to seven species in flower during every month of the year, and a species may bloom twice in the same year.

In Australia a part of the species bloom very irregularly, as white box gives a flow of nectar in some seasons during May, June, July, and August; yellow box in

Fig. 55.—Blue Gum (Eucalyptus globulus). Photographed by Richter.

Fig. 56.—Forest Gray Gum (Eucalyptus tereticornis). Photographed by Richter.

Victoria blooms from October to March, while Gippsland stringy-bark may bloom throughout the year, though the crop comes from a burst of bloom in autumn. The honeys obtained from the different species differ widely in color and flavor. The yellow box yields a pale, almost white honey, with an excellent flavor, while that of brown stringy-bark is dark, with a disagreeable odor and flavor. Chemical analysis shows that the honeys also differ in the percentage of sugar, moisture, and ash. Some granulate within a week, others remain liquid indefinitely. When the conditions are right the flow of nectar is astonishing; from red gum of western Australia (*F. calophylla*) Rayment gathered a teaspoonful of nectar from three flowers. The flow does not always cease at night, and bees have been observed working during bright moonlight nights.

The characteristic honeys yielded by this remarkable genus of trees are of great interest; but, as few of them except blue gum are common enough in California to produce a surplus, we are compelled to go to Australia for descriptions of their qualities (*Australian Honey Plants* by Tarlton-Rayment). Most of the following species are found in California.

Blue gum (*E. globulus*) does not yield nectar as heavily as some other species, the best record being 56 pounds per colony. The honey is amber, thin, and granulates in a few months. In California it has an acid flavor, and is in little demand for general use. Blue gum produces great quantities of cream-colored pollen. (Fig. 55.)

Yellow box (*E. melliodora*) blooms profusely every other year, but yields little pollen. It is a most popular honey-tree, and a surplus of 350 pounds is not considered remarkable. The honey is pale yellow, or nearly white, and has a delicious flavor. It remains liquid almost indefinitely.

Red box (*E. polyanthemos*) yields a pale, dense honey of fine flavor, which rarely granulates. Every second year the trees are white with bloom. Very little pollen, but a splendid nectar-producer.

Gippsland stringy-bark (*E. eugenioides*). Honey amber, with a very rich flavor, but so viscous that it clings to the extractor in thick ropes and effectually checks its revolutions. An abundance of cream-colored pollen.

Red iron bark (*E. sideroxylon*). The flowers vary in color from white to pale pink. An abundance of pollen. The pale yellow honey is not so dense as that from red gum, but the flavor is good. It blooms late, but bees do not winter well on it, probably because it has ripened imperfectly.

Jarrah (*E. marginata*) yields a dark strong honey unfit for table use.

Red gum (*E. rostrata*) is considered the finest honey-producing tree in Australia. A surplus of 150 pounds per colony is not rare. A nearly white honey, which granulates within a week with a fine grain. The tree blossoms every year, but is not everywhere reliable.

Red stringy-bark (*E. macrorhyncha*) honey has at first a strong flavor, but when fully ripe is excellent. It does not granulate.

Brown stringy-bark (*E. capitellata*) produces a dark-colored honey, which has a disagreeable odor and flavor.

Sugar gum (*E. corynocalyx*) secrets nectar copiously and yields an abundance of pollen. A mild, bright amber-colored honey; the tree blooms profusely and exhales the fragrance of ripe cantaloupes. A handsome shade tree. Five hundred trees would be a good investment for every beekeeper.

Apple box (*E. staurtiana*) in New South Wales yields a dark inferior honey with a sharp flavor. It granulates with a fine grain.

Gray box (*E. hemiphloia*) yields a honey with a "tallowy" flavor, whence it has been imagined that bees gathered tallow from sheep skins hung up to dry.

Mahogany gum (*E. botryoides*) yields a pale yellow acid honey.

There are many other species, each having its own distinctive characteristics, but they do not differ essentially from those described.

By far the most widely planted and probably the best adapted to the climatic conditions of California is the blue gum, or *E. globulus,* which is found in almost every town in the state from San Francisco to San Diego, and inland as far as the edge of the Imperial Desert region. It is apparently as vigorous in California as in its native Australia. It is claimed to be the fastest-growing tree in the world. Seedlings will average a growth of 50 feet in height in six years and 100 feet in ten years; and under favorable conditions a seedling may reach a height of 35 feet in eight months, and in three years a height of 70 feet. The wood is heavy, hard, and strong, but is not durable in contact with the soil, and is therefore not suitable for fence-posts or telegraph-poles. It is used for cabinet work as well as for fuel; the leaves yield large quantities of medicinal oil. Windbreaks of two or three rows of blue gum afford excellent protection to orchards.

The bark of the blue gum is smooth and pale brown. The leaves are sword-shaped, 6 to 12 inches long, tough, leathery, and bluish-green on young trees, but dark green on older trees. The flowers are solitary (in most other species they are in small clusters) in the axils of the leaves and appear from December to June. When the flower-bud expands the top of the calyx drops off, and there is a "veritable starburst" of some 100 creamy-white stamens. A flower consists of the cup-shaped lower portion of the calyx, which is well adapted to hold the very abundant supply of nectar, and a ring of stamens, with the pistil in the center. There are no petals. The seed-cases are round, top-shaped, or in the blue gum angular, and a pound of seed will produce 10,000 plants. *E. globulus* was introduced into California in 1856.

Other species of Eucalytus which are promising commercially are the sugar gum (*E. corynocalyx*), the red gum (*E. rostrata*), and the gray gum (*E tereticornis*); but none of these are comparable to blue gum (*E. globulus*) in rapid growth and ability to flourish over a wide range of conditions in California. The sugar gum is much used in southern California as a street tree and for windbreaks. It strongly resists drought, but succumbs easily to frost. The red gum has been largely planted in the Sacramento and San Joaquin valleys, California. The gray gum endures drought and cold better than many species, and can, therefore, be planted over a wide range of the state. The timber of all three species is strong and valuable. (Fig. 56.)

ESPARCET.—See Sainfoin.

EYSENHARDTIA.—See Rockbrush.

FALSE INDIGO (*Amorpha fruticosa*).—In Iowa this genus is ranked in importance with *Trifolium* (clover) and *Melilotus* (sweet clover). In Nebraska it grows throughout the state in wet valleys, deep ravines, and along streams. In broad ravines it occurs as a tall treelike shrub, forming rather close thickets. It blooms in early spring, and the dense masses of violet-purple and yellow flowers stand out prominently in the landscape. Honeybees visit the flowers in great numbers and gather both pollen and nectar. Unlike most members of the pulse family, the flowers have no keel or wings and the pollen is easily obtained.

FENNEL, DOG.—See May-weed.

FETTER-BUSH (*Lyonia nitida*).—An evergreen shrub with oblong-oval, leathery, entire leaves, growing 3 to 6 feet tall, and found in wet land from Virginia to Florida and Louisiana. The white urn-shaped flowers are in clusters in the axils of the leaves. They bloom in April or May, are nectariferous, and helpful in building up the colonies, but do not furnish a surplus.

FIGWORT (*Scrophularia marilandica*).—Also called heal-all, square-stalk, and carpenter's square. A strong-smelling herb, 3 to 6 feet tall, with square stems, opposite leaves, and small greenish-purple flowers, growing in woodlands from Massachusetts to Kansas and Louisiana. The abundant nectar is secreted in two large drops at the base of the ovary. In 1879 a small field of figwort at Medina, Ohio, made a remarkable showing. Honeybees visited the flowers from morning until night during the entire period of blooming. After the nectar was removed other drops would exude in about two minutes. At one time this plant excited a considerable *furore* among beekeepers, as it was thought that for artificial pasturage it would exceed anything then known. The honey obtained, however, would not warrant the large expense of its cultivation. The flowers are very frequently visited by wasps, and are often called wasp-flowers. In California *S. californica* occurs on the mountains, and would be of great value if it were more common. This species was formerly known as *S. nodosa*, var. *marilandica*.

FILAREE—See Alfilerilla.

FLOWERING CURRANT.—See Currant.

FOG FRUIT.—See Carpet Grass.

FORGET-ME-NOT (*Myosotis macrosperma*).—In Oregon this species is reported to yield a light-amber honey of thin body with the aroma of the blossoms.

FRUIT BLOOM.—See Apple, Pear, Plum, and Cherry.

GALLBERRY (*Ilex glabra*).—Inkberry. Evergreen winterberry. An evergreen shrub, 2 to 6 feet tall, with oval or elliptic leathery leaves, smooth, shining, deep green above, and paler and dull beneath. The diffusely branched bushes form dense thickets which withstand the encroachments of all other plant growths, and can be passed through with difficulty. The gallberry multiplies both by offshoots and by seed, and in the southeastern states is rapidly extending over land which has been recently cleared of forest. (Fig. 57.)

The blooming period lasts for about a month, beginning with May and closing early in June. The small flowers, in a multitude of little clusters, are produced in the greatest profusion, and 3000 have been counted on a bush with a stem only half an inch in diameter. The flowers are largely diœcious; that is, a part of the bushes produce chiefly staminate and a part mostly pistillate flowers; and they are, therefore, dependent on insects for pollination. The sterile flowers are in clusters of 3 to 6, while the fertile are solitary. The bushes begin to bloom the second year. The berries (drupes) are shining black, and are sometimes used for dyeing wool or in making a substitute for ink, whence the name inkberry. They are also called winterberries, because they remain on the bushes in great numbers during the winter and afford a never-failing food supply for birds. Ripe fruit can be found on the bushes every month in the year. In the spring when they are in full bloom there still remain on the branches a part of the fruit of the previous year. As the name indicates, the fruit is very bitter; but to some extent the gallberries are eaten by birds.

The gallberry grows in sandy soil along or near the coast from eastern Massachusetts to Florida and Louisiana. It is the most valuable honey plant in the southeastern section of the United States, rivalling or surpassing the gum trees in the amount of honey produced. It covers thousands of acres of the sour or acid soil of the Coastal Plain where sweet clover, white clover, and tobacco will not flourish. The limestone soils so necessary to the prosperity of the clovers are unfavorable to this shrub. Without it there would be an immense area of sour soils with no honey plant well adapted to the needs of bee culture. Upon the surplus secured from this species the South is largely dependent for a good table honey. In the swampland are found the gum trees, or tupelos, while on the higher ground grow the gallberry, blackberry, and huckleberry. Much of this land can never be drained or reclaimed, and will thus always remain inviting territory to the beekeeper. So abundantly is the nectar produced that in southeastern Georgia the little drops can be plainly seen glistening in the flowers. So much better and so abundant is this source of nectar, that honeybees were observed by E. R. Root actually to desert the tupelos in the height of the honey flow, and devote their attention wholly to the newly opening bloom of the gallberry.

The gallberry first becomes important as a honey plant in the swamps around Norfolk, in southeast Virginia, where with the gum trees it seldom fails to yield a surplus. In Tidewater North Carolina there is a vast area of low land comprising 20,000 square miles, the eastern half of which is not more than 20 feet above sea level. Gallberry covers thousands of acres along the rivers and bays, blooming from May 10 to June 1, and yielding an excellent but rather thin honey, which is liable to ferment unless well ripened. The Coastal Plain is far superior to any other part of the state for beekeeping, and "there is practically," writes the State Specialist in Beekeeping, "no limit to the extent beekeeping can be developed in this section." The average size of the apiaries of 55 beekeepers in 19 counties is 88 colonies, but there are 20 apiaries which contain 100 or more colonies. The

Fig. 57.—Gallberry (Ilex glabra). Photographed by J. J. Wilder.

largest number of colonies is found in the southeast counties, where the tupelos, or gum trees, are also abundant. There is much unoccupied territory. In the near future it is probable that gallberry honey will be sold by the carload.

Owing to the much smaller area of swampland in South Carolina the gallberry is less important than in North Carolina; but in southeast Georgia it is a very common honey plant, and a great yielder of nectar. It blooms ten days later than in North Carolina. We have never failed, says J. J. Wilder, to get a surplus from it even during the most unfavorable weather conditions. In over half a century there is no record of its ever once disappointing the beekeeper. The largest surplus that has been obtained from a single colony is 147 pounds. During the honey flow the bees disregard all other bloom, working for pollen until about eight o'clock in the morning, when the flow begins and continues for the remainder of the day. The honey is light amber, very heavy, and very mild and pleasant in flavor. When it is pure and well ripened it has never been known to granulate. Although the honey is produced by the hundred tons it is rarely shipped north, as it is impossible to fill the demand for it in the southern markets. Wilder declares that he has never known a gallberry section to be overstocked, and in one location 362 colonies did nearly as well as 100. Good gallberry locations in Georgia are, in his opinion, nearly numberless, and large quantities of this fine honey are annually lost for want of bees to collect it. Beekeepers often regret that fires started by the men engaged in the production of turpentine burn the gallbery thickets, but a much better growth is thus secured the following year. Wilder reports that he makes a practice of burning over one-half of the gallberry lands in his location once in two or three years. On a burned-over section the bushes make a rank growth, while on a section not burned over they are thin and scattering.

But a warning is given by E. R. Root to the northern beekeeper who proposes "to go southeast." He must not suppose that there are no obstacles to overcome or no failures. A large part of this remarkable bee country is swampland and will always remain a wilderness. There are venomous snakes and hosts of mosquitos and redbugs; the population is sparse; the villages are small and primitive; the country roads are very poor; the winters and springs are damp and chilly and the summers are very hot; and there are few modern conveniences. But there is reported to be very little malaria, and from most of the dangers and difficulties enumerated the adventurous apiarist can protect himself.

In Hamilton County, on the north border of Florida, the larger part of the surplus comes from gallberry; and, on low sandy soil and in the vicinity of swamplands, it ranks as one of the major honey plants of this state. The coast region of Alabama is a nearly level sandy plain, and is covered largely with a forest of long-leaved pine. In the spring colonies may average 60 pounds of honey from titi and gallberry; but in many localities there is a scarcity of honey plants later in the season, and colonies are likely to perish from want of stores unless fed. The soil of southern Mississippi is sand with little or no clay subsoil. The trees are almost entirely long-leaved yellow pine, and wherever the forest has been cut gallberry springs up in great profusion. It extends as far north in the state as the thirty-second degree of latitude. Gallberry is found in Louisiana only in the extreme southeastern section of the state.

Pure gallberry honey has nearly the flavor of white clover honey mixed with that from basswood; but it differs from this blend in that it has a slightly tart reaction ten to fifteen seconds after it has been tasted. Its flavor is often injured by an admixture of honey from black titi (*Cliftonia monophylla*), which is abundant in the swamps and blooms a little earlier.

Swamp gallberry (*Ilex lucida*). This species is also an evergreen shrub, resembling the common gallberry in leaf, flower, and fruit; but it is a little larger, blooms a little earlier, and grows in swamps. It extends from Virginia to Florida and Louisiana. The honey is very similar to that of *I. glabra,* but is reported to be a little milder. Other species of *Ilex* of value to the beekeeper are holly (*I. opaca*), dahoon (*I. Cassine*), yaupon (*I. vomitoria*), possum haw (*I. decidua*) and black alder (*I. verticillata*).

GAURA (*Gaura biennis*).—A herbaceous plant with spikes of white flowers, which in sunshine fade by ten o'clock and turn pink. It is in some seasons visited by bees very eagerly. At Grenola, Kansas, it blooms early in September. Minnesota to Georgia.

GERMANDER (*Teucrium canadense*).—American germander, or wood sage, is a perennial herb, 2 to 3 feet tall, belonging to the mint family. It extends from New England to Nebraska, and in the central states is common in low ground. The conspicuous purple flowers are in racemes, and bloom from the last of June to the middle of August. They are frequently visited by honeybees. The corolla tube of this species is only 6 mm. long, so that honeybees easily reach the nectar; but a part of the species of this genus have a longer floral tube and are adapted to bumblebees. Wood sage is a bee-flower, but as a honey plant it is only of secondary importance.

GILIA (*Gilia floccosa*).—This species is listed as a honey plant in southern Arizona, where it is very common. *Gilia virgata,* according to Alice Merritt, in California secretes nectar freely, and is visited by both honeybees and butterflies. In Bear Valley, California, bees gather both pollen and nectar from *Gilia densiflora.* There are many other species of *Gilia* which are valuable for nectar and pollen on the dry hills and mesas of New Mexico, Arizona, and Nevada.

GILL-OVER-THE-GROUND (*Nepeta hederacea*).—Also called ground ivy and field balm. A creeping vine with tubular blue flowers which bloom from May to June. Introduced from Europe, it has become common in waste places. In Illinois both honeybees and bumblebees are abundant on the bloom. A considerable amount of extracted honey has been obtained from this plant, according to A. I. Root. The honey is rather dark and strong flavored, but is valuable for brood-rearing.

GLOBE ARTICHOKE.—See Artichoke.

GLOBE THISTLE (*Echninops sphaerocephalus*).—Chapman honey plant. Thistlelike herbs, 3 to 6 feet tall, with spinose leaves and globular heads of whitish or bluish flowers. Introduced from Europe, and frequently cultivated for ornament and rarely growing wild in waste places. About 1886, Hiram Chapman of Versailles, N. Y., planted some three acres of globe thistle and described it as a most promising honey plant. It was also cultivated and warmly praised by many other beekeepers; and even as late as 1918 a writer in the British Bee Journal declared: "No bee plant that I have ever grown was so attractive to bees. Whenever the weather was favorable the heads were crowded." Of this species C. C. Miller writes: "Upon its introduction I planted quite a patch of it and I never saw bees so thick on any other honey plant. But close observation showed that the bees were not in eager haste in their usual way when getting a big yield, but were in large part idle. I should not take the trouble to plant it now if land and seed were furnished free." As a honey plant globe thistle is almost unknown to the present generation of beekeepers; but it has considerable historical interest, as at one period its value as a honey plant was frequently a subject of discussion in the bee journals.

GOLDENROD (*Solidago*).—Goldenrod honey is very thick and heavy, with the golden-yellow color of the blossoms. The quality is poor when first stored; but, when capped and thoroughly ripened, the flavor is rich and pleasant. It is the general testimony of New England beekeepers that many persons prefer this honey to any other. They regard its color, body, and flavor as the qualities of an ideal honey. When granulated and cut up into cubes for table use, it is hardly less attractive than white clover honey. Its genuineness is never questioned. But the flavor is stronger than that of white clover, which would be given the preference by the great majority as the great universal staple to be used with bread and butter. Extracted goldenrod honey granulates with a coarse grain in about two months.

In a large part of New England goldenrod never fails to yield freely even in cold wet weather, but it does exceptionally well during a warm dry fall. The honey has always proven an excellent winter food for bees, and is the main reliance of many beekeepers for this purpose.

While the bees are bringing in the nectar the whole apiary is filled with a disagreeable sour smell, which on a calm evening can easily be perceived at a distance

Fig. 58.—Flat-topped Goldenrod (Solidago graminifolia). Photographed by Lovell.

of one hundred feet. The odor observed during a goldenrod honey flow has sometimes been likened to that of decaying carrion, but this is a mistake. When such an odor is present in the apiary, it is caused in most cases by one or more stink-horn fungi (*Phallus impudicus*). Where there is decaying organic matter, as near old stumps, these fungi frequently spring up in the fall. They exhale a strong fetid scent similar to that of dead carrion, which may easily mislead the beekeeper, as we have learned by experience. The stem is hollow, and the caplike top deliquesces into a sticky, semi-liquid mass, filled with spores, which is very attractive to carrion-flies. The flies feed on the thick syrup and thus become an agent in distributing the spores. As soon as the fungi are removed the carrionlike odor disappears. The sour smell of the nectar also vanishes in a few days.

Goldenrod is an important honey plant in Nova Scotia and New Brunswick, but is of less value in southern Quebec. On the sandy plains of Gatinau Valley, about 40 miles north of Ottawa, *S. puberula* and *S. squarrosa* occasionally yield 40 pounds of honey per hive. In Manitoba, east of Winnipeg, from 50 to 100 pounds per colony are obtained. Goldenrod in southern Maine is a very reliable honey plant, never failing to yield a surplus, but northward in the state is reported to be of less value. In Vermont, Massachusetts, and Connecticut a marketable surplus is often obtained in September, and once in three or four years strong colonies on Cape Cod will store nearly a hundred pounds from fall flowers. Various species of goldenrod are also common and valuable in New York, New Jersey, Georgia, Kentucky, Texas, and other eastern and southern states. In California, western goldenrod (*S. occidentalis*) is common in marshes and on river banks, and *S. californica* occurs on dry plains and in the mountains throughout the state; they yield an amber-colored honey. But in Illinois and Maryland, and in general throughout the

Fig. 59.—Worker bees on the Tall Hairy Goldenrod (Solidago rugosa) and White Clover (Trifolium repens). The single flower-cluster in the upper left-hand corner is Alsike Clover (Trifolium hybridum). Photographed by Lovell.

Fig. 60.—Tall Hairy Goldenrod (Solidago rugosa). Photographed by Lovell.

southwestern states, it is not regarded as important. A surplus of goldenrod honey is occasionally obtained in southern Georgia. It has a deep golden color, but is rather strong in flavor and odor. *S. rigida* is worthless for bees in Iowa, but in southwest Wisconsin attracts great numbers of them.

All the goldenrods, so far as known, yield nectar; but the species vary widely in different localities in the quantity of nectar secreted. Even when growing side by side two species may yield nectar very unequally. In the former case the variability in nectar secretion may be largely explained by external factors, as soil and climate; but in the latter case the differences depend upon internal conditions, as variations in the chemical changes or functions of the plants themselves. Of the four very common species of goldenrod in Maine, the bushy goldenrod (*S. graminifolia*) and the tall hairy goldenrod (*S. rugosa*) are the best honey plants, while the early goldenrod (*S. juncea*) and the field goldenrod (*S. nemoralis*) are much less often visited by bees. Nectar is most freely secreted in hot dry autumns, when the flow lasts well into October. (Figs. 59 and 60.)

Fig. 61.—Salt-Marsh Goldenrod (Solidago sempervirens). An excellent honey plant. Abundant in salt marshes. Photographed by Lovell.

The stately and beautiful genus of goldenrods begins to bloom at midsummer, or earlier in the case of the early goldenrod, and in November there are still visible the flower clusters of the Canada goldenrod and the tall hairy goldenrod, while the salt-marsh goldenrod may prolong the season until December. There are about 85 described species, confined chiefly to North America, with a few in South America and Europe. Fifty species occur north of Tennessee and east of the Rocky Mountains. They are closely allied, often hybridize, and are difficult to distinguish. There is a form adapted to almost every kind of location. The woodland goldenrod (*S. caesia*) is found in open woodlands; the field goldenrod (*S. nemoralis*) is very common in dry fields; the rock goldenrod (*S. rupestris*) prefers rocky situations; the swamp goldenrod (*S. neglecta*) lives in swamps; while the seaside goldenrod (*S. sempervirens*) thrives in salt marshes. (Fig. 61.)

Although the individual heads are so small, conspicuousness is gained by massing them in great plumelike clusters or panicles. Their bright yellow color renders them visible both by day and evening; and, as the temperature at night is several degrees above the surrounding air, they sometimes serve as a temporary refuge for insects. The floral tube is very short, seldom over one millimeter in length, so that there are few insects which are unable to gather the nectar. In Wisconsin Graenicher has taken on the early goldenrod (*S. juncea*) 182 different species of bees, wasps, flies, butterflies, and beetles either sucking nectar, or collecting or feeding on pollen; and on the Canada goldenrod (*S. canadensis*) 144 visitors. The honeybee visits the florets so rapidly that the number of visits per minute can not be counted. A large amount of pollen is gathered both by the domestic bee and by wild bees. So abundant, indeed, are the flowers, and so ample the stores of pollen and nectar, that four or five of our native wild bees, which fly only in autumn, never visit any other plants. Some of the goldenrods are pleasantly scented, others are odorless.

GOOSEBERRY.—See Currant.

GORSE.—See Furze under Pollen Plants.

GRANJENO.—See Hackberry.

GRAPE (*Vitis*).—It has been claimed that the grape is wind-pollinated, and occasionally the pollen is carried by the wind; but the whole structure of the flower is adapted to insect pollination. The small greenish yellow flowers depend on their delightful odor to attract insects. The fragrance resembles that of mignonette and can be perceived for a long distance. Kerner relates that in a journey up the Danube, in the valley called the Wachan, the air was so filled with the scent of the flowers that it seemed impossible that they could be far off, and yet the nearest vines were 900 feet away from the boat. There are five nectaries, which in central Europe are functionless and secrete no nectar; but in warmer regions Delpino states that nectar is secreted freely. At Dulac, La., a good flow from wild vines is reported, and *Vitis californica* in the Coast Ranges of California is said to yield nectar; but these observations should be confirmed. There is only a small amount of pollen. The flowers are visited by honeybees, bumblebees, and smaller bees. Beetles are common on the flower-clusters and sometimes do great injury. The vines remain in bloom for about ten days.

GREASEWOOD.—See Chamise.

GUM-PLANT (*Grindelia squarrosa*).—Many acres of the dry plains of Manitoba and Minnesota are covered by its yellow flowers, which honeybees visit in great numbers. A coarse herb, blooming from July to October.

HACKBERRY (*Celtis*).—Also called sugarberry and nettle-tree. The hackberries are shrubs, or small trees, on which the leaves and flowers appear at the same time. The small greenish flowers are borne in the axils of the leaves, and are partly staminate and partly pistillate. The species are more numerous in the South than in the North, and are valuable for both pollen and nectar. The common hackberry (*C. occidentalis*) extends from Canada to western Texas, and in Illinois

Fig. 63.—Heartsease or Smartweed (Polygonum Persicaria.)

The extracted honey has a good flavor and varies in color from a light to a dark amber. Heartsease comb honey, in point of color, is almost as white as that of clover. The extracted granulates in very fine crystals, and looks very much like the candied product of any white honey. Care should be taken in liquefying, as heartsease honey is injured more easily, and to a greater extent, by overheating than any other honey.

In New England heartsease is of no value as a honey plant, and is apparently of little importance throughout the eastern United States. The genus *Polygonum* also contains many other species, some of which are known to be of value locally. Water smartweed (*P. acre*), common in wet places, in Yolo and Colusa counties, along the Sacramento River, California, is reported to yield large quantities of dark honey of a poor quality. Another species (*P. Bolanderi*), found only on rocky outcroppings in the Napa and Mt. Hood ranges in the same state, is said to yield every year about 20 pounds per colony of an amber-colored honey. It blooms during September and October. The fields also are often white with the flowers of the common knotweed (*P. lapathifolium*), found throughout temperate North America and also in Europe and Asia.

HEATHER (*Calluna vulgaris*).—The heathers or heaths are not indigenous to America, although three species occur locally in eastern New England; but in northern and western Europe heather or ling (*Calluna vulgaris*) covers vast areas of waste or sterile lands called moors. When it grows a yard tall the fine evergreen leaves, the purple stems, and profusion of pink flowers present an expanse of color long to be remembered. Its uses among the peasants are numberless, being employed for brooms, brushes, fuel, brewing, roofing, beds, dyeing, and fodder. Another beautiful heath, the purple heath (*Erica cinerea*) is also common on the lower moors of Great Britain. Both secrete nectar plentifully and furnish a generous surplus of amber-colored honey, with an aromatic flavor and a pungent aroma, but so thick that it is difficult to extract. It is highly prized as a fine honey throughout Great Britain. In southwestern Africa the heaths reach their maximum, and the 500 species are a prominent element in the vegetation of that region, reaching the height of 12 feet and being covered with white or pink blossoms for a large part of the year. (Fig. 64.)

On the Bayard Thayer estate at South Lancaster, Mass., there have been planted on the hillsides in irregular masses two or three acres of heather (*Calluna vulgaris*). The plant is hardy and seeds itself, but it is necessary to keep the land free from underbrush. It is propagated by means of cuttings rooted in a greenhouse. The original seed came from Scotland and was planted at Townsend, Mass., by a Scotch woman, homesick for her native heather hills. No surplus has thus far been obtained from the bloom.

HEDGE NETTLE (*Stachys*).—There are many species of hedge nettle, which are found in both the East and West. They belong to the mint family, and, owing to the long corolla tubes, are adapted to the long-tongued bees, especially bumblebees. But honeybees also visit the flowers, which are doubtless entitled to a place among the honey plants of secondary importance.

HELIOTROPE (*Heliotropium curassavicum*).—Listed as a honey plant in California, where it is common along the seashore and in low alkaline land.

HERON'S-BILL.—See Alfilerilla.

HILL VERVENIA.—See Phacelia.

HOLLY (*Ilex opaca*).—American holly. White holly. A small evergreen tree, 20 to 50 feet tall, with a trunk sometimes six feet in circumference. Bark smooth and grayish white, twigs light brown. The leaves are elliptical, leathery, spiny-toothed, dark green, shiny above and dull beneath, with bright red berries in the axils. As in the common gallberry, the flowers are small, white, and a part are pistillate and a part staminate, the staminate being clustered and the pistillate solitary. Only one kind of flower occurs usually on an individual tree. Holly extends throughout the southern states from Florida to Texas, and in the Mississippi Valley northward to Missouri, and along the coast to Massachusetts, but is not abundant north of Virginia. (Fig. 65.)

American holly is widely distributed in Georgia, but is seldom very common in any one locality. The flowers expand in April, and, although the honey is never obtained pure, it is undoubtedly excellent. In Florida it is confined to the northern part of the state, where it blooms a little earlier than in Georgia. The honey is always mixed with that from other early spring flowers; for example, on the eastern coast it forms a fine blend with the honey of the saw palmetto. In South Carolina holly is also considered a valuable honey-producer, and the odor of the flowers is very noticeable in the apiary when the trees are in bloom. In Massachusetts the holly does not flower until June. There is in this state a variety with yellow fruit.

But locally throughout western Mississippi and southern Arkansas holly is an important honey plant, and the source of a large amount of surplus. At Grayspport, Grenada County, Mississippi, it is the only honey plant yielding a surplus. The honey is almost white or a very light amber in color, heavy, excellent in flavor and when pure will not candy for years. "I would go out of business," writes a beekeeper from this town, "were it not for holly. It is always reliable except when

Fig. 64.—Heather (Calluna vulgaris). Photographed by Lovell.

Fig. 65.—American Holly (Ilex opaca). Staminate flowers in bud. Photographed by Lovell.

Fig. 66.—Black Alder (Ilex verticillata). Solitary pistillate flowers in the axils of the leaves. A cluster of staminate flowers on the left. Photographed by Lovell.

it rains constantly during the blooming period. I have had strong colonies store 17 pounds per day from holly." Holly begins blooming from April 25 to May 2, and the blooming-period lasts from two to three weeks, according to the season. The secretion of nectar seems to be stimulated by a warm damp atmosphere. Colonies are usually not strong enough to make the most of the holly flow, as it is difficult to build them up sufficiently so early in the season. A large quantity of holly honey is also secured in the Yazoo Delta, Mississippi. In southern Arkansas holly is one of the major honey plants. It blooms in May, and the flow here, as in Mississippi, is not greatly influenced by the weather. At Buchner colonies store about 40 pounds of honey in spring, most of the surplus coming from holly and black gum. Along the Ouachita River in the southern part of the state holly is the principal honey plant.

Holly is the most widely known of all the species of *Ilex,* as great quantities of the branches with their bright green leaves and red berries are used for decoration during the Christmas holidays. But there are several species of *Ilex* with red berries which are liable to be confused with American holly, as they are valuable as honey plants and may attain the size of small trees; but they may be easily distinguished by their smaller leaves, which are never spiny-toothed. The myrtle-leaved Ilex (*I. myrtifolia*) is fairly common in cypress swamps in the wire-grass region of Georgia. It blooms at the same time and usually in the vicinity of the common gallberry, and the honey is believed to be equally good, for the bees gather the nectar most eagerly. The yaupon (*I. vomitoria*) and the dahoon (*I. Cassine*) may grow twenty feet or more tall, and in the southern states are helpful in building up colonies in early spring. At Bay City, Texas, yaupon yields a clear honey with a heavy body and a greenish sulphur-yellow color. It has a decidedly bitter taste when first gathered, but this disappears with age. There is a fair yield every other year. In New England the black alder (*I. verticillata*) is a common shrub in swamps and is much visited by bees. (Fig. 66.)

HOLLYHOCK (*Sidalcea malvaeflora*).—Checkerbloom. Wild hollyhock grows in profusion along the roadsides and irrigating-ditches and on the plains of southern California, and is an excellent source of nectar in winter, sometimes yielding a small extracting. It also stimulates brood-rearing, if there is warm sunny weather. There are two kinds of flowers, one perfect with a large corolla, the other pistillate with a small corolla.

HONEY LOCUST (*Gleditsia triacanthos*).—The bloom of the honey locust has the odor of honey and secretes nectar freely; but it does not, as a rule, yield a surplus, and is of far less value than black locust. A honeylike odor is apt to prove misleading as to the quantity of nectar a flower yields, both to bees and beekeepers. A thorny tree with rough bark, pinnate leaves, and small greenish flowers. It grows in rich woods from New York southward to Texas and westward to Nebraska, but it is not abundant. It is cultivated for ornament as an avenue tree, and as a high windbreak.

HONEY PLANT, ROCKY MOUNTAIN.—See Rocky Mountain Bee Plant.

HONEYSUCKLE (*Lonicera*).—The flowers of the different species of honeysuckle are nectariferous, but are adapted chiefly to bumblebees and humming-birds. The bush honeysuckles, among which the Tartarian honeysuckle is common in cultivation, bear bumblebee-flowers, from which honeybees are able to gather little except pollen. The blossoms of the familiar climbing honeysuckle of the garden (*L. Periclymenum*) are adapted to hawk-moths, and those of the trumpet honeysuckle (*L. sempervirens*) to humming-birds. Both have the nectar far beyond the reach of honeybees.

HOP-TREE (*Ptelea trifoliata*).—A shrub with small greenish white flowers in level-topped clusters. The odor is disagreeable. Honeybees are frequent visitors for nectar, and many other bees and insects resort to the bloom. It is reported of value in Texas. In the absence of insects the hop-tree is self-sterile. The bitter fruit is used as a substitute for hops. Also called shrubby trefoil, wingseed, and water-ash.

HORSE-CHESTNUT (*Æsculus Hippocastanum*).—A bumblebee flower, but honeybees gather both pollen and nectar. From the California buckeye (*Æsculus californica*) as much as 25 pounds of honey per colony has been obtained; but under certain conditions this honey may be poisonous to the bees, as told in Circular 301, Agricultural Experiment Station, Berkeley, Calif. See Buckeye.

Fig. 67.—Common Locust (Robinia Pseudo-Acacia). Photographed by Lovell.

HOREHOUND (*Marrubium vulgare*).—A woolly herb with small white flowers in whorls. Naturalized from Europe, horehound is widely distributed throughout the United States, blooming in spring and summer. In Texas it yields nectar freely. The honey is very sweet, with a peculiar, rather nauseating flavor. It is common in the foothills of the Coast Ranges and Sierra Nevada, in the Sacramento and San Joaquin valleys, and in southern California, where it yields a dark amber honey, too strong for table use, but largely used in medicine. Horehound is a pest on a sage range, for if only a small quantity of its nectar is gathered, the color and flavor of the sage honey are impaired. (Fig. 68.)

Horehound was introduced into Australia by the early settlers, and in many localities has become very common. Bees visit the flowers throughout the day, and horehound honey is valued for its medicinal qualities. It has a greenish shade and too often flavors the honey gathered from other plants.

Fig. 68.—Horehound (Marrubium vulgare). Photographed by E. R. Root.

HORSE-BEAN (*Parkinsonia aculeata*).—Retama. A graceful small tree or shrub, growing in warm regions and found from Florida to the Pacific Coast. The fragrant light yellow flowers open in spring, and after rain may appear at any time throughout the year. It is visited by bees but does not yield a surplus.

HORSEMINT (*Monarda punctata*).—The genus *Monarda,* which was named for Nicolas Monardes, a Spanish physician and botanist who lived in the 16th century, contains about 15 species, all natives of North America. The most valuable species to the beekeeper is the common horsemint *M. punctata.* It is a perennial herb with lance-shaped leaves and two-lipped yellowish flowers spotted with purple, which grows in sandy fields and prairies from New York to Wisconsin and southward to Florida and Texas. In western Wisconsin and eastern Minnesota it is common on the sandy jackpine lands and oak barrens, where it yields nectar abundantly and is very attractive to honeybees. In Iowa it occurs only in the sandy section along the Mississippi, Wapsipinicon, Cedar, and Iowa rivers. It is likewise listed as a honey plant in western Mississippi and northwestern Louisiana. In Texas it is an important source of surplus.

Horsemint was first brought into notice a few years ago, when it was highly recommended to beekeepers and the seeds sold quite extensively. Subsequently it was almost forgotten until large crops of honey from this source obtained on the low alluvial lands bordering on the Mississippi River attracted attention. Afterwards wonderful reports came from different parts of Texas. While horsemint is found in nearly every county in eastern Texas, it is most abundant on the Black and Grand Prairies. According to the reports of hundreds of Texas beekeepers, it ranks second in importance among the honey plants of that state, and it is estimated that about 20 per cent. of the total surplus comes from this source. Beginning in June, or a little earlier, it blooms from four to six weeks, or if there is much rainy weather for a much longer time. The surplus in the cotton belt is largely dependent on the horsemint, and the average per colony in commercial apiaries ranges from 20 to 100 pounds. But it is not reliable every year, and in hot dry seasons the flow greatly decreases. The extracted honey is clear light amber in color, a little darker than the comb, and of good body. It has a pronounced flavor, and has been compared with the basswood honey of the North. Horsemint honey is preferred to white clover honey by many persons, but it is the general opinion that it has a little too strong a flavor.

In the Black Prairie region of Texas lemon mint (*Monarda citriodora*), an annual with pink or nearly white flowers, is abundant. In localities where there is much lime in the soil the flowers are bright red. The corolla tube is about the same length as that of the preceding species and the nectar is readily reached by honeybees. *M. clinopodioides* is also very common on the dry plains of Texas. The form of the flower and the time of blooming are nearly the same as in *M. punctata.* There are also several other species of horsemint growing in Texas, the nectar of which can be easily gathered by honeybees.

But a part of the species of this genus have long corolla-tubes and are adapted to insects with a much longer tongue than the honeybee, so that even if common they are of little value to the apiarist. Bradbury's Monarda (*M. Bradburiana*), occurs on dry hills in Illinois, Missouri, and Kansas. The pink flowers have a tube 18 mm. long and are adapted to female bumblebees, but butterflies and hummingbirds also are common visitors. The tube at times fills with nectar for more than half of its length, and, as its mouth flares sufficiently to admit the head of the honeybee for 5 mm., worker bees in large numbers have been seen gathering the nectar. Wild bergamont (*M. fistulosa*) is another bumblebee-flower. It is found abundantly by the roadside and on dry hills from Maine and Minnesota to Florida and Louisiana. The corolla-tube is 18 mm. long, or a little less than three-quarters of an inch. The pink flowers attract many butterflies. Wasps bite holes in the base of the tubes in order to obtain the nectar, and honeybees very often make use of these perforations to gain a part of the sweet liquid. Bee balm or Oswego tea (*M. didyma*), with scarlet flowers, also has very long corolla-tubes. The horsemints belong to the mint family, or Labiatae.

HUAJILLA (*Acacia Berlandieri*).—Southern Texas is a semi-arid country with many days of intense sunshine and an average annual rainfall which ranges from 16 to 26 inches, according to the locality. The soil is sandy, gravelly, or rocky, and the surface is partly level and partly broken by ridges and hills. The dry uplands are about one-half covered with a scrubby, thorny growth, 2 to 12 feet tall, often forming impenetrable thickets. Mesquite, white brush, prickly pear, Texas ebony, catsclaw, retama, and huajilla are abundant and yield a large surplus of honey. The three most important honey plants are mesquite, catsclaw, and huajilla; but the main surplus, according to Scholl, comes from huajilla. Huajilla occupies the rocky ridges and hills of this section extending from the Nueces River to the Rio Grande. It is not found in Cameron County in the extreme south, but is common in Uvalde and many other southern counties. It is a nearly unarmed shrub, 2 to 10 feet tall, with bipinnate leaves and small cream-colored flowers in globular heads. It blooms from the last of March to the middle of May, and yields nectar in such abundance that it is impossible for the bees to gather all of it in favorable seasons; but a sudden rain will bring the flow to an abrupt close. Many carloads of this honey have been shipped from Uvalde County. The honey is white or a very light amber, and is probably the lightest-colored honey produced in the state. It has a very mild rich flavor, and is famous for its excellent quality and pleasing aroma. It granulates early with a coarser grain than catsclaw honey. The flow is not reliable every year. Huajilla honey unmixed in commercial quantities is difficult to get, as this species and catsclaw usually bloom at the same time.

HUCKLEBERRY (*Gaylussacia*).—There are seven species of huckleberry in the southern states, which grow on a great variety of soils, as in deep forests, on the mountains, rocky hillsides, sandy pine lands, deep swamps, and along the coast. In Georgia huckleberry bloom has been reported to yield occasionally a surplus of thin honey, with a pink hue and a cherry flavor. The blooming-period extends from February to May. The first supers of honey nearly always contain more or less huckleberry honey. A common species is *G. baccata.*

On Cape Cod, Massachusetts, without *Gaylussacia baccata* a crop of honey would not be certain oftener than every other year. Beekeepers do not distinguish always between the huckleberries and the high-bush blueberries. In the State of Washington the high-bush "huckleberry" (*Vaccinium ovatum*) covers large areas of the poorest sandy soil. It blooms in April and may yield a small surplus of excellent honey. In Oregon the scarlet "huckleberry" or red bilberry (*V. parvifolium*) is very abundant on land that has been burned over after it has been cleared of forest. The honey is light amber, very mild, with a characteristic flavor. When ripe it is so thick and viscous that it is difficult to extract. The moist climate of the Coastal region is necessary for the secretion of nectar by this shrub. The berries are bright red. See Blueberry.

HUISACHE (*Vachellia farnesiana*).—Also called yellow opopanax and cassie. A shrub, or small tree, sometimes 30 feet tall, which occurs from Florida to southern California. The flowers are fragrant, bright yellow, and are valuable for pollen in early spring, but yield very little nectar. (*Acacia farnesiana*).

HYSSOP (*Agastache nepetoides*).—A perennial herb with yellowish flowers found in woods and thickets from Vermont to Nebraska. It is in bloom for about six weeks and attracts many honeybees.

INDIAN CURRANT.—See Coralberry.

INDIAN SAGE.—See Boneset.

INKBERRY.—See Gallberry.

JACKASS CLOVER (*Wislizenia refracta*).—A rank-scented annual usually called stinkweed. It is known as "jackass clover" chiefly among California beekeepers. It is very abundant on the alkaline land of the San Joaquin Valley, and is so rapidly spreading that it promises to become a honey plant of great importance. It yields a mild water-white honey, which granulates in three or four months into a pastelike solid. According to Richter it secretes nectar so copiously that a Fresno beekeeper, in 1919, extracted 30 pounds per colony each week for six weeks from this source. It blooms in autumn, but heavily only every other year. One to three feet tall; flowers yellow in dense racemes; leaves of three leaflets like a clover leaf.

JAPAN PLUM.—A variety of plum introduced from Japan. See Plum. The loquat is also sometimes called Japan plum. See Loquat.

JERUSALEM ARTICHOKE.—See Artichoke.

JEWEL-WEED (*Impatiens biflora*).—A surplus of honey has been reported to be gathered from this plant in Minnesota. When honeybees enter the pendulous flowers their backs are dusted with the white pollen, and the beekeeper knows that they are working on the bloom of jewel-weed. Bumblebees often bite holes in the spurs, and then both bumblebees and honeybees rob the nectaries without entering the flowers. Also called touch-me-not from the elastic seed vessels.

JOE-PYE WEED.—See Boneset.

KIDNEY ROOT.—See Boneset.

KINNIKINNIK (*Rhus virens*).—The green sumac in Texas is found in hilly woodlands from the Colorado River to the Rio Grande. It is reported to yield a strong-flavored, greenish-colored honey, which does not granulate. Kinnikinnik is also one of the vernacular names of the red bearberry *Arctostaphylos Uva-Ursi.*

KNOCKAWAY.—See Anaqua.

KNOTWEED.—A variety of *Polygonum lepathifolium* called *P. incarnatum,* with large leaves and long drooping spikes of flowers, at Hebron, Ohio, is abundant and yields much fine honey. It grows best in moist lands and in wet seasons. The pink blossoms present a beautiful appearance. During drought there is less bloom; but if there is much rain it blooms in September, and the combs are filled nearly to the bottom-bars. See Heartsease.

LAUREL.—See Mountain Laurel.

LEMON (*Citrus Limonium*).—At Corona, California, a six-acre lemon grove was covered with tobacco cloth, a strong kind of mosquito netting. The purpose was to secure a more even temperature — warmer in winter, cooler in summer. Two colonies of bees were placed in the grove, but they dwindled rapidly and died in a few months. The honey gathered was a very light yellow and possessed a slightly tart flavor. The fruit increased fourfold, but unfortunately a heavy wind wrecked the structure. Near Pasadena, California, there is a lemon orchard 50 acres in extent. Three colonies of bees placed in this orchard gathered a small amount of lemon honey, the qualities of which are described as excellent. It had a mild acid flavor suggestive of the lemon. The lemon blooms more or less at all seasons of the year. It is cultivated in southern Florida, and in southern California near the coast; but it is by no means as important as the orange as a honey plant.

LIME (*Citrus Limetta*).—The lime, a native of India, is cultivated occasionally in California. The fruit is similar in shape to that of the lemon, but the flesh is greenish and the juice very acid. The flowers yield nectar.

LIME-TREE.—See Basswood.

LIMA BEAN (*Phaseolus lunatus*).—Seventy-five per cent. of all the beans harvested in the United States are grown in California, and more than 50 per cent. of the entire crop comes from the southwestern counties of Ventura, Orange, Santa Barbara, and San Diego. In 1920 the four counties reported 115,000 tons of beans. Of the various varieties of beans grown in California only the Lima bean is of value to the beekeeper, although the black-eyed bean has been erroneously stated to yield an amber-colored honey. In 1920 there were under cultivation in southern California 149,837 acres of Lima beans, and in 1919 the crop was 900,000 bags. California produces 85 per cent. of the Lima beans grown in the world. The only other region in which they are grown on a commercial scale is in the Island of Madagascar. In California they are adapted to a coastal strip 20 miles in width extending from Santa Barbara County southward to San Diego County, which is subject to heavy ocean fogs. Cool sea fogs and the absence of protracted hot spells are required for the maturing of the plant, which otherwise is apt to blight; but the dense fogs often retard the flight of the bees.

The Lima bean is a twining vine with racemes of small white flowers, and compound leaves of three pointed, ovate, entire leaflets. The pods are scimitar-shaped with a few large flat seeds. A bush variety has been very extensively planted during the past few years. It is grown a little farther away from the ocean under irrigation. In 1920 thousands of acres of this variety were planted in the San Fernando Valley, which was the haven of many migratory beekeepers. Nectar was yielded in abundance by the bush Lima bean fields, while bees dependent on pole varieties were starving. The vines bloom in July and August and yield a heavy, white, mild honey which has a pleasant flavor. Most of the honey is secured during the first two weeks of bloom. It granulates quickly. The crop of honey from this source is rather uncertain, as it is greatly influenced by weather conditions. If there are many days of hot sunshine little nectar is secreted, and too much fog prevents the flight of the bees.

In Ventura County whole farms comprising more than a thousand acres each are devoted to growing Lima beans, and there are rows a mile in length. In no other county is there so large an acreage under cultivation. The crop of bean honey is usually fairly reliable, and an average of 50 pounds per colony is secured in a good season, but as much as 150 pounds per colony has been obtained. Twice the crop has been a failure. Some years ago a hot wind literally withered the bloom, and again in 1920 the bloom was reported as nectarless. After the honey flow from the sages is over, many beekeepers move their colonies to the bean fields. There have been as many as 2000 colonies in a radius of three miles, and it has been estimated that nearly 500 beekeepers migrate to the bean fields.

As a rule the beans never receive a drop of rain from the time of planting to harvesting. The ground water and the ocean fogs furnish all the moisture they require. All weeds are destroyed by cultivation and hand-weeding. As there is no rain, the vines do not require poles but can lie on the ground without rotting. After the bean crop is removed the fields are so bare that a sheep would starve on a hundred acres, and there is no pasturage left for bees.

In reply to a special request M. H. Mendleson, one of the most prominent and successful beekeepers in California, describes his experience in the production of lima bean honey as follows:

"I have had over forty years' experience in the bean fields, and was the pioneer in moving on an extensive scale colonies of bees among the beans in southern California. In Ventura County, following wet winters, there are over 100,000 acres planted with beans, mostly with the vine Lima bean. In the San Fernando Valley, in Los Angeles County, the Henderson bush Lima bean, which also yields a white honey, is largely grown; but in California the surplus of bean honey is obtained chiefly from the tall or vine variety. Ventura's rich valley lands formerly produced as much as fifty sacks of beans of eighty pounds each to the acre. One ranch alone of 1900 acres furnished a big trainload of beans each year. A warm damp atmosphere causes the best secretion of nectar. On a Lima bean range in a frostless belt, at an elevation of some twenty-five feet, about three miles from the ocean, where the heavy fogs hung low, I secured 140 pounds per colony of Lima-bean honey of superior quality. Irrigation does not produce results as good as those which follow rainy winters. When irrigated, the twining Lima bean grows too much to vine, and, as it is difficult for the bees to visit the limited bloom, they store much less honey. Sometimes a hot east wind, continuing for three days, will cause the bloom to drop and bring the honey flow to an abrupt end; but fortunately we seldom have these winds from the deserts during blooming-time.

"The honey flow from the bloom of Lima bean follows so closely the flow from the mountain sages that bees must be moved hastily to the bean fields. It is better to move the bees a couple of weeks before the bean flow begins; but I have moved colonies about ten miles and had them fill up in six days with water-white honey. As a rule, bees should be moved to the bean fields from July 1 to 10—the earlier the better. Sage bloom ends usually between June 20 and the first of July, but in wet seasons it may continue longer. With sufficient rainfall and favorable weather I have secured from 40 to 140 pounds per colony of choice Lima bean honey. If it has been properly ripened, extracted Lima bean honey compares favorably with the white clover honey of the eastern states. It is water-white, delicious in flavor, and excellent when candied. It absorbs moisture much more than any other honey, and must be extracted as soon as it is completely capped over and ripened or otherwise it loses its fine flavor. Extra-strong colonies are required to ripen this honey thoroughly. Weak colonies can not do this as well, unless the hive space is contracted proportionately to the strength of the colony. Even then strong colonies give the best results, as near the coast the heavy ocean fogs are very penetrating. The honey can be ripened much better farther inland where the air is dryer. A number of beekeepers extract Lima bean honey when it is only partially capped over, and then it ferments.

"Honeybees are believed to be a great benefit in the pollination of Lima bean bloom. A ranchman, to whose bean fields I moved a large number of colonies, declared that he harvested a much heavier crop of beans from the vines in the vicinity of my bees than elsewhere."

LINDEN.—See Basswood.

LOCUST, BLACK (*Robinia Pseudo-Acacia*).—This is one of the finest honey trees of the eastern and southern states. It belongs to the great family of the Leguminosae, which includes many of the best honey plants, as the clovers, Acacias, and locusts. It is a native of the mountains from Pennsylvania to Georgia, and westward to Missouri and Arkansas; but it has become extensively naturalized in Canada, New England, and the eastern states. Large plantations of it have been made for timber. The wood is hard and very durable, and is much used for posts. There is a saying that stone will crumble before locust will rot. The tree grows to great size, and is long-lived except when attacked by borers. It spreads rapidly by sprouts rising from the roots, which run for long distances near the surface of the ground. When the trees are cut, or killed by borers, the roots send up a great many sprouts, which grow very rapidly and flower within two or three years. See Honey Locust.

The white fragrant flowers resemble pea-blossoms, but are in pendent clusters like those of Wisteria. The tree blossoms in May or June, remains in bloom for about ten days, and yields a large amount of water-white honey of heavy body and mild flavor. It is, however, an unreliable yielder, not blooming at all in some seasons, and in others the bloom secreting nectar very sparingly. The black locust is reported to yield a surplus in Sacramento and Marin counties, California. In many localities it is becoming more abundant. Also called white locust, red locust, yellow locust, green locust, false acacia, post locust, and locust tree. (Fig. 67.)

LOGWOOD (*Haematoxylon campechianum*).—A tree common in the West Indies and Central America. There are large areas in the tropics where this tree is the predominating growth. It yields an almost water-white, heavy honey, with a pleasant flavor and the fragrance of the bloom. In Jamaica, logwood is the principal source of honey and one of the heaviest yielders of honey in the world. Logwood usually blooms twice, once in November and again about Christmas time. The earlier flow is usually light, and the honey is mixed with that from other plants in bloom at the same time. This early flow is chiefly valuable for increasing the number of bees. The second flow, which begins about Christmas time, may last through January. With strong colonies and sufficient rain a large surplus has been secured — in one instance 500 pounds from a single colony. But continuous dry weather will cause the buds to droop and wither under a torrid sun, and defeat the hopes of the apiarist. A single shower has been known to cause hundreds of acres to burst into bloom and set the apiary a-roar. The main bloom lasts for about six weeks. Logwood is the source of dyes extracted from the heart of the tree.

LOQUAT (*Eriobotrya japonica*).—The loquat is a pear-shaped fruit, orange-red or lemon yellow in color, with a soft, juicy, sweet flesh. Many commercial orchards of this semi-tropical fruit are found in Orange and Ventura Counties in southern California. The loquat blooms from October to February; and the white fragrant flowers are always covered with bees, when the weather permits. The loquat also finds a congenial home in Louisiana, Mississippi, Alabama, Georgia, Florida, and as far north as North Carolina. It was brought from Japan, in 1889, to California, under the name of "Japan plum;" but it should not be confounded with the true Japan plum. It is now generally called loquat.

LUCERNE.—See Alfalfa.

LUPINE (*Lupinus*).—The lupines are pollen-flowers and honeybees often gather pollen from them, but they are wholly nectarless, although often reported to yield a "little nectar." See Lupine under Pollen-Flowers.

MADROÑA (*Arbutus Menziessii*).—A widely branching tree, 20 to 125 feet tall, with dark red or crimson bark and evergreen leaves shining above and whitish beneath. Its range extends from British Columbia to southern California. It grows on mountain slopes and in gravelly valleys of the Coast Ranges and reaches its highest development in Mendocino and Humboldt counties. The small, white, globular flowers in terminal clusters open in early spring and attract a great company of honeybees, bumblebees, and other insects. There are ten nearly transparent nectaries in a circle at the base of the corolla. The honey is a light amber, with a very heavy body and a pleasant slightly aromatic flavor. At Melrose, Oregon, madroña is a common and reliable honey plant, yielding an excellent light-amber honey. The red or orange-colored berries are ripe about Christmas time, when great flocks of wild pigeons feast upon them. Like manzanita, madroña belongs to the heath family. It is the handsomest ornamental tree in this family, and never fails to excite admiration.

MAGNOLIA (*Magnolia grandiflora*).—English names are laurel-bay and bull-bay. The noblest tree of the Magnolia family and the largest evergreen tree in our flora. The leaves are dark green, smooth and shining above, and rusty-red, finely tomentose beneath. The creamy-white flowers are broadly bell-shaped, lemon-scented, gradually turning brown as they fade. They produce large quantities of pollen and also secrete nectar. A surplus of honey has been reported to have been obtained from magnolia during damp weather, but there is little evidence that it is of much value as a honey plant. In hammocks and along rivers from North Carolina to Florida and westward to Texas. (*Magnolia foetida.*)

MAGUEY.—See Agave.

MALLOW (*Malva*).—Often called cheese-flower, cheeses, and cheese-cake. The mallows are common in waste places and along roadsides. Musk mallow (*M. moschata*) is listed as an important honey plant in New Jersey. The large red flowers of the high mallow (*M. sylvestris*) are very attractive to a great many bees, which eagerly gather the nectar but pay little attention to the pollen. In the absence of insects it is self-sterile. In California cheese-weed (*M. parviflora*) is very common in the interior valleys near dwellings, and is considered a desirable honey plant.

MAMMOTH CLOVER.—See Clover.

MANDARIN.—See Orange.

MANCHINEEL (*Hippomane Mancinella*).—An evergreen tree, with smooth leaves, resembling a pear tree, found on sandy beaches in southeast Florida, on the Keys, in the West Indies and tropical America. It has an acid milky juice, which is very poisonous and was used by the Caribs to poison their arrows. "Probably the most poisonous member of our arborescent flora. The juice and the smoke from the burning wood are very injurious to the eyes." The yellowish-green flowers are very small and open from February to April. Occasionally it is reported to yield heavily.

MANGO (*Mangifera indica*).—The mango tree grows in the southern part of Florida in hammocks and on some of the Keys. The small whitish flowers are in large clusters or panicles composed of 200 to 4000 flowers. Some varieties remain in bloom but 10 days, others for nearly two months. Nectar is secreted in small drops on a white disc, collecting in larger quantities at the base of the petals. Honeybees are important visitors and are also fond of the overripe dropped fruit.

Fig. 72.—Black mangrove on the left; red mangrove on the right.

MANGROVE, BLACK (*Avicennia nitida*).—There are in southern Florida three different trees called mangrove — the red mangrove, the white mangrove, and the black mangrove. The red mangrove (*Rhizophora Mangle*), an evergreen shrub or tree, belongs to the mangrove family. Along the coast of southern Florida and the Keys it forms dense tidal swamps, advancing into the water where the shores are flat and mucky. The seeds germinate while still on the tree, and, falling into the mud in an upright position, immediately strike root. The trunk and branches send out aerial roots, which, descending in an arched fashion, give the tree the appearance of stepping forward. The flowers are small and yellowish.

The white mangrove (*Laguncularia racemosa*), also called white buttonwood, is a member of the white mangrove family. It is common on the seashore of peninsular Florida and in the West Indies and tropical America. In Florida it is commonly a shrub with leathery oval leaves and small greenish flowers. Neither the red mangrove nor the white mangrove is of value to the beekeeper.

Fig. 73.—Black Mangrove (*Avicennia nitida*). Photographed by J. J. Wilder.

The black mangrove (*Avicennia nitida*), also called blackwood, and black tree, belongs to the verbena family, most of the species of which in the North are herbaceous plants. It grows on the seashore of southern Florida, the Keys, and eastern Texas, and also in tropical America. In Florida it is not found to much extent north of Ormond on the east coast. It usually grows back of the red mangrove, and in localities where both grow together the red mangrove fringes the shore and makes new land, while the black mangrove is a soil-former. Both are valuable in catching drift and lodging humus and gradually transforming the shallows into reefs and islands, and finally into solid land. But the black mangrove does not actually grow in the water. (Fig. 72.)

The black mangrove, when it grows to the size of a tree, resembles a scraggly old oak with a rough brown bark. It may be 25 to 50 feet tall, with a trunk diameter of four feet, or on the Keys it may attain even greater size. Northward it is seldom more than a shrub. The leaves are leathery, oblong, with very short stems, and when they unfold are somewhat hairy, but later become bright green and shining above, paler or nearly white beneath. The flowers are small, inconspicuous, in terminal clusters, appearing at all seasons of the year. The wood is dark brown and very durable in contact with the soil. When used as fuel it burns with intense heat. (Fig. 73.)

As a source of honey the black mangrove has attracted more attention than any other tree in Florida. Up to the year of the "big freeze," in 1894, phenomenal yields were reported. As much as 400 pounds of honey from one hive in a single season has been recorded. In those earlier days migratory beekeeping was in practice, and many colonies of bees were moved to the vicinity of Hawks Park from points up and down the coast and from inland localities 50 miles distant. It was hardly possible then to overstock a mangrove section in a favorable season. But the severe winter of 1894 froze and killed the mangrove to the ground. It did not recover from this check for 18 years, and not until 1909 did it again yield nectar, and then only in small quantities. Since that year the bushes have gradually grown in size and the yields have increased also, but as yet they can not even be compared with those preceding 1894.

On the numerous small islands of Indian River and along the east shore of Florida southward from Ormond, there are thousands of acres of black mangrove from six to fifteen feet tall. There are a few beekeepers located in the mangrove swamps of southwestern Florida, but not so many as on the east coast, as at Ariel and near New Smyrna. At Cocoanut Grove, Dade County, a mixture of mangrove and cocoanut honey is secured, which is much lighter than the mangrove honey alone, owing to the cocoanut honey. There are also a few colonies of bees in the vicinity of Everglade, which is about 70 miles south of Fort Myers. This is a promising section, but it is wholly undeveloped and the country is as wild as it was 40 years ago. It is the home of the Seminole Indians, and few white people live there.

At Punta Gorda, on the west coast, black mangrove begins to bloom from May 1 to 15, according to the season, and remains in bloom until July 15 or a little later. When atmospheric conditions are favorable the nectar can be seen in large drops shining in the little cups, and a bee can obtain a load from a single blossom. According to Frank Stirling, of the State Plant Board of Florida, the honey is dark-colored and is used very largely in the manufacture of sweet cakes. On the east coast it is usually blended with the honey from cabbage palmetto, which blooms at the same time, and is in consequence lighter colored. At Punta Gorda, says Ward Lambkin, when there is a heavy flow the honey is light colored but thin and not very sweet, with a salty or brackish taste, as the trees grow on the sand flats, which are often flooded with salt water by the tide. A sample of pure extracted mangrove honey from Punta Gorda has granulated with a fine grain, is yellowish brown in color, not very sweet, and has a poor flavor with an after-taste which is slightly salty.

The secretion of nectar is greatly influenced by the weather. In 1911 near New Smyrna it yielded well early in the season, and the bees left their hives for the

Fig. 74.—Manzanita (Arctostaphylos manzanita). Photographed by Richter.

mangrove swamps almost before dawn, hurrying across the coves of salt water the entire day; but after two weeks the weather suddenly changed and hardly a bee was seen again on the blossoms, although they still continued to open. At Punta Gorda in 1919 the crop of mangrove honey was very small, but in 1918 it probably exceeded 100 pounds per colony. In this same year a beekeeper below Ft. Myers reported the crop a failure.

MANZANITA (*Arctostaphylos manzanita*).—An evergreen shrub or low bushy tree, 3 to 18 feet tall, with wide-spreading, very crooked branches. Associated with other spiny shrubs it forms dense thickets called chaparral. The clustered flowers are small, urn-shaped, white or tinged with pink, and are produced in great profusion. Frosty nights stimulate the flow of nectar and do not injure the flowers. Nectar may often be shaken from the flowers like dew. In Mendocino County, California, there are in the foothills hundreds of acres of manzanita forming an almost impenetrable growth. It begins blooming early in January and the honey flow lasts until the middle of March. Colonies in good condition have built the combs and stored two standard ten-frame supers with honey. At Applegate several colonies packed their hives in February with manzanita honey. The honey is light amber, has a good body, a delicious flavor, and the fragrance of the bloom. While this early flow starts brood-rearing it is followed in localities by freezing weather and an absence of other nectar-secreting flowers, so that it may be all consumed by the bees. Special care is required in order to have colonies strong enough to store a surplus from this source. The field force of bees must be raised during the preceding fall, and well protected during the winter. To prevent loss from cold winds the apiary should be placed in the midst of the manzanita. This is the most common species of manzanita. (Fig. 74.)

Other species of manzanita in California, which are common and yield a similar honey, are the hairy manzanita (*A. tomentosa*) and the big-berried manzanita (*A. glauca*), both abundant in the Coast Ranges.

MAPLE (*Acer*).—The maples bloom so early in the season that their value as honey plants is usually greatly underestimated. In early spring the colonies are so weak that a surplus from this source is seldom obtained, and the maples are regarded as important only for brood-rearing. There are about 100 species in the genus *Acer,* which are confined chiefly to the northern hemisphere. Many of the trees are very common, and the rock maple forms extensive forests. In the states east of the Rocky Mountains a small surplus of maple honey has been reported in Iowa and Alabama. (Fig. 75.)

The red maple (*Acer rubrum*) is a well-known tree in the eastern United States extending from Canada to Georgia and westward to Missouri. The scarlet flowers appear in early spring before the leaves, and yield large quantities of pollen and considerable nectar; but the weather is often so cold and stormy that it prevents the bees from flying freely. (Fig. 76.) In New England and in the region of the Great Lakes the forest is in many sections almost exclusively made up of the rock or sugar maple (*A. saccharum*). The trees are completely covered with yellowish-green, pendulous flowers, which are attractive to great numbers of honeybees. Their contented hum is audible at a long distance. Strong colonies in many localities should store a small surplus from this source. The flowers of the silver maple (*A. saccharinum*) appear in earliest spring in advance of the leaves. As in red maple, the stamens and pistils are in different flowers and usually on different trees. It is widely distributed throughout the eastern states. The box elder, or ash-leaved maple (*A. Negundo*), grows from Manitoba to Texas, but is not found near the coast. The small green flowers appear before the leaves and are a valuable source of nectar. (Fig. 77.)

In Washington and Oregon broadleaf or Oregon maple (*A. macrophyllum*) is an important spring honey and pollen plant blooming in April and May. It is found mainly west of the Cascades and below an elevation of 3500 feet. Vine maple (*A. circinatum*) is a much more important honey plant than broadleaf maple. It grows below an altitude of 5000 feet, mainly west of the Cascades, and blooms a little later than the preceding species. The honey has a fine flavor and is white or

Fig. 75.—Rock or Sugar Maple (Acer saccharum). Photographed by Lovell.

amber-colored with a faint pinkish tinge. The flowers of both of the Oregon trees appear a little later than the leaves.

MARSH FLEABANE (*Pluchea petiolata*).—The marsh fleabane is an erect herb with a simple or branched stem, oblong leaves, and purplish flowers in small heads. It is abundant on swampy or low land in the eastern and southern states. At Moorhead, Minnesota, a surplus of 160 pounds was obtained from this species in 1922 in about two weeks. The honey was bright amber-colored with a disagreeable taste and odor. It granulated quickly after it was extracted. In southern California arrow-weed (*P sericea*) is also a good honey plant. See Arrow-weed.

MARSH FLOWER (*Floerkea Douglasii*).—Also meadow foam. A spreading annual with yellowish green leaves and white flowers, yellowish at base. On low ground in the valleys of the Coast Ranges it colors large patches white and yellow. According to Jepson it is cultivated in England as a honey plant.

MARIGOLD (*Gaillardia pulchella*).—A diffusely branching annual found in dry soil throughout the southwest from Nebraska and Missouri to Texas and Arizona. In Texas marigold is an important honey plant: When there are ample rains in spring it covers the Black Prairies, and about the middle of May it produces a profusion of large showy heads of fragrant yellow flowers. On hundreds of acres in a favorable season the flowers are so abundant that they almost touch each other. But it is seldom seen on sandy or gravelly ridges where oaks grow. It continues in bloom for three weeks, or for a month if there is sufficient moisture in the soil.

The honey is a light transparent yellow in color with a golden-yellow comb. It is seldom obtained entirely pure, as mesquite and horsemint bloom at the same time. The honey is not highly regarded by large commercial buyers. But it is seldom se-

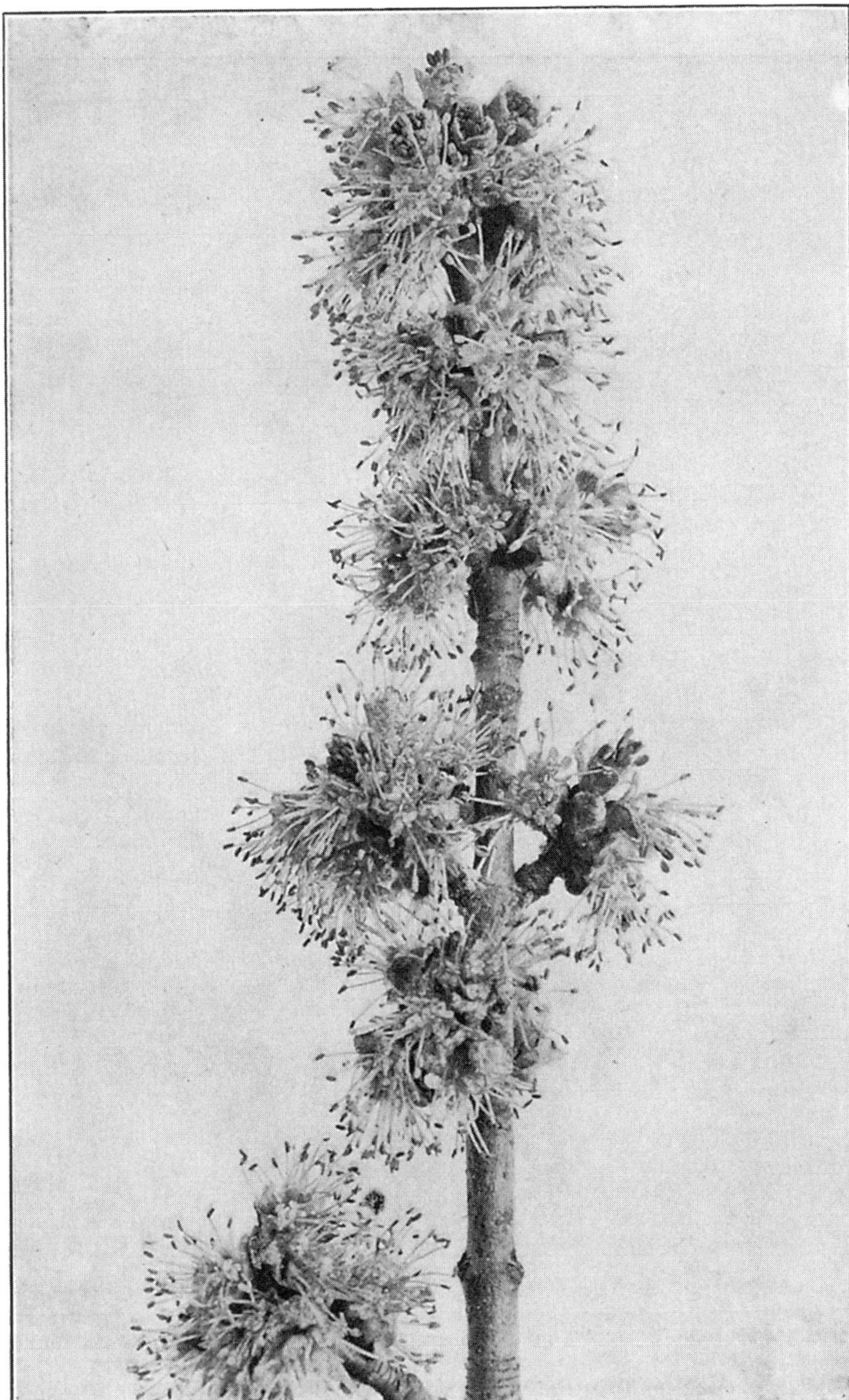

Fig. 76.—Red Maple (Acer rubrum). Staminate flowers. Photographed by Lovell.

Fig. 77.—Red Maple (Acer rubrum). Pistillate flowers. Photographed by Lovell.

cured more than 85 to 90 per cent. pure, and the blend is in ready demand among local buyers. It has a mild flavor, a medium body, and in cool weather granulates almost solidly. In dry seasons the flow is nearly a failure. In 1918 a single colony of bees at Levita stored 150 pounds of marigold honey, but a fair average is about 30 pounds. A hive on scales has recorded a gain of 10 pounds in

Fig. 78.—Marigold (Gaillardia pulchella). Photographed by E. R. Root.

a day when the flow was at its height. A large amount of pollen is also gathered from the flowers. Also called blanket flower. (Fig. 78.)

MATGRASS.—See Carpet Grass.

MARJORAM (*Origanum vulgare*).—A perennial herb, naturalized from Europe, and found in fields and waste land from Ontario to Pennsylvania. Leaves small, entire or crenate, with pink two-lipped flowers in dense terminal clusters. It is so abundant in Bennington County in southwest Vermont that a surplus of 50 pounds per hive has been obtained in August. The color is as light as that of clover honey, but has a pinkish tinge. It is locally called "horsemint." Self-sterile in the absence of insects.

MATRIMONY VINE (*Lycium vulgare*).—A climbing woody vine, escaped from cultivation in the eastern and central states. The greenish purple, funnelform flowers are visited very diligently by bees throughout the entire day.

MAY-WEED (*Anthemis Cotula*).—Dog's camomile, dillweed, dog-fennel. A smooth annual, very common by the roadsides and in waste places all over North

Fig. 79.—May-Weed (Anthemis Cotula). Photographed by Lovell.

America, also widely distributed in Europe, Asia, Africa, and Australia. Flowers yellow with white rays, appearing in summer and autumn, yielding a light-yellow, very bitter honey. Reported of value in the Sacramento Valley, California. (Fig. 79.)

MELILOT.—See Sweet Clover.

MESQUITE (*Prosopis glandulosa*).—This genus includes about 15 species of trees or shrubs growing in the warm arid regions of both the Old and New Worlds, but most abundant in America. In the United States there are two species, *P. glandulosa* and *P. velutina*. Texas mesquite (*P. glandulosa*) is also called algaroba and honey-pod. In the lower Rio Grande Plain it is a large tree attaining a height of 40 feet and a diameter of 2 feet; but on the dryer soil south and southwest of San Antonio there is a vast mesquite forest consisting of trees 10 to 15 feet tall. On arid land the mesquite becomes a straggling shrub with crooked branches. It is found from Kansas to Texas, Arizona, and southern Nevada, and southward to Mexico. (Fig. 80.)

It is probable that mesquite first invaded Texas from Mexico near Matamoras about one hundred and fifty years ago. Until comparatively recent times the Rio Grande Plain was a grassland, but mesquite and various shrubs have spread over it very rapidly. There are many who remember when hundreds of acres now brush-

Fig. 80.—Mesquite (Prosopis glandulosa). Photographed by Louis Scholl.

land were in grassland. At present a scattered growth of mesquite is found over the larger part of Texas except in the northwest corner of the Panhandle or Staked Plains and that portion of eastern Texas east of the lower Brazos River, the Navasota River, and a line extending northward to Hunt County, and thence running westward to Montague County. The heaviest growth of mesquite is in the twelve counties of Atascosa, Bexar, Dimmit, Frio, La Salle, Live Oak, McMullen, Medina, Nueces, San Patricio, Uvalde, and Zavalla, embracing an area of 14,915 square miles southwest of San Antonio. The annual rainfall averages between 20 and 30 inches. The greater portion of this area is covered with a shrublike growth of mesquite, which stools out at the ground into a number of slender, crooked stems. On the higher and dryer land huisache and catsclaw are associated with it. The mesquite trees are only about 10 feet tall, and average less than 2 inches in diameter. Along the streams it is crowded out by elm, ash, hickory, and live oak. Only about 9000 acres, located in San Patricio, Uvalde, and Live Oak counties, are commercially valuable for logging. The logs average only 3 feet straight length and 3 to 4 inches in diameter.

Throughout the central denuded region from Hardeman and Wilbarger counties on the Red River to Valverde County on the Rio Grande mesquite is generally distributed, but throughout its northern range the trees become more dwarfed and the stands more scattering. Short, crooked trunks with long, irregularly curved branches produce scraggly trees, suggestive of long-neglected orchards. The trees have a large taproot which extends to a great depth, and it is not uncommon to find the larger part of the tree under rather than above the ground.

The rapid spread of mesquite is largely due to the abundance of seed, and to its wide dissemination by live stock, but its distribution is determined chiefly by the character of the soil. It can not compete with the native trees, and is forced to occupy low or level areas where the soil is fine and compact. It does not occur on very moist soils along the streams because it is crowded out by the native hardwood trees. It occupies the level areas with fine silty soils, which are less porous, known as "mesquite flats." Large areas of the mesquite prairies have been reclaimed for agricultural purposes, but fortunately the greater part of the land is planted with cotton, which is as valuable as mesquite as a source of honey. In parts of the Trans-Pecos region mesquite is one of the surplus honey plants, although the main dependence of the beekeeper is irrigated alfalfa.

The leaves are bipinnate, composed of two branches or pinnae, each of which bears from 6 to 25 pairs of narrow leaflets. At the point of union of the two branches of the leaf there is a gland. The small fragrant flowers are in yellowish cylindric spikes, 3 to 5 inches long. There are 5 sepals, 5 petals, and 10 stamens. The fruit is a round pod, 6 to 8 inches in length, which is greedily eaten by cattle. Even human beings find that these beans have a good flavor, and children particularly relish them each season as they ripen. They vary considerably in taste, however, some being so bitter that they can not be eaten, while others are very sweet and agreeable in flavor. Their chief value is as food for cattle.

Mesquite has usually two separate and distinct blooming periods during the year, although in some seasons there is no interval. The first comes during April, and the second during the last of June or in July. These periods are sometimes a week or more earlier or later, according to the season, the occurrence of cold weather, and the rainfall of the preceding fall and winter. If rain has been abundant during the winter, no matter how dry the following spring, there will be a profusion of bloom and a heavy flow of nectar. The long tap root penetrates the soil to a great depth and is thus able to obtain water, which is beyond the reach of many other shrubs and trees. (Fig. 81.)

According to H. B. Parks mesquite may bloom at any time from late spring to early autumn. The number of bloomings is governed by the number of rainfalls, and may be as many as five or six. One inch and over of rain will cause a new blooming, but a very small rainfall, as one-eighth of an inch, will cause the bloom to fall. There is an interval of about five weeks between the falling of the rain and the opening of the induced bloom. A rainy December and January and a dry March and April are favorable to a honey flow in the spring, while a wet spring and a dry June and July favor a summer honey flow.

From 25 to 100 pounds of honey per colony are stored from the bloom, according to the locality and weather conditions. The honey is light amber in color and of good quality. It is considered a better table honey than any other of the Texas honeys, since one does not tire of it as quickly as in the case of a honey with a more pronounced flavor. The honey, although ranked very high in Texas, would in the North probably be classed with the amber honeys. Nectar secretion is more reliable on light sandy soils than on heavy land.

In New Mexico, in the valley of the Rio Grande River, beekeepers formerly depended on mesquite and other desert plants, but now pay attention only to al-

Fig. 81.—Algaroba tree. Mesquite (Prosopis juliflora). Algaroba trees growing near the sea on the lee shore of the Island of Oahu, Hawaiian Islands. Photographed by Leslie Burr.

falfa and sweet clover. At La Mesa in some seasons the mesquite flow is excellent, but it is often a failure. Texas mesquite is also valuable in Otero County, N. M., and in southern Nevada.

Arizona Mesquite (*Prosopis velutina*). This tree is the largest of the mesquites, often attaining a height of 45 feet or more. It grows in the hot dry deserts of southern Arizona, southern California, and Sonora. The leaves, flowers, and pods are very similar to those of Texas mesquite. Arizona mesquite blooms for the first time in May, and again in July. On the arid deserts it is often the chief dependence of the beekeeper. At Wellton, Arizona, it is the most reliable honey plant for surplus. At Polonias, in the eastern part of Yuma County, there is an apiary of 65 colonies which stores about a 60-pound can of honey per colony from mesquite. In mid-summer it becomes very hot and dry on the deserts; and, as there are no other honey plants and no honey flow of any kind, the bees must depend on the stores gathered earlier in the season or die of starvation. Along the Colorado River in the eastern portion of Imperial and Riverside counties, California, there is an extensive growth of mesquite which yields a large honey flow. The Liguanea Plain on the south side of the Island of Jamaica is largely covered with another species of mesquite (*Prosopis juliflora*).

MESQUITE IN THE HAWAIIAN ISLANDS

In the Hawaiian Islands the mesquite (*P. juliflora*) is not only the chief but almost the only source of floral honey. Here it is called algaroba, or in the native language *keawe.* The history of honey plants offers no more interesting chapter than that describing the introduction of this tree and its rapid increase, until to-day it yields annually large quantities of algaroba honey and has rendered beekeeping profitable in sections of the islands where previously little honey was stored. In earlier times the apiaries seldom exceeded 50 colonies in number, and were located near forest trees growing in the mountains, which yielded comparatively little nectar.

About 1828 the seed was brought from the Royal Gardens of Paris by Father Bachelot, founder of the Roman Catholic Mission; and until a few years ago a tree raised from this seed was still standing on Fort Street in Honolulu. Once introduced, the mesquite increased with remarkable rapidity. It thrives from the level of the sea-coast, where the spray of the waves falls upon the foliage, up to an altitude of 2000 feet; but it succeeds best at a slight elevation in the semi-arid climate. The algaroba forests are confined chiefly to the lee or western side of the islands. The reason for this is that the windward, or eastern side, is exposed to trade winds, which blow with few exceptions during the entire year. As a result of these winds the climate on one side of the islands is entirely different from that on the other side, even in the case of an island that is only a few miles across. On the windward side there is a heavy rainfall, in some places in excess of 200 inches annually, and at times it may reach 400. On the lee side there is less rain, or the climate may be so dry that the land is little better than a barren desert. Where there is much rain the mesquite is entirely absent, or does not thrive.

On the western side of the islands there are vast forests of algaroba trees, covering thousands of acres of land. In the island of Oahu alone there are not far from 17,000 acres. Cattle are continually disseminating the seed, and the number of trees is also largely increased by systematic planting. Prior to October of the year 1916 there were planted over 100,000 trees in Oahu. It is estimated that a tree with a 30-foot spread of branches will produce 2½ pounds of honey in a normal year. One strip of algaroba forest in Molokai supports nearly 2000 colonies of bees. Of the 600 tons of honey produced in the Hawaiian Islands, more than 200 tons come from the flowers of the algaroba. The trees begin to bloom when they are from four to six years old. There are two periods of blooming: the first period begins in March or later, according to the locality of the island, and lasts until August; in Hawaii the second period of blooming ends about the first of October.

The honey is water-white, about as thick as that from white clover, and possesses an agreeable though peculiar flavor; but it is suitable for a table honey. It granulates soon after it is gathered. Honey which has granulated in the combs is placed in huge solar extractors, which will hold several hundred combs at a time. The sun's heat liquefies the honey without darkening it; and it also melts most of the wax, which is extracted from the "slum gum" by the usual methods.

The trees grow rapidly and attain a height of 45 to 50 feet with a diameter of 2 feet or more. The flowers and pods are similar to those of the Texas mesquite. The yellow pods are eagerly eaten by cattle, and the crop in Oahu is estimated at 25,000 tons.

Fig. 82.—Mexican Clover (Richardia scabra). Photographed by E. R. Root.

MEXICAN CLOVER (*Richardia scabra*).—Also called Florida clover and Spanish clover. An annual herb which springs up in fields as soon as cultivation stops. It also grows along railroad tracks and public roads, but it is often found in pastures in southern Alabama. It blooms from May until late in the season, or in Georgia until frost. In Volusia County, Florida, it fills in the gap between orange bloom and partridge-pea. The honey is light amber in color, and has a peculiar tart flavor. Mexican clover grows in sandy soil in the Gulf States from Florida to Texas, and also in Mexico and South America. (Fig. 82.)

A smaller species (*Richardia braziliensis*) is found in the pine woods as well as in the fields of Florida. While the bees visit it more or less, it does not secrete nectar as well as the larger species. Both species belong to the madder family (Rubiaceae) and are not allied to the true clovers (*Trifolium*).

MIGNONETTE (*Reseda odorata*).—Honeybees visit frequently the very fragrant flowers of the garden mignonette both for nectar and pollen. This genus contains about 55 species, but they are all natives of the eastern continent. Besides the garden mignonette, three species have been introduced into the United States and grow in waste places. If more common they would be valuable. (Fig. 83.) In California mignonette has been reported to be visited by thousands of bees, and to be a valuable honey plant giving a great abundance of flowers and a very long period of bloom. The honey has a most pleasant odor and flavor. Tested at the Michigan Agricultural Station it proved a failure, as it was visited by very few bees. In Maine likewise it is not attractive to bees.

MINT (*Mentha spicata*).—Spearmint. In Sacramento County, California, it yields an amber-colored honey. Naturalized from Europe and common throughout eastern North America. Small, odorous, purplish flowers in close clusters or whorls. The nectar is sheltered completely by a ring of hairs; but, as the floral tube is short, it can be gathered by flies as well as bees.

There are many other species of mint, as peppermint (*M. piperita*), water mint, and field mint, all of which are of more or less value to the beekeeper. Peppermint was formerly cultivated chiefly in Massachusetts, New York, and Ohio; but more than 2000 acres are now grown in southwest Michigan. The oil is used in flavoring candy, and in medicine as a carminative.

MILKWEED (*Asclepias syriaca*).—The extracted honey is so light in color that it is usually classed as white, but it not unfrequently has a yellowish tinge. The cappings of the comb honey are pearly white. The flavor is excellent with a slight tang, but it becomes milder with age. In hot dry weather it is so thick and heavy that it can not be extracted until the combs are warmed. In most localities it is mixed with the honey of other flowers blooming at the same time.

In Michigan the honey flow lasts for about thirty days, between July 1 and August 15, varying somewhat with the season. The nectar is secreted very rapidly, and a large colony has gathered from 13 to 17 pounds in a single day. An average of 11 pounds for 10 successive days has been obtained. At Bellaire an apiary will average 50 pounds per colony annually. The pollen grains are bound together in waxy masses called pollinia, and are useless to bees.

As a honey plant milkweed is of most importance in the northern part of the Southern Peninsula of Michigan, in the counties of Emmet, Cheboygan, Charlevoix, Antrim, and Grand Traverse. It grows on any kind of soil from white shore sand to heavy clay; but, as with clover, the clay soil gives the most nectar. In localities it has taken almost complete possession of the land, and is steadily spreading over a larger area. It is regarded as a noxious weed; but it is difficult to eradicate, as the roots go down into the soil for a depth of six to ten inches. The chief honey plants of this section of Michigan are fireweed, raspberry, and milkweed. After the forests have been lumbered, large areas are soon covered with dense thickets of raspberries; and in sections which have been burned over there springs up a rank growth of fireweed or willow-herb. But in a few years these plants become less vigorous and other forms of vegetation begin to take their place; thus the time must come when the raspberry and the fireweed will no longer be the chief reliance

Fig. 83.—Mignonette (Resedà odorata). Photographed by Lovell.

of northern Michigan. On the other hand, milkweed is annually increasing in abundance, and bee culture here is likely to be largely dependent on it in the future. Swamp milkweed (*A. incarnata*) is abundant in swamps throughout the northern states and extends southward to Tennessee and Louisiana. It yields a large amount of honey. In California two species of milkweed (*A. mexicana* and *A. speciosa*) are of immense value.

The common milkweed, or silkweed, is a stout plant with milky juice, 3 to 5 feet tall, with oblong opposite leaves, which are very hairy beneath but smooth above. The small brown-purple flowers are in terminal or lateral clusters, called

Fig. 84.—Common Milkweed (Asclepias syriaca). Photographed by Lovell.

umbels. Their structure is very intricate, rivalling that of the orchids. The fruit is a pair of pods containing many flat seeds, tipped with tufts of long silky hair or down. (Fig. 84.)

Milkweed flowers are called pinch-trap flowers because they possess a remarkable clip-mechanism found in no other family of plants. Two club-shaped masses of pollen are attached by flexible bands to a small, dry, triangular disc placed midway between them. In this membraneous disc there is a wedge-shaped slit at one end. In its efforts to obtain a foothold on the smooth flowers an insect is likely to thrust a claw, leg, antenna, or tongue into one of the slits. If one of these organs is drawn upward in the slit, the dry disc becomes tightly clamped to it. When the

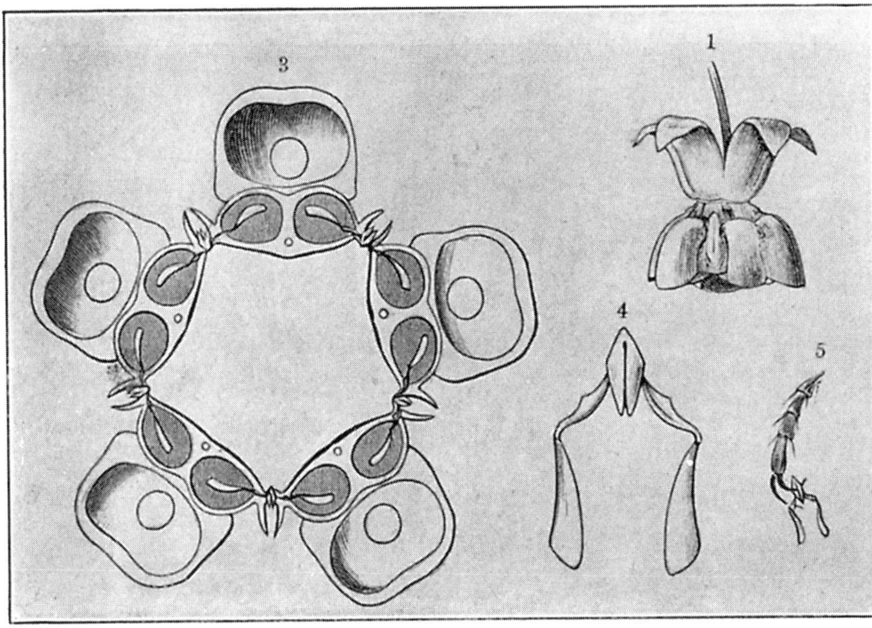

Fig. 85.—Clip mechanism of milkweed flower. 1. Flower of common milkweed (Asclepias syriaca) seen from the side. 3. Transverse section through the flower (enlarged). 4. Corpusculum (clip) with two pollinia (enlarged). 5. Foot of an insect with pollinia fastened to it by the clip. (After Kerner.)

insect flies away it carries with it the disc and the two masses of pollen strapped to it. Exposed to the air, the strap-like stalks dry and draw the pollinia close together. As the bee alights on another flower, they are easily thrust between two anther wings, where they come in contact with the stigma; but, once inserted and pulled upward, they can not again be withdrawn. The insect can obtain its liberty only by breaking the connecting bands. If it can not do this, it perishes slowly of starvation. Disc after disc may thus become attached to an insect, until it is crippled or helpless. (Fig. 85.)

One season an English beekeeper lost thousands of his bees from the effects of strings of these clips, but in America not much loss has been reported. Bees thus afflicted are sometimes expelled from the hive. Not a season passes that inquiries are not received from beekeepers requesting information in regard to these peculiar appendages; and many different explanations of them have been given by persons not familiar with the flowers of the milkweed. Some regard them as a fungus, others as a protuberance growing on the bee's leg, and still others as a winged insect-enemy of the bee. The dry membranes are often described as sticky or glutinous glands, but this is never the case.

MISTLETOE (*Phoradendron flavescens*).—A leafy yellowish-green shrub parasitic on deciduous-leaved trees, as the red maple, tupelo, oak, apple, mesquite, and elm. The small greenish flowers are sunk in the joints of the short, jointed spikes. The stamens and pistils are in different flowers, which are borne on different plants. The flowers have an orange odor, and nectar is secreted by a nectary which forms a ring at the base of the calyx on the inner side. Both kinds of flowers secrete nectar; but the pistillate yield the greater amount and are sometimes filled to the tips of the calyx lobes. Under certain conditions the flowers are nectarless. There are only three anthers, which produce a small amount of pollen which is thickly beset with short spines. The tongue of a bee probing for nectar becomes covered with pollen, which is carried away to another flower. The insects active in pollination are flies and bees. In Texas mistletoe blooms in February and March

Fig. 86.—Motherwort (Leonurus Cardiaca). Photographed by Lovell.

Fig. 87.—Sheep Laurel (Kalmia angustifolia). Photographed by Lovell.

and is much visited by honeybees for both pollen and nectar. In 1923 a small surplus of honey was reported to have been stored from this source. Honeybees sometimes visit only the staminate flowers and fail to fly to the smaller pistillate blossoms, which of course contain no pollen. Common in the southern states.

MOTHERWORT (*Leonurus Cardiaca*).—An erect perennial herb, belonging to the mint family, with pale purplish, two-lipped flowers in dense clusters in the axils of the leaves, common in waste places around dwellings. Honeybees and bumblebees frequently gather nectar, but seldom pollen, from the bloom. (Fig. 86.)

MOUNTAIN LAUREL (*Kalmia latifolia*).—Calico bush. Poison laurel. Wood laurel. Mountain ivy. One of the handsomest of North American shrubs, common from Massachusetts to Florida, but especially abundant in western North Carolina, where on the mountain slopes it often presents an unbroken sheet of bloom. In localities in this section a poisonous honey is gathered year after year. Hundreds of pounds of this poisonous honey have been thrown into the streams, and in many instances beekeepers have retired from the business. When eaten freely it produces acute nausea, which lasts for several hours. The bees themselves apparently suffer no ill effects from eating the honey. The honey is commonly reported to be gathered from the flowers of the mountain laurel, although other genera of the heath family, as *Leucothoe* and *Andromeda*, are believed to secrete poisonous nectar. The source of this honey is open to great doubt, as a number of beekeepers have reported that they have never seen a honeybee on the flowers of mountain laurel, or very rarely.

Kalm, the Swedish traveler, after whom the genus *Kalmia* is named, says that if domestic animals eat the leaves they fall sick or die, but that they are harmless to wild animals. The belief that the leaves are poisonous seems to have extended to the honey. But Dr. Bigelow states in his Medical Botany that he has repeatedly chewed and swallowed a green leaf of the largest size, without perceiving the least ill effect in consequence. The taste of the leaves is mild and mucilaginous. Bigelow believed that the noxious effect of the leaves on young domestic animals was due to their indigestible quality. Sheep laurel, lambkill, or calfkill (*K. angustifolia*), common in New England, has been supposed to be a noxious shrub, but apparently on insufficient grounds. (Fig. 87.)

MOUNTAIN LILAC (*Ceanothus*).—Many species of these free-blooming shrubs occur in the Coast Ranges and in the Sierra Nevada, where with manzanita, pea chaparral, and scrub oak they often form extensive and almost impenetrable thickets on the foothills and higher slopes. The small white, pink, or blue flowers are in clusters and are sought by many insects for both pollen and nectar. Snow brush (*C. velutinus*), so-called from its numerous white flowers, has a very sweet odor and yields a delicious white honey. It blooms in May and June. Blue blossom (*C. thyrsiflorus*) is very abundant on logged redwood lands. The blue flowers open from February to April and yield a white honey which is chiefly valuable for brood-rearing. Deer brush (*C. integerrimus*) is also a common species blooming in July and August, yielding an amber-colored honey of good flavor. Mahala mats (*C. prostrata*) is a low shrub with prostrate spreading branches, thickly matting the ground on the mountain slopes. The blue flowers yield an excellent white honey in May. All of the species are honey plants of more or less importance. See Ceanothus.

MOUNTAIN MINT.—See Basil.

MUSKMELON (*Cucumis Melo*).—The staminate flowers open first and are more numerous than the pistillate. Valuable for pollen. Muskmelons are grown chiefly in New Jersey, Delaware, Florida, Georgia, Indiana, Illinois, Colorado, and the Imperial Valley, California. In the absence of bees no fruit is produced.

MUSTANG.—See Blue Curls.

MUSTANG MINT (*Monardella lanceolata*).—A fragrant herb common in the foothills of the Sierra Nevada. The rose-colored flowers yield a moderate amount of nectar.

Fig. 88.—Black Mustard (Brassica nigra). Photographed by Lovell.

MUSTARD (*Brassica*).—Mustard, turnip, rape, cabbage, and charlock all belong to the same genus. Charlock (*B. arvensis*), naturalized from Europe, is often very abundant in grain fields, presenting an unbroken expanse of yellow. Almost equally common is black mustard (*B. nigra*), common everywhere in waste places. Both the black and the white mustard (*B. alba*) are extensively cultivated for their pungent seeds. In the Lompoc Valley, California, mustard seed is an important crop, and in a single year the farmers have realized a quarter of a million dollars from its sale. The average yield per acre is from 800 to 1000 pounds of seed. The period of bloom lasts for nearly a month, and when the sowings are made at different intervals it can be prolonged for a period of ten weeks. The honey is light in color, but the flavor is generally regarded as inferior in the West. Although not as heavy-bodied as alfalfa, it has the same tendency to candy quickly. It may

Fig. 89.—Common Yellow Mustard (Brassica campestris). Photographed by Richter.

granulate in the tank in four or five days, but this may be prevented by using a tank with a glass top exposing the honey to sunlight. Honey gathered from wild black mustard is stated to be amber-colored with an odor characteristic of the plant. In 1884 M. H. Mendleson, of Ventura, California, had one colony which gathered from an abundance of mustard bloom, while 199 gathered from the sages. This was an exceptional case, and would indicate that the bees preferred the nectar from sage. (Figs. 88 and 89.)

NAPA THISTLE (*Centaurea melitensis*).—Tocalote. First introduced at Napa, California, it has become common in grain fields and arable land throughout the state. The yellow flowers appear during the last of May, and in Sacramento County yield a light amber honey of good quality.

NETTLE, HEDGE.—See Hedge Nettle.

OLIVE.—See Pollen Plants.

OAK.—See Pollen Plants.

ONION (*Allium Cepa*).—Valuable for honey when grown for seed. In California most of the onion seed is raised in the Santa Clara and Sacramento Valleys and in San Benito County. The honey is amber-colored, and the peculiar onion odor and flavor almost disappear as it ripens. The flowers are white.

ORANGE (*Citrus Aurantium*).—Orange honey is nearly white or light yellow in color, has the fragrance of the bloom, and a most pleasant flavor. In Florida it is not as heavy as the honey from scrub palmetto, and it granulates in a few months after extracting. It is considered one of the finest honeys produced in the United States, but in Florida it is seldom obtained pure. In California it is very easy to obtain orange honey unmixed with any other honey, as sage ordinarily does not blossom until the orange trees have nearly ceased to bloom. The nectar is frequently very thin when first gathered, and naturally is thinner for a few days after irrigation; but toward the end of the flow, if the weather is warm, it becomes much thicker. When well ripened the honey is heavy, white, and excellent in flavor; but, as it usually granulates in a few months, many dealers prefer to buy sage honey. While orange bloom yields nectar in Florida, a greater amount per tree is secreted in California. The climate of California is more favorable for orange nectar secretion.

The commercial citrus fruits include the orange, lemon, grapefruit, and lime. Large areas are under cultivation in southern Europe, Asia, North America, and in the tropics. The trees flourish on a great variety of soils, providing they are well drained and the region is free from severe frosts. In the United States the

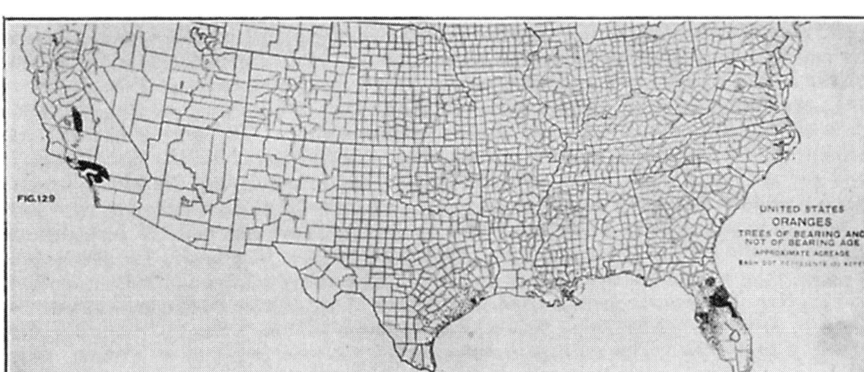

Fig. 90.—Acreage of citrus fruits in the United States. California has nearly two-thirds of the trees and produced in 1909 about three-fourths of the crop. (After Baker. Geog. World's Agr.)

commercial orange crop is confined to Florida and California, but there is a belt of orchards along the coast of the Gulf of Mexico from Florida to Texas, and also an area in southern Arizona.

In northwestern Florida, west of the Suwannee River, and northeastern Florida north of St. Augustine only a very small area of citrus fruits is under cultivation. In the central lake region, comprising the eight counties of Alachua, Marion, Putnam, St. John, Volusia, Orange, Lake, and Sumter, there are over a million and a half of trees. Along the west coast in Citrus, Hernando, Pasco, Hillsboro, and Manatee counties there are also numerous orange groves, Hillsboro County reporting a larger acreage than any other county in the state. Farther south Polk, Brevard, and DeSoto counties each contain over 200,000 trees. Lee County, the seat of the Big Cypress Swamp, commonly regarded as a wilderness, has also many orange groves. Lemons are produced principally in southern California, but there is a small acreage chiefly in Monroe County in the southwest corner of Florida. The grapefruit orchards are found in western Florida, in Lee, Manatee, and Hillsboro counties. Limes are grown commercially only in the southern tip of Florida (Monroe County) and the adjoining islands. It has been estimated that there are in this state 10,000 square miles adapted to orange culture (Fig. 90.)

Fig. 91.—Orange blossoms.

The date of the beginning and the length of the period of blooming vary greatly in Florida according to the variety of the fruit, the extent of cool weather in winter and early spring, and differences in rainfall and soil. For instance, in the middle of the state flowers have been seen on the round orange as early as Feb. 6; but in 1912 the first bloom did not appear until March 15. On an average Feb. 20 is the date on which blossoms begin to appear in this section. The spring of 1915 was unusually cold and rainy, and the flowers did not open until about the first of March, and did not yield nectar well until the end of the month, when the bloom became very abundant. Usually nectar is not gathered later than April 10; but this year a hive on scales showed a gain of three pounds on April 20, and the flow did not cease entirely until a few days later. The trees remain in blossom for about four weeks, if the weather is not too hot and dry. As a rule the later the bloom appears, the shorter the time it lasts. Cool and frosty weather will prolong it unless the frost is so severe, as in 1911, that it injures the blossoms, when it brings the flow speedily to a close. The average surplus in a good year is about 40 pounds. (Fig. 91.)

A large acreage of oranges, mostly of the Satsuma variety, has been planted along the coast of the Gulf of Mexico from Florida to Texas, but the bloom does not yield much nectar. About 40 miles below New Orleans, La., in the Delta of the Mississippi River in Plaquemines and Jefferson counties, there are many miles of almost continuous orange groves. The first orange trees were planted in this section more than one hundred years ago. In Texas the orange orchards are largely restricted to Galveston and Brazoria counties on the Gulf Coast, but citrus-growing is in course of development in the Rio Grande Valley.

In the Salt River Valley in Maricopa County, Arizona, where there are 2500 acres of groves, orange culture is an important industry, as it is also in the Imperial Valley, California. The orchards are confined to the slopes, which are free from orange-killing frosts. Oranges in this section ripen early, and the first shipments often reach the eastern markets in time for the Thanksgiving trade. The culture of the orange and grapefruit will expand considerably in those parts of the Salt River Valley where winter temperatures permit and there is an average water supply.

The orange was introduced into California by the early Catholic missionaries, but its cultivation on a commercial scale began 45 years ago. To-day the state has two-thirds of the trees and produces three-fourths of the crop in the United States, devoting 234,600 acres of its fertile soil to growing oranges and lemons.

As with many other honey plants, the secretion of nectar varies in different localities and is greatly influenced by weather conditions. In the cool regions near the coast there is little nectar. Fog also often interferes with the flight of the bees so that there may be very few days which are ideal for field work. In the foothills it is occasionally very cold. An apiary at an elevation of a few hundred feet has been snowed under for a few hours, while in the valley below the orange trees were also white — but with flowers, not snow. At Redlands the weather is very warm and there is little fog, with the result that in four years out of five orange bloom yields a fair crop, in proof of which may be cited the experience of a beekeeper who states that he has shipped one or more carloads of pure orange honey every year except 1904. Even here, when the weather is cool, very little nectar is gathered. But when the conditions are suitable there is probably no other plant in the United States which secretes nectar more copiously. At times the clothing of pickers and pruners is wet by the dripping nectar, the horses and harness require washing at the close of the day's cultivation among the trees, and even the ground is dampened by the many falling drops. In southern California the trees begin to bloom the last of March, or early in April, and the blooming period lasts until the middle of May. The flowers of the navel orange open first, followed by those of Sweets, Valencias, and Seedlings. It would be an advantage if the honey flow were later, for the weather is sometimes so cold that tons of nectar are lost because the bees are forced to remain in the hives. The colonies are also not sufficiently strong to bring in all the nectar. With large colonies and clear warm weather it comes in very rapidly. At Redlands a hive on scales showed a gain of 119 pounds in 17 days from April 7 to 23. The honey was secured in about five hours of each day from 11 to 4 o'clock. During the morning the bees brought in pollen from various flowers, but before noon they were all at work on the orange bloom. A surplus of from 60 to 120 pounds per colony has been obtained. At Pomona the land for miles is entirely occupied by groves, and it is difficult to obtain room for an apiary. Here, after the flow from orange is over, the bees bring in nothing for the rest of the season except a dribble of dark honey from pepper and horehound.

ORANGE HAWKWEED (*Hieracium aurantiacum*).—Devil's paint-brush. Introduced from Europe about 1875, this plant has spread from Quebec to Pennsylvania. In localities it completely covers the fields. Commonly bees do not visit it to any great extent, but in Aroostook County, Maine, in some seasons, when there is sufficient moisture and a bright hot sun, it yields well. The honey is light-greenish yellow. (Fig. 92.)

OREGON GRAPE (*Berberis aquifolium*).—A low, trailing shrub with yellow flowers and blue or purple berries resembling grapes. It is listed in Oregon as a minor honey plant. Barberry.

OREOCARYA.—Coarse rough-hairy plants with white or yellow flowers, of

Fig. 92.—Orange Hawkweed (Hieracium aurantiacum). Photographed by Lovell.

which there are nearly a score of species growing on dry hills and sandy plains in Colorado and Wyoming. According to Pellett, in Colorado they yield nectar freely and an average of 40 pounds per colony has been obtained from this source. The honey is light amber-colored and poor flavored. *O. virgata* grows on the eastern slopes of the Rocky Mountains. A minor honey plant.

PALM.—The palm family (*Palmaceae*) is represented in Florida by 15 species, not including the cultivated date palm. There are in the world 130 genera and over 1000 species of palms which are nearly equally divided between the tropics of both hemispheres. Palm trees, with their slender, unbranched columnar trunks surmounted with a crown of immense fernlike leaves, are among the most stately and graceful of trees, and occupy an important place in both story and history. The individual flowers are small, stemless, and usually white or greenish-colored; but they are borne in enormous branched spikes or flower-clusters. The largest flower-cluster in the world, which is 40 feet in length, is produced by the Talipot palm of Ceylon. The natives of the tropics utilize every portion of the palm for food, wine, clothing, medicine, and dwellings. A part of the palms are pollinated by the wind and a part by insects, although in some species both methods occur. In Florida the cabbage, scrub, and saw palmettos, and the royal palm are valuable sources of nectar. The cocoanut palm is reported to yield a surplus, but further observation is desirable. The date palm is wind-pollinated and the flowers are nectarless, but honeybees often in large numbers gather the pollen.

The cocoanut palm (*Cocos nucifera*) grows in southern Florida, and throughout the West Indies and the tropical regions of both worlds. It has been reported to yield an amber-colored honey with a flavor resembling horehound honey. On Key Biscayne, on the east coast of Florida, the cocoanut is said to be practically the only source of honey, but in Porto Rico it is not considered a good honey plant. When the stalks of the great flower-cluster, 3 to 6 feet long, are wounded, a sweet sap flows freely, which in the East Indies is collected and evaporated into a crude sugar.

The royal palm (*Oreodoxa regia*) grows in southern Florida, Cuba, and Porto Rico, is pollinated by insects and is nectariferous. (Fig. 93.) The tree has no regular time of blooming, but the flowers appear at intervals throughout the year, and there may be fruit of four different ages on the trees at one time. The tough buds open with a sharp cracking sound, exposing the clusters of flowers, which are 3 or 4 feet long and consist of hundreds of yellowish-white, strong-scented blossoms. It is not usual to get a surplus from this palm, although once in a while a strong colony will store a pound a day; but it is a valuable honey plant since it yields during the summer when there is no other honey. In localities where the royal palm covers large areas a small amount of honey may be placed on the market. The honey is light amber, very thin, and has a strong flavor. (*Roystonea regia.*)

PALMETTO, CABBAGE (*Sabal Palmetto*).—The cabbage palmetto, so called from the cabbagelike terminal bud, which is boiled and eaten like a cabbage, is found in the sandy coast regions from North Carolina to Florida, and also occurs in Cuba and the Bahamas. It grows from 20 to 50 feet tall, and is abundant along the east and west coasts, on the banks of rivers, and in hammocks throughout southern Florida. The erect trunk is gray-colored and bears a crown of fan-shaped leaves, about 5 feet in length and almost equally broad. The flowers have three sepals, three petals, six stamens, and a three-celled ovary. It is a picturesque tree and is widely planted for ornament. The drooping flower-cluster, which is three or more feet in length, consists of a central, much branched axis, bearing on the ultimate smaller branches hundreds of small, white, stalkless flowers. They exhale a strong fragrance as pronounced as that of apple blossom. (Fig. 94.)

In the extreme southern part of Florida the cabbage palmetto begins to bloom about the first of July, but in the northern portion of the state not until August. The flowers are very sensitive to the weather: too much dampness blights, and a dry hot atmosphere blasts the bloom. According to Baldwin it is on an average a good yielder only one year in three—for example, 1907, 1909, and 1912. In a

Fig. 93.—Royal Palm (Oreodoxa regia). The royal palm is an important honey plant in Porto Rico and Cuba, where it forms large forests, and in southern Florida. It blooms throughout the year and yields nectar when there are no other sources of honey.

good year it secretes nectar very freely, and on the St. Lucie River 65 colonies gathered 3500 pounds of extracted honey in two weeks. (Fig. 95.)

The honey is nearly white, or light amber-colored, and has a characteristic aroma, which does not resemble at all that of scrub palmetto. It is very thin, and in warm weather runs almost like water; and even in cold weather it never thickens. The flavor is extremely mild, but it is inferior to that of scrub palmetto. Gas bubbles may frequently be seen under the cappings of the sealed cells, and during extracting the honey foams considerably, as though it were fermenting, but after it has stood for a few days the bubbles wholly disappear. But honey from unsealed cells will ferment enough to deprive it of its flavor. As it is a mild honey it blends well with other honeys. In the vicinity of Hawks Park, Fla., it blooms almost simultaneously with mangrove, so that the two honeys are always secured together. Farther south they are obtained separately. (Fig. 95.)

PALMETTO, SAW (*Serenoa serrulata*).—This species, which closely resembles the scrub palmetto in flower and fruit, also gives a large honey flow. The honey is similar to that of the scrub palmetto, with which it is usually mixed, as it blooms at the same time. The saw palmetto has a wider distribution, extending from North Carolina to Florida and Texas, and becoming a small tree in southern Florida. Unlike the scrub palmetto, the edges of the leaf-stalks are armed with rigid, spinelike teeth, and the margins of the leaves have no threadlike filaments. No doubt the two species are often confused.

PALMETTO, SCRUB (*Sabal megacarpa*).—A low shrub with long, creeping, crooked stems, which are partly subterranean. At intervals the stems root, and send up clusters of light-green, fan-shaped leaves four to seven feet tall. The

Fig. 94.—Cabbage Palmetto (Sabal palmetto). Tree in bloom.

foliage is not injured by frost, and, when burned to the ground, a new growth requires only a year. The scrub palmetto grows well over the southern two-thirds of the peninsula of Florida, becoming rarer and smaller toward the northern boundary of the state. It reaches the largest size south of a line extending from Tampa to the east coast. On the west coast north and south of Tampa it forms an unbroken sea of green. The traveler on the Seaboard Airline Railroad may ride for miles without losing sight of the scrub palmetto, which offers an impressive appearance in such large masses. It grows over all of the "flatwoods," or low pine lands, which overflow more or less during the rainy season, also along all the water courses, and on the edges of the heavy hammocks. The latter places are the most suitable, for there these shrubs grow 8 feet or more tall and yield the most honey. The stunted plants in the flatwoods do not yield as well. There are still large areas of scrub-palmetto hammocks and flatwoods as yet unoccupied by beemen, which offer an attractive field for bee culture. An objection to these localities is that the palmetto is usually the only nectar-secreting plant in the region, and colonies must be watched closely or they will run out of stores in seasons when it does not bloom.

Fig. 95.—Cabbage Palmetto (Sabal Palmetto). Flower-cluster. (Florida Photo. Concern.)

But year in and year out there is no more reliable honey plant in Florida than the scrub palmetto. (Fig. 96.)

It begins to bloom at Fort Myers and Miami in April, and farther north in May. The small, white, stalkless, fragrant flowers are borne in a great, many-branched flower-cluster, which is 2 to 4 feet in length. Too much rain during the

Fig. 96.—Scrub Palmetto (Sabal megacarpa.) (Florida Photo. Concern.)

blooming-period causes the flowers to mildew, and too much heat, to wither. In the flatwoods the plants are often stunted and do not yield as well as where they grow larger. When both soil and weather conditions are favorable there is a profuse secretion of nectar. From 6 to 8 pounds daily per hive have been reported from this source, and 8 or 9 are not unknown; an average of 100 pounds per colony has been secured. The honey is lemon yellow, thick and heavy, with an aromatic flavor and fragrance. It is considered one of the finest honeys of Florida, but possibly is surpassed by tupelo honey. It granulates, but not as quickly as orange honey. (Fig. 97.)

PALO VERDE (*Parkinsonia torreyana*).—Also called green-barked acacia. A small tree growing in the desert regions of Arizona, southern California, and Lower California, with bright-green branches and large, showy golden flowers. It blooms in spring and is abundant on the hills of Arizona, where it is attractive to bees, and is listed as a honey plant. When the water supply fails the trees drop their leaves and depend on the green bark for the making of food. If there are autumn rains a new crop of leaves may unfold.

PARSNIP (*Pastinaca sativa*).—A common weed in eastern North America, growing by the roadsides and in waste places. The yellow flowers are in large

Fig. 97.—Scrub Palmetto (Sabal megacarpa). In bud. Photographed by Lovell.

flat-topped clusters or umbels. The nectar is unprotected and the bloom is visited by a great number of insects, including honeybees. Where abundant, this plant should prove of value. Near the great truck gardens in the Sacramento River Valley where large areas of parsnip and celery are grown for seed, hives five or six stories high are packed with honey from these plants. The honey is light-colored and medium in quality.

PARTRIDGE-PEA (*Cassia Chamaechrista*).—Also called sensitive pea. The genus *Cassia* contains more than 275 species, which are very abundant in tropical America. It is of interest to note that the flowers of this great genus are both nectarless and odorless. They are pollen-flowers. While all the species may be cross-

Fig. 98.—Partridge-Pea (Cassia Chamaechrista). Photographed by Lovell.

pollinated by insects visiting them for pollen, they all retain the power of self-fertilization. The species consist of herbs, shrubs, and, in tropical regions, of trees, with evenly pinnate leaves and yellow or white flowers. There are 5 species in the northeastern states, and 25 species in the southeastern states, a part of the species being common in both areas.

Partridge-pea is a herbaceous, much branched, spreading annual with pinnate leaves, and showy yellow flowers which often have the petals purple-spotted at base. It extends from Maine to Florida and westward to Indiana and Texas, but it is valuable as a honey plant chiefly in Florida and Georgia. In the north-central part of Florida there are thousands of acres in bloom during July and August, and for miles the ground is covered with a yellow carpet of flowers. It is also common in Georgia; in many dry sandy sections of the South, indeed, it is the main dependence of the beekeeper, making beekeeping possible in otherwise very unfavorable localities (Fig. 98.)

The blooming-period is long, beginning the last of June and closing late in September. The flowers are wholly nectarless, and are pollinated by bumblebees, which visit them for pollen. The nectar is secreted by extra-floral glands located on the upper side of the leafstalk (petiole) near its base. The gland is saucer-shaped and there is usually only one to each leaf. Unless the summer rains are very heavy, nectar may be gathered continuously for 100 days or more. In rainy weather the nectar is easily washed away. It is very thin and contains a large percentage of water.

From one to three supers, or 100 pounds of honey per colony, have been obtained from partridge-pea. The honey is medium light amber, exceptionally thin, with a poor flavor. At Fort White, Florida, the surplus comes from partridge-pea and chinquapin. Inferior as is the flavor of this honey, its fine appearance has caused it to sell at a high price. The extracted honey is bought by bakers and the large quantity obtained partly atones for the poor quality.

PEACH (*Prunus persica*).—The acreage of peach trees is restricted chiefly to the southern states and to Fresno County, California; but there are dense areas in Delaware, in New York south of Lake Erie, and on the eastern shore of Lake Michigan. In west-central and northwestern Georgia there are miles of peach trees, but they bloom while the weather is yet cold and frosts are frequent. An apiary in a peach orchard failed to gather a pound of surplus. In Fresno County, California, a small surplus is obtained in favorable seasons. The honey is dark-colored and poor-flavored.

PEANUT (*Arachis hypogaea*).—The small yellow flowers are borne in the axils of the leaves, that is, in the angle formed by the leaf-stem with the stalk of the plant. After pollination the corolla falls off and the flower stem elongates and bends downward, pushing the seed vessel into the ground where it ripens. The peanut belongs to the same family as the pea and bean, and the fruit is a pod or legume, not a nut. Until recent years the peanut has been chiefly cultivated as a commercial crop in Virginia, Tennessee, North Carolina, South Carolina, and Georgia, but to-day it is grown in all the Gulf States and westward to California. In 1919 the acreage in the United States was 1,738,400 acres.

In Florida the peanut begins to bloom the latter part of June, and remains in bloom until the latter part of September, or even until the first of November, thus being in bloom for three or four months. The rainy season in Florida occurs during the months of July, August, and September, and may often for days prevent the harvesting of the crop; but there are frequently dry periods during which the bees gather large quantities of nectar. According to Frank Stirling, the secretion of nectar is not affected to any extent by the weather. The color of the honey is a trifle darker than that of white clover, resembling somewhat the honey of the purple sage of California. It is as thick as orange honey, but not so clear. It has a characteristic flavor which is very mild, and does not resemble the flavor of peanut butter or peanut oil. The honey which took the blue-ribbon prize at the St. Louis Fair in 1918 came from peanut bloom.

PEAR (*Pyrus communis*).—Nearly one-half of the pear trees of the United States are in the five states of California, New York, Michigan, Oregon, and New Jersey. A surplus of honey is obtained in good seasons. Cross-pollination is essential. The odor of the white flowers is like that of the thornbush. Waite states that at times they secrete nectar so copiously that it falls in drops to the ground.

PECTIS (*Pectis papposa*).—A strong-scented herb with yellow flowers in small heads, growing on semi-arid plains from New Mexico to southern California. In southern Arizona, if there is sufficient rain in July, this weed springs up very quickly, covering hundreds of acres. It blooms in 6 to 8 days. Bees visit it in multitudes, and continue their visits until frost, unless the secretion of nectar is checked by dry weather. If there is very little rain, it does not appear.

PENNYROYAL (*Satureja rigida*).—A perennial shrubby plant, 2 to 3 feet tall, growing on sandy barrens and pine lands throughout southern Florida; but it is of little value to the beekeeper south of Lake Apopka. The stems branch diffusely and bear headlike clusters of light purple, two-lipped flowers. It is abundant in the southwestern part of the state below Tampa on the west coast, and near Stewart on the east side. The honey flow comes in January, at a time unfortunately when the colonies in this section are usually weak; but the plant does not entirely cease to bloom until March. During the flow colonies build up very rapidly and fill the hives with stores. While good crops have been obtained three years in succession, rainy weather is very likely to interfere with the flow.

The honey is light-colored and has a minty flavor and odor. The aromatic taste might not be agreeable to everyone, but only a small quantity of the honey is placed on the market. A beekeeper at Hansford writes under date of January 17: "I examined 15 colonies yesterday and found them full of bees, with from 4 to 6 frames of brood in each hive and an average of 50 pounds of surplus in the supers. Pennyroyal is now just at its best and has yet 60 days to yield. Don't call it erratic. This is the third season I have secured a surplus." See Purple-Flowered Mint.

PEPPERBUSH.—See Clethra.

PEPPERMINT.—See Mint.

PEPPER TREE (*Schinus molle*).—A small evergreen tree, with fragrant compound leaves, and greenish white flowers in feathery clusters. The fruit is a small red drupe with a pungent flavor, whence the name. It is not the source of pepper, which comes from plants belonging to the genus *Piper*. The honey in California is reported to be dark with a strong flavor. It blooms at all seasons.

PERSIMMON (*Diospyros virginiana*).—When grown in a dense forest the persimmon may reach a height of 100 feet, but in open land it is rarely 50 feet tall. The small, urn-shaped, greenish yellow flowers are four-parted, and appear from the last of April in the extreme South to the middle of June at the northern limit of the species. The clustered staminate and the solitary pistillate flowers are on different trees, and are largely dependent on bees for pollination; but the pollen, which is light and powdery, is also carried by the wind. Throughout the southern states beekeepers very widely report that the nectar and pollen gathered from the persimmon are valuable for building up the colonies in the spring. At Chandler, Oklahoma, a small surplus is reported to have been obtained one year.

The persimmon is most abundant in the fields and woodlands of the southeastern United States, but it extends westward to central Kansas. Northward a few trees are found in Connecticut, New York, and Michigan. It is most productive in Virginia and the Carolinas, and westward to Missouri and Arkansas. The fruit has been described as "good for dogs, hogs, and 'possums," but the early Spanish explorer, De Soto, called it "a delicious little plum."

PHACELIA (*Phacelia tanacetifolia*).—Fiddle-neck. A hairy herbaceous annual, 6 inches to 2 feet tall, with bluish flowers in scorpioid racemes, 3 to 4 inches long, common from Sacramento Valley to southern California. It blooms in about six weeks from seed and furnishes an excellent bee pasturage for about the same length of time. The honey is amber-colored with a mild aromatic flavor. The color of the pollen is blue. It was formerly very abundant in California, but owing to the over-pasturing of the cattle-ranges it has almost disappeared from thousands of acres of wild land. It is now found chiefly in the underbrush where it can not be reached by cattle. It has been introduced into Europe, where it has been highly praised as a honey plant. In Sweden, on a trial plot of 500 square meters one kilogram of Phacelia seed was broadcasted and harrowed into the soil. The plants came up in 8 or 10 days, and by the middle of August had reached maturity, averaging a half meter in height. Eight weeks after planting they began to bloom, and the blooming period lasted for about four weeks. Throughout the entire time the field was constantly visited by bees from 3 o'clock in the morning to 9:30 at night. The field is estimated to have yielded 50 to 60 kilos of honey, and if the summer had not been dry the crop would undoubtedly have been larger. The honey was clear and thin, and excellent in quality.

The racemes of the caterpillar Phacelia (*Phacelia hispida*) are coiled and covered with slender white hairs, whence the common name. Its range is from Santa Barbara County to San Diego County. It was formerly very abundant in Ventura County, but it has been largely destroyed by mountain fires. It grows luxuriantly if there have been abundant winter and spring rains, and attains a height of 2 to 4 feet. One season M. H. Mendleson, of Ventura, secured from this plant a large surplus. The honey was extracted into a seven-ton tank, and before the tank was full it granulated at the bottom. A few feet of liquid honey were drawn off and about three feet of solid candied honey was then shoveled out. "The honey," writes Mendleson, "is water-white and has a fine flavor; in the candied condition it is like a fine flour paste. I have never had a surplus before nor since that season from this source." Of the 50 other species of *Phacelia* which occur in the United States few appear to be of much value. (Fig. 99.)

At Terra Bella in order to raise farm crops they practice summer fallowing every alternate year, and the land remains unsown for that season. On an area about two by eight miles in extent caterpillar Phacelia springs up and grows four to five feet tall. The honey flow lasts not far from ten days. The honey candies very quickly into a pastelike solid. It is sometimes discolored by honey from other plants.

PIGEON CHERRY (*Prunus pennsylvanica*).—Wild red cherry. Bird cherry. Pin cherry. Common in open woods, clearings, and thickets, from Labrador to Pennsylvania and southward along the mountains. Blooms early, at a time when there are few flowers. Pollen and nectar.

PIN CLOVER.—See Alfilerilla.

PINK-VINE (*Antigonon leptopus*).—Mexican vine. Also called Spanish vine. A vigorous vine with large leaves and showy clusters of pink flowers, which is planted extensively in Florida for ornament. It grows quickly and is used for covering arbors, fences, and trellises. It is found wild in the counties of Pasco, Sumter, and Hernando. The blooming-period lasts from early summer until late fall. Around Tampa it is so common that it adds materially to the stores of the apiary, and a small surplus may be obtained. In the vicinity of Dade City the bees gather quantities of pollen and nectar from the flowers, and forty per cent. of the flow is secured from it. The honey is white with a flavor resembling that of aster honey. Also called coral vine and rosa de montana.

PLUM (*Prunus*).—The plums bloom profusely, and in small orchards the air is filled with wild solitary bees hovering about the trees. Honeybees are frequent vistors, gathering both nectar and pollen. From cultivated plums a surplus has been reported in California. About one-third of the acreage of plums and prunes in the United States is in California, centering in the Santa Clara Valley, which has over 3,000,000 trees. There is a large area also in Oregon and Washington. In the Pacific States there are more prunes than plums.

There are many native plums, widely distributed over eastern North America, which are of value to the beekeeper, as the Canada plum (*Prunus nigra*), which is a mass of flowers before the leaves expand; the wild goose plum (*P. hortulana*); the chickasaw plum; the sand plum and the beach plum. In river swamps and hammocks from South Carolina to Florida and Louisiana the hog plum (*P. umbellata*) is common, and stimulates early brood-rearing. The cultivated plums may

Fig. 99.—Phacelia (Phacelia tanacetifolia). Photographed by Lovell.

be divided into three groups according to their origin—the European, the Japanese, and the American. They are all largely self-sterile in the absence of cross-pollination. Colonies of bees should be placed in all orchards.

POINSETTIA (*Euphorbia pulcherrima*).—The poinsettia, a native of Mexico, with its brilliant scarlet bracts, stands first among decorative plants. The yellow extra-floral nectaries secrete nectar so richly that it falls down to the ground in drops. Widely cultivated for ornament, but of no importance as a honey plant in the United States.

POISON OAK OR IVY.—See Sumac.

POPLAR.—See Tulip tree. Also poplar under Pollen-Flowers.

PRAIRIE CLOVER (*Petalostemum*).—There are half a dozen species, which are common on the dry plains east of the Rocky Mountains. Bees visit the flowers eagerly.

PRICKLY ASH (*Xanthoxylum*).—There are in the United States five species of prickly ash; shrubs or small trees, with small clustered white or greenish flowers. They bloom in spring, are nectariferous, and are frequently visited by bees. *X. americanum* occurs from Canada to Virginia, west to Nebraska. *X. Clava-Herculis* is a very spiny tree or shrub, abundant along the coast from Florida to Texas. In Texas east of the Brazos River it is very abundant, blooming from the middle of April to the middle of May. A surplus is frequently obtained. The honey is light in color and pungent in flavor. It is also common throughout Florida, except on the Keys. The abundant supply of pollen and nectar attracts thousands of bees to the bloom, which appears from April to June. Toothache-tree, pepperwood, and Hercules' club. Colima or wild lime (*X. Fagara*) is common in southern Florida and along the Gulf coast of Texas. It occasionally yields a small surplus. Yellow-wood (*X. flavum*) is found chiefly in the West Indies. The staminate and pistillate flowers of the above four species are on different trees.

PRICKLY PEAR.—See Cactus.

PRIVET (*Ligustrum vulgare*).—A shrub with fragrant, small white flowers in terminal clusters. It is used for hedges and has escaped from cultivation in New York, Pennsylvania, and other states. Bees and many other insects gather the nectar. The Japan or California privet (*L. japonicum*), which is widely used for hedges, produces an abundance of white flowers in panicles. The honey is ill-flavored; and, where the hedges are neglected, so that they bloom, the honey becomes mixed with the major flow, lowering the grade of the whole. Privet is listed as a honey plant in New Jersey, Texas, and Oregon.

PRUNE.—See Plum.

PSORALEA (*Psoralea tenuiflora*).—The few-flowered Psoralea belongs to the Pulse family. It is a slender herb two or more feet tall, with purple flowers. In the country around Okmulgee, Oklahoma, it is considered a remarkable honey plant. It springs up after the grass has been cut and the wheat harvested.

PUMPKIN (*Cucurbita Pepo*).—The large yellow flowers yield an amber-colored honey that candies quickly. The pumpkin is believed to have been known to the Indians at the time of the discovery of America.

PURPLE SAGE.—See Sage.

PURPLE-FLOWERED MINT (*Mesosphaerum spicatum*).—An annual herb considered very valuable by beekeepers in Alachua, Polk, Lake, DeSoto, Pinellas, Hillsboro, and several other counties in Florida. It is abundant in sandy soil along the roadsides, railroads, and in waste places. The honey is medium amber, mild in flavor, and does not crystallize. The small purple flowers are in dense clusters in the axils of the leaves. The plant blooms continuously from the time it is 8 inches high in March until it is 8 feet high in November. Purple-flowered mint is also found in Alabama and in tropical America.

PURPLE THISTLE (*Eryngium Leavenworthii*).—A smooth herb, 1 to 2 feet tall, with purplish flowers in dense heads and spiny-toothed sessile leaves, resembling a thistle; but it belongs to the carrot family (Umbelliferae). At Bay City, Texas, there is often a honey flow from this plant in July. It furnishes the most nectar during extremely hot dry weather. The honey is dark-colored with a poor flavor. In dry soil from Kansas to Texas.

QUEEN OF THE MEADOW.—See Boneset.

RABBIT-BRUSH (*Chrysothamnus*).—Rayless goldenrod. Chico. Bees gather a moderately large amount of honey from the flowers, which appear in the fall. The honey is a deep yellow color, thin, and poor in quality. It granulates quickly even in the comb, and when it is present in a section of alfalfa or sweet clover it granulates before either of these honeys. The intense yellow color of the pollen stains the surface of the combs. At Independence, Inyo County, California, according to Wm. Muth Rasmussen, *C. nauseosus,* a perennial plant, is fairly abundant in waste places. It bears small yellow flowers in clusters at the ends of the stems, which are ash-colored or white. The flowers appear in September and October, and bees work vigorously on them until they fade. The honey is dark and has so disagreeable an odor, and tastes so nauseous, that even the Indians will not eat it. Many beekeepers remove their sections when rabbit-brush begins to bloom. (Fig. 100.)

There are 18 species of rabbit-brush, or rayless goldenrod, in the Rocky Mountain Highlands, several of which are very common in the dry hills and plains of Colorado and Wyoming, as *C. lanceolatus, C. pumilus, C. frigidus,* and *C. plattensis.* They are shrubby plants with narrow entire leaves and yellow flowers, which open in the fall and resemble the goldenrod. The white, tufted appearance of the inflorescence after blooming suggested the name rabbit-brush.

RADISH (*Raphanus Rhaphanistrum*).—Wild radish, or jointed charlock, is a troublesome weed in fields, which yields a small amount of nectar and pollen.

RAMONA.—See Sage.

RASPBERRY (*Rubus idaeus* variety *aculeatissimus*).—The wild red raspberry is a valuable honey plant in the northern part of the Southern Peninsula of Michigan, in the Adirondack Mountain region of New York, in north-central Pennsylvania, and to a smaller extent in northern New England. Probably no other raspberry location is so well known as that of the Lower Michigan Peninsula. The northern portion of this section of the state was once covered by an extensive forest of white and Norway pine, in which there were belts of magnificent hardwood timber consisting largely of beech, maple, and elm. Nearly all of the pine has in recent years been cut for lumber. During the first dry season following the cutting, fire burns over the stumpland, leaving a blackened, desolate, almost weird pine barren. Two or three years later willow-herb may spring up, but raspberries are either entirely absent from the pine barrens, or are so short and stunted as to be of little value as honey producers.

It is upon tracts from which the hardwood lumber has been cut that the wild red raspberry offers as reliable a bee pasturage as is to be found anywhere. So luxuriant is the growth that it is possible while riding along a wood road, to pick the luscious ripe berries from the tall bushes bending with the fruit. If the land is not burned over, the rich loam, mulched with brush, produces large thrifty bushes, which yield great crops of honey for several years. A hive on scales showed in fair weather a daily gain of 6 to 13 pounds. But the rapidly growing young trees soon smother the bushes, and the beekeeper is forced to seek a new location. If, however, the land is occasionally burned over, the average annual surplus is less, but the bee pasturage lasts much longer. Thus the raspberry district is constantly changing, and this shrub is not a permanent source of nectar in one locality,

Fig. 100.—Rabbit-Brush (Chrysothamnus). Photographed by Lovell.

like white clover. As new areas are lumbered off "new pastures are offered to new comers." There is only one way, according to Hutchinson, to find a desirable location, and that is to hunt for it. (Fig. 101.)

Wild raspberry honey is described by Hutchinson, who produced it in large quantities, as having a delicious raspberry flavor, and, although not as white as white clover honey, it is classed as a white honey of the very finest quality. In a warm season it begins to bloom about the first of June; but, if the spring is cold and backward, the flow does not start until the middle of June. "I think that I can safely say," wrote Hutchinson, "that the wild raspberry never fails to produce nectar. It does not winter-kill as does clover, and nectar secretion is less affected by the weather. I have seen bees doing well working upon it when the weather was so cool that clover would not yield a drop of nectar." A luxuriant growth of

Fig. 101.—Raspberry (Rubus idaeus, variety aculeatissimus). Photographed by Lovell.

bushes yields more nectar than a stunted one. The bloom lasts fully as long as that of white clover, and bees have been seen gathering nectar from it during the latter part of July. If there is a drought in August, followed by rains and warm weather in September, it sometimes blossoms again and furnishes a second crop of honey. The blossoms are inverted, a provision for protecting the nectar from rain.

Across the Straits of Mackinac in Upper Michigan raspberry is abundant on the cultivated hardwood land, and is second in importance only to alsike clover. In the acid sandy soils north of the Adirondack region in New York the clovers do not thrive, but raspberry is abundant and yields well. Raspberry honey is produced in commercial quantities year after year near Massena Springs. It is the only source of surplus and about 50 pounds of extracted honey per colony are obtained. Considerable of Franklin County is to be included in the raspberry area, as from Malone southward.

Where the raspberry is cultivated on a large scale for market it is also an important honey plant. The largest acreages of bush fruits are located on the eastern shore of Lake Michigan, on the southern shore of Lake Ontario, in southern New Jersey, and around Cincinnati, Ohio; St. Louis, Missouri; Los Angeles and San Francisco, California; and Salem and Portland, Oregon. The red varieties, especially the Cuthbert, are believed to furnish the most honey. Bees work on the flowers closely, and the honey is excellent in quality. The raspberry blooms between fruit trees and white clover, so that large fields of it are a great acquisition.

RATTAN (*Berchemia scandens*).—Where abundant, it yields a surplus of dark amber honey used by bakeries. A much branched climbing shrub bearing small greenish flowers, which open in April. It grows in damp soil or swampy land from Virginia to Florida and Texas. This species is abundant in the swamplands of southeastern Texas, where it blooms from April to July; and sometimes if there are summer rains there is a second blooming period. J. D. Yancy, of Bay City, reports the honey as a light lemon yellow with a mild agreeable flavor.

RED-BAY (*Persea Borbonia*).—Florida mahogany. Tisswood. Laurel tree. A tree reaching a height of 50 feet, with leathery leaves and yellow-green flowers, which open in spring. It is reported as valuable in many localities in the Gulf States, as Tasmania, Florida; Waycross, Georgia, and in eastern Texas. The honey is dark and poor in quality.

REDBUD (*Cercis canadensis*).—Judas-tree. A rapid-growing, beautiful tree, often cultivated. The pink-purple flowers appear from March to May, and yield both nectar and pollen. Widely distributed in rich soil throughout the states east of the Mississippi River, and westward in the river valleys to Arkansas. The Texas redbud and the California redbud have smaller flowers, which also bloom in early spring, and are valuable for brood-rearing.

RED CLOVER.—See Clover.

RED GUM (*Eucalyptus rostrata*).—Nectar abundant; a promising species. See Eucalyptus.

RETAMA.—See Horsebean.

RHODODENDRON.—The Rhododendrons are bumblebee-flowers. The following report from a beekeeper at Divide, West Virginia, is noteworthy: "I am reasonably sure that honeybees never visit the Rhododendrons in this locality. Last season tons of nectar dropped from the flowers near my apiary, but I was unable to find a single bee on the bloom. The nectar fell in large drops, was pleasant to the taste, and very sweet. The Rhododendrons cover the land for mile after mile, and when in blossom present a most beautiful appearance." A number of observers have reported visits to the bloom by several species of the larger bees, but the Rhododendrons are not listed as honey plants in any state. The flame-colored azalea (*R. calendulaceum*) is exceedingly abundant in the mountains of North Carolina. See Mountain Laurel.

ROCK BRUSH (*Eysenhardtia amorphoides*).—A branched shrub with glandular pinnate leaves and white flowers, common on the limestone hills of southern Texas and northern Mexico, blooming in spring and yielding an excellent honey. Nine hundred pounds from 12 hives have been secured, and an abundance left for the bees. It belongs to the pulse family, or Leguminosae.

ROCKY MOUNTAIN BEE PLANT (*Cleome serrulata*).—A smooth annual, with trifoliolate leaves and showy rose-colored flowers in racemes. Its habitat is from Minnesota southwest to New Mexico and Arizona. In Colorado the importance of Cleome has been greatly overestimated, although formerly it was much more important than to-day. In exceptionally favorable seasons it still yields considerable surplus. It grows widely over the plains east of the mountains, while beekeepers are found chiefly in the irrigated areas. The honey has been reported as of light color and fair flavor, and also as dark and strong. It begins blooming in July, and bees at times work freely on it. About 1890 the Michigan Agricultural College experimented with several acres of this plant for the purpose of testing its value. A good stand of plants was not secured, and the honey obtained was far from paying expenses.

Two other species of this genus also deserve description. Spider-flower (*C. spinosa*) is cultivated for its handsome white or rose-colored flowers. It grows in waste places from Illinois southward to Florida and Louisiana, and often yields nectar very copiously. With a medicine dropper a teaspoonful of nectar has been drawn from thirteen flowers. It has indeed been gathered with a spoon in sufficient quantities to permit its flavor to be tasted, and a single bee can not gather all there is in one blossom. There are several hundred blossoms to each plant, and they yield nectar until late in the fall. An acre of these plants in bloom presents a most beautiful sight.

Yellow Cleome (*C. lutea*). The yellow Cleome has yellow flowers and blooms in June. It is found in the western highlands from Nebraska to Washington and Arizona. Unlike the purple Cleome, which seems to prefer cool well-watered locations in the creek bottoms and upper mountain valleys, the yellow species is seldom found anywhere except in the desert and in the cultivated land of the warmer valleys. If the winters are dry the seed does not germinate, but lies dormant in the soil until there is sufficient moisture, so that there may be few or no plants for several years. But after a winter with a sufficient precipitation of rain or snow it springs up so quickly that the desert for miles looks as though it were covered with a carpet of gold. In moist or irrigated land it grows to the height of two feet or more, and blooms nearly all summer. Usually it grows only 12 to 18 inches tall, and dies in two or three weeks after it begins to bloom.

Under favorable conditions it yields a moderately large amount of honey. Nearly a super per colony has been obtained. The honey is rather dark-colored, but the flavor is good. Cleome fills in the gap between fruit-bloom and alfalfa.

ROSE.—See Pollen Plants.

ROSIN-WEED (*Silphium laciniatum*).—A tall rough-leaved herb with numerous large heads of bright yellow flowers growing on the prairies from Ohio to Texas, but not forming dense masses. It blooms from July to September, and the honeybee and many other bees have been observed sucking the nectar. Another species, the cup-plant (*S. perfoliatum*), is common on the prairies, and honeybees gather from the flowers both nectar and pollen. July to September.

SACALINE (*Polygonum sachalinense*).—This hardy perennial from the island of Sakhalin is closely allied to buckwheat. The stems are over 6 feet tall, the leaves are heart-shaped, and the small greenish-white flowers are borne in clusters in the axils of the leaves. It blooms profusely in August and is said to be a great favorite with bees. Tests of sacaline have been made in many states, but most of the reports are adverse to the plant.

SAGE (*Salvia*).—Sage honey, which is widely known for its delicious flavor in Europe as well as in America, is a product peculiar to California. The crops of honey secured from black sage during the past 25 years have been so immense that fine sage honey is now offered for sale in many of the principal cities of the world. During the past 40 years there have been two or three exceptionally heavy flows when 200 pounds per colony were secured. At Ventura, Mendleson has obtained an average of 300 pounds per colony in a single season from the black and purple sages. While black sage occurs to a limited extent on Mt. Diablo, near San Francisco, and in localities in San Mateo County, practically the entire sage region of this state is restricted to the Coast Ranges, extending from the foothills in the northern part of San Benito and Monterey counties to San Diego County in the southwest corner. The largest amount of sage honey comes from Ventura and San Diego counties, while only a small surplus relatively is secured in Riverside and San Bernardino counties. The sages belong to the genus *Salvia* (from the Latin salveo, to save, from the supposed medicinal value of some of the species) and to the mint family, or Labiatae. *Salvia* is a large genus comprising nearly 500 species widely distributed in both temperate and tropical regions.

The three species most valuable as honey plants in California are the black, white, and purple sages. Black sage is so called because the foliage is very dark green, and also because the flowers after blooming turn black and adhere to the bush until the next season. Purple sage has purple blossoms and the foliage has a very striking grayish-purple appearance on the hillsides. When the two shrubs are seen side by side in the distance on the foothills, the contrast is very marked, the one looking dark or black and the other purple. The foliage of the white sage is grayish-white and the flowers are also white. The black and purple sages are bushy shrubs very leafy at base, but the white sage has longer stems and is less bushy. The purple sage is much larger than the black sage and is sometimes six feet tall. The white sage grows on the flat mesa lands, while the black and purple sages are abundant on the foothills and the sunny slopes of the canyons.

Black Sage (*Salvia mellifera*). Also called button sage and blue sage. The black sage is a shrubby plant, 3 to 6 feet tall, with oblong leaves, dark green above and woolly beneath, and numerous flowering branches, which bear about five dense whorls or "buttons" of flowers. The corolla is two-lipped, white or pale purple, and rather small. The whorls, the larger of which are about an inch across, diminish in size toward the tip of the stalk, and in fading the flowers turn dark, but do not fall from the bushes. (Fig. 102.)

The honey flow lasts from the middle of March or the first of April until about the first of July. The crop is unreliable every other year, and there is a total failure once in three or four years. Every fifth year a large crop may be expected, and if the rainfall has been ample a fair surplus is sometimes obtained three years in succession. The black sage does not yield nectar freely unless there has been a sufficient rainfall during the winter, followed by a clear warm spring. The rainfall varies greatly in different years, presenting great extremes; but frequently it is less than 12 inches. In 1882 there was only 2.94 inches, while in 1905 it amounted to 22.12 inches. Although the plants are well adapted to live in semi-arid regions, if there is a drought they dry up and become valueless to the beekeeper. The flowers are often injured by the sage worm, and the foliage by rust. The honey is water-white, thick and heavy, and does not granulate.

Purple Sage (*Salvia leucophylla*). This species is a much larger shrub than the black sage, and the whorls of light purple flowers are nearly twice the size of those of the latter species. The honey is water-white, does not granulate readily, and its flavor is considered a little superior to that of the other sage honeys. It begins to bloom usually a little later than the black sage, but the honey flows from both species nearly coincide. Purple sage is most abundant in Ventura County, where it is a characteristic feature of the vegetation of the foothills. (Fig. 103.)

White Sage (*Salvia apiana*). A shrub, 3 to 5 feet tall, which is less bushy than the black sage, the branches being long, straight, and slender. The two-lipped white flowers are produced in great profusion in lateral racemes and the leaves are grayish-white on both sides. The lower flowers open first. It begins to bloom the latter half of May, and the blooming period lasts from six to eight weeks. On the dry plains or mesa lands and foothills of southern California there are thousands of acres of this beautiful shrub, and one may ride through avenues of it for miles. One range is described as a mile wide and two miles long, consisting practically of unbroken white sage.

The white sage secretes much less nectar than does either the black or purple sage. In districts where both the black and white sages are abundant, beekeepers have estimated that the black yielded ten pounds of honey to one from the white species. To produce a vigorous growth and a profusion of flowers there must be a sufficient rainfall. The honey is white and heavy, and does not candy, while the quality is nearly as fine as that of the black sage. Much of the white-sage honey, so-called, comes from the black sage. At Caldwell, Idaho, white sage is reported to yield a surplus of honey. A colony of bees carried eight or nine miles away from

Fig. 102.—Black Sage (Salvia mellifera). Photographed by E. R. Root.

the alfalfa fields to a sage range gathered fully as large a crop as colonies near alfalfa. The honey was water-white and its flavor was excellent. (Gleanings in Bee Culture, Sept. 15, 1908.)

The larger portion of the surplus from sage comes from the three species described above. The black and purple sages yield almost equally well, but the white sage is not a heavy yielder, although the surplus varies in different years. Black sage is practically the only species of *Salvia* found in Monterey County. It grows plentifully along the coast and on southeast and southwest slopes, decreasing inland as ridge after ridge is passed, and disappearing in eastern San Benito

County. In the Salinas Valley it blooms from April 15 to June 5; in the northern part of the county it remains in bloom until July 1, and along the coast until July 15. In the Gavilan Mountains northeast of Gonzales seed of the white sage was sown in 1885, and this species still perpetuates itself in this locality. Purple sage has been reported near Monterey. In San Luis Obispo and Santa Barbara counties all three species occur, but the white sage is not common in San Luis Obispo County. (Fig. 104.)

The distribution of the sages in Ventura County is described by M. H. Mendleson as follows: "Purple sage is most abundant in this county, although there is a large amount of black sage scattered through the interior; but there is only a small amount along the coast. The white sage is the rarest in this county. Following a wet winter, the black and purple sages are about equally valuable for the production of honey. The black sage blooms first, then the purple, and last of all the white sage, the blooming periods usually overlapping. The white sage can seldom be depended upon for a crop, and the honey is inferior to the flavor of that stored from the other two species. The white sage honey invariably granulates, while the black and purple, when well ripened and gathered in the interior, remain liquid; but on the coast they invariably candy. The time of blooming varies from March 1 to June 1, depending on climatic conditions, as rain and a cold or warm winter and spring; but usually it begins from April 15 to 30, and closes between June 20 and July 10.

Fig. 103.—A bush of purple sage. Photographed by E. R. Root.

"According to my experience of 41 years, most of the sage honey comes from the black and purple species, and only a small percentage from the white. All of the sage honeys darken with age, and become the color of molasses, although still remaining translucent. A bottle of sage honey 36 years old has lost its sage flavor, but has the flavor of a delicious syrup.

"Many thousands of acres of the sages are destroyed annually by stockmen, and in the near future good sage ranges will be at a premium. The great destruction of brush by fire is lessening our rainfall. The government should take steps to reforest the state promptly, as the good soil washes away during the heavy rains, leaving a soil so poor that it will not maintain a new growth."

In Los Angeles County the white sage is most abundant, and the black sage the rarest. White sage is found on the mesas adjacent to the foothills, and as high as 2000 feet above sea level. The black and purple sages occur in the canyons and high hills up to an elevation of 4000 feet. The sages bloom in May, June, and July, and the duration of the honey flow depends upon rainfall and temperature. The

Fig. 104.—White Sage (Salvia apiana). Shrub without the blossom. Photographed by Richter.

black and purple sages bloom at nearly the same time, but the white sage blooms several weeks later, and remains in bloom for a month or six weeks.

In San Diego County white sage is estimated to be three times as abundant as black sage. Purple sage is less common than northward. The black and white sages occur in San Bernardino County on the foothills in the southwest portion. The black sage is the better honey plant, blooming several weeks before the white.

In Riverside County, according to T. O. Andrews, the black sage, which is found on the foothills and mountain slopes up to 3500 feet, is most abundant and the best yielder of nectar. It often begins to bloom in February and in some seasons continues to bloom until July 20. The white sage grows on more level land and on better soil at the foot of the hills. It begins to bloom in May and may last through June. There is a little purple sage. This species is central in Ventura County near the mountains, and is more widely distributed than either the black or white sage. In 1920 there was in Riverside County the best flow from black sage in 25 years. Strong colonies averaged over four pounds per day for 15 days. The honey was water-white and very heavy. The purple and white sages also yielded well.

There are several other species of sage which deserve mention. The creeping sage (*S. sonomensis*), or ramona, covers the ground with a matlike growth, from which arise flowering stems four or five inches tall bearing light violet flowers. This species is rare, but it is widely distributed in the Sierra Nevada and Coast Ranges and in localities is important. The honey is like that of the other sages. Annual sage (*S. Columbariae*), or chia, is common in the foothills and mountains of the Coast Ranges, and in some districts yields a surplus. The seeds were formerly used for food, and were also considered of medicinal value in cases of fever. It blooms in April and May. Thistle sage (*S. carduacea*) likewise yields a white honey of fine flavor. Two or three stems rise from a rosette of root-leaves, and bear from 1 to 4 whorls, or "buttons," of light blue flowers. The leaves are more or less spiny-toothed, whence the English name common in southern California and the San Joaquin Valley. It blooms in June.

The common garden sage (*S. officinalis*) is reported to yield a white honey, where it is extensively cultivated as a culinary herb. Lance-leaved sage (*S. lanceolata*) grows in bare fields and on dry plains from Nebraska and Kansas to Texas and Arizona. It is about a foot tall, with lance-shaped leaves and blue flowers, and yields nectar from early July until frost. It is helpful in maintaining the strength of the colonies. The blue sage (*S. azurea*) is listed as a honey plant in Texas. In Australia the verbena sage (*S. verbenacea*), introduced from Europe, yields a little honey during the dry months of the year. But many species of *Salvia* have a corolla tube so long that they are adapted to bumblebees. Other species, as the crimson sage of California, the cardinal sage of Mexico, and various Brazilian species with scarlet or bright red corollas, are humming-bird-flowers, the nectar being secreted at the bottom of a tube 2 inches long, far beyond the reach of bees.

SALMON-BERRY (*Rubus parviflorus*).—In Oregon the large white flowers of the salmon-berry bloom about the 10th of March, and yield a thin red honey, which is very sweet. Michigan to Oregon and California. A raspberry.

SAINFOIN (*Onobrychis sativa*).—Sainfoin has long been extensively grown in England, France, and Belgium for hay. In America it has been successfully cultivated in a few places throughout Ontario and in various parts of the United States. It succeeds best on a limestone soil, or where lime is used as a fertilizer. It is not suited to a semi-arid country; and, as the stems are shorter and smaller than those of alfalfa, it does not produce as many tons of fodder per acre. Like the clovers, alfalfa, and sweet clover, it belongs to the pulse family. The spikes of light pink flowers appear in summer. (Fig. 105.)

Bees gather the nectar very eagerly. The honey is pale yellow, makes a handsome section, and is said to be almost as clear as spring water. It does not granulate readily; the quality is excellent and compares well with that from white clover. Some retailers prefer it to alfalfa honey. A field of sainfoin is pink with bloom for a long time. Also called crocette and esparcette.

"The chief honey plant on the chalky uplands of southeast England," says Sladen, "is giant sainfoin, a variety of *Onobrychis sativa*. It is sometimes cut twice in a single season. It blooms first about June 10 and yields a crop of honey, if the weather is not too wet and it is fairly warm. It blooms again in July when the main honey crop is gathered. It often blooms for a third time in middle August; but it yields no nectar, although the fields are pink with the bloom and the conditions are as favorable as in July."

In the Department of Loiret, south of Paris, France, at Outerville sainfoin occupies about one-third of the land. It remains in bloom for about 28 days before it is cut for hay. A month after the first cutting, it is in bloom again and later in the season there is a third crop. Few plants yield as much nectar, and the secretion continues abundant even in windy, cloudy, or rainy weather. During a very unfavorable season a hive of bees on scales showed a daily increase in weight of five pounds. The honey is famed throughout France. It is a pale amber or straw color and very sweet. As the bees come into the hive covered with pollen the new combs are quite yellow, and the cappings are of a pretty saffron shade.

SAVORY (*Satureja hortensis*).—Summer savory. A garden annual growing wild sparingly from New Brunswick to Kentucky. Honeybees visit the flowers in great numbers in some localities, but in Maine under cultivation the bloom was entirely neglected.

SEA GRAPE (*Coccolobis uvifera*).—An unattractive small tree or shrub with twisted and crooked branches, which grows along the beach, frequently in the water. It is common in southern Florida and on the Keys. The round leaves are so leathery that they are not blighted by the stinging force of the salt spray driven by severe storms. The flowers are small, green, and in clusters; the fruit resembles the common grape in size and appearance and is fairly palatable. At Sarasota, Florida, it is reported to be an important honey plant. It belongs to the buckwheat family.

SENSITIVE PEA.—See Partridge-pea.

SHADBUSH (*Amelanchier canadensis*).—There are some 23 species in North America, mostly shrubs, which are abundant in localities. The large white flowers open in April or May, and, as they furnish both pollen and nectar, are helpful to the apiary. The Canadian shadbush is the most common species in the eastern states. (Fig. 106.)

SHOESTRING-VINE.—See Blue Vine.

SIDA (*Sida spinosa*).—There are some 17 species in the South, growing in sandy soil from Virginia to Florida and westward to Arizona. Also found in the warmer parts of South America, Asia, and Africa. Sida is listed as a honey plant in Texas, Porto Rico, and Hawaii (where the native name is *ilima*). The prickly sida, a common species, has solitary yellow flowers which contain nectar.

SILKWEED.—See Milkweed.

SIMPSON'S HONEY PLANT.—See Figwort.

SILVER WATTLE.—See Acacia.

SILVER-BERRY (*Elaeagnus argentea*).—A shrub with twigs covered with brown scurf becoming silvery, found from Utah to British Columbia and Minnesota. The fragrant, pale yellow flowers secrete nectar freely, and are visited by bees.

SMARTWEED.—See Heartsease.

SNEEZEWEED (*Helenium autumnale*).—Also called false sunflower. A perennial plant with bright yellow flowers, growing in swamps and wet land throughout a large part of the United States. Honeybees gather both pollen and nectar. It belongs to the same genus as bitterweed.

SNOWBERRY.—See Buckbush.

SNOW-ON-THE-MOUNTAIN (*Euphorbia marginata*).—An annual, the upper leaves of which have white margins. Grows wild from Minnesota to Texas, where a surplus of 15 pounds per colony has been reported. The honey is a very dark amber with a peculiar but not unpleasant taste. Of interest chiefly because the honey has been reported to be poisonous.

SNOWVINE (*Cissus arborea*).—Pepper-vine. A climbing woody vine belong-

Fig. 105.—Common Sainfoin (Onobrychis sativa). Photographed by Lovell.

Fig. 106.—Shadbush (Amelanchier canadensis). Photographed by Lovell.

ing to the grape family. In rich soil from Virginia to Florida, Texas, and Mexico. Also in Cuba. The flowers are small and greenish. The honey flow in Georgia lasts sometimes from the middle of June to September. The honey is dark amber and of fair quality.

SOAPBERRY (*Sapindus*).—There are three small trees belonging to this genus in the southern states, the sumac-leaved soapberry (*S. Saponaria*), the Florida soapberry (*S. marginatus*), and Drummond's soapberry (*S. Drummondii*). In Texas Drummond's soapberry sometimes yields a small surplus, as at Vance. It grows in river valleys from Louisiana to Mexico. Also called Chinaberry and wild China. The small white flowers are in dense panicles. This species must not be confused with China tree (*Melia Azedarach*), cultivated and growing wild in the Gulf states.

SOAP-BUSH (*Porliera angustifolia*).—At Uvalde, Texas, bees store honey from soap-bush only in very dry seasons in April. The honey is white, has a good flavor, and granulates quickly. It is usually a shrub, but occasionally a small tree. The large violet-purple flowers are borne at the end of small spreading branches. (*Guaiacum angustifolium*).

SOURWOOD (*Oxydendrum arboreum*).—Also called sorrel tree, lily-of-the-valley tree, and elk tree. This is a very important source of honey in the South. It is a fine tree, belonging to the heath family, or Ericaceae, growing 40 feet to 60 feet tall and a foot in diameter. The smooth bark is brownish red, and the young twigs are light green. The leaves are oblong, pointed at the apex, smooth on both sides, and have a sour taste. The numerous white urn-shaped flowers are in slender one-sided racemes, 5 to 6 inches long, which hang in clusters at the ends of the branches. From the resemblance of the blossoms to those of the little perennial herb of the garden, sourwood is often called lily-of-the-valley tree. The popular names sourwood and sorrel tree are derived from the sour odor and flavor of the leaves and twigs. (Fig. 107.)

Sourwood grows in rich woods from southern Pennsylvania to western Florida and southern Alabama, westward to southern Indiana, the Arkansas mountains, and western Louisiana. It is most abundant in the mountainous tract of country occupied by the Alleghenies and the Blue Ridge, but eastward it extends in places as far as tidewater, and westward to central Tennessee. It is planted for ornament as far north as Massachusetts, but as a source of nectar it is chiefly valuable in the mountainous regions of North Carolina, South Carolina, and Tennessee. It flourishes on high dry soil, and is common on poor woodland ridges, but in the forests along the rivers, where the soil is rich and deep, it makes a much larger growth and they are often beautifully checkered in July by the white blossoms.

Sourwood begins blooming about June 20, and the harvest from this source usually lasts until the latter part of July. The urn-shaped corolla is pendulous and contracted at the mouth, so that the bountiful supply of nectar is protected from both rain and injurious insects. Sourwood is considered one of the most important honey-producing trees of the South. The nectar is secreted in such abundance that it may be shaken in small drops from the bloom. The honey flow is usually dependable; and in localities where it is abundant the beekeeper seldom misses a harvest. In northwest North Carolina the surplus comes largely from this source, and the flow is reliable three years in five. At Brookneal, Virginia, there is a total failure about one year in four. As the honey flow comes so late, the beekeeper has ample opportunity to build up strong colonies which can gather nectar very rapidly during the short honey flow.

Under favorable conditions, sourwood honey is produced in enormous quantities, but it is seldom found in the markets outside of the region in which it is gathered. It is nearly all consumed in the localities where it is produced, as it is regarded as one of the finest flavored honeys in the United States, and often commands in local markets a premium of a few cents per pound. The honey is white

Fig. 107.—Sourwood (Oxydendrum arboreum). 1, sourwood bloom in the forest; 2, a single flower-cluster. Photographed by L. E. Webb.

or light colored, with a delicious, slightly aromatic flavor, and is very slow to granulate. It is, however, often mixed with basswood honey, or with persimmon honey gathered earlier in the season.

Few are acquainted with the merits of sourwood outside of the region where it is an important source of honey, and it is sometimes confused with black gum and sour gum, much to its disparagement. A beekeeper familiar with the honeys from basswood, tulip tree, clover, buckwheat, goldenrod, and aster, declares that it has no superior among the honey-producing trees of the United States either in its beautiful appearance or in the amount of nectar secreted.

SOUR CLOVER (*Trifolium fucatum*).—The cream-colored flowers, which turn pink with age, yield a moderate amount of nectar. It grows rankly in low alkaline soil throughout California.

SPANISH NEEDLES (*Bidens aristosa*).—The honey has a golden color, excellent flavor, and good body, weighing full 12 pounds to the gallon. It is so thick that there is little water to evaporate, and the cells can be sealed soon after they are filled. This plant has showy, large yellow-rayed heads, and yields immense quantities of honey along the bottom-lands of the Mississippi and Illinois rivers. It is found in swamps from Illinois to Louisiana, blooming from August to October, and yielding a honey which is superior to, or is unsurpassed by, that from any other fall flowers.

A typical Spanish-needles swamp is located at the foot of the bluffs of the Illinois River where there is a broad expanse of low marshy land from 3 to 5 miles wide. This land is subject to an overflow from the river once a year, which usually occurs in early spring. This renders a large portion of the soil unfit for tilling purposes, and in consequence Spanish needles has secured a permanent foothold to the exclusion of nearly all other plants. Early in September the bright yellow rays begin to appear, and in a short time the whole district is aglow, and its dazzling brilliancy reminds one of a burnished sheet of gold. The bees revel in this great field of flowers, so rich in nectar, and rapidly store a surplus. A single colony stored 63 pounds of honey in six days, and 43 colonies produced 2021 pounds in 10 days, an average of 47 pounds per colony.

There are many other species of *Bidens* widely distributed throughout America, all of which are of more or less value to bees. The common beggar-ticks (*Bidens frondosa*) is one of the most abundant. They are all fall flowers, and usually grow in wet places, one species being aquatic.

SPEARMINT.—See Mint.

SPIKEWEED (*Centromadia pungens*).—A branching annual with spinescent, sweet-scented leaves. The yellow flowers yield an amber-colored honey of good quality, which granulates quickly. Much spikeweed honey has been produced in Fresno County, California. Jepson says that spikeweed is "abundant on the plains of the lower San Joaquin, southward to southern California and westward to Walnut Creek and Alameda. On the alkaline plains of the upper San Joaquin this species covers tens of thousands of acres, and often forms thickets 4 to 5 feet high." It is also abundant in low, more or less alkaline land on the plains of Solano County, and forms extensive colonies in summer fields. It is a valued bee plant.

SPIDER FLOWER.—See Rocky Mountain Bee Plant.

SQUASH (*Cucurbita maxima*).—The stamens and pistils are in separate flowers on the same plant. The squash is dependent on bees for pollination. The nectar is abundant.

STINKWEED.—See Jackass Clover.

STRAWBERRY (*Fragaria virginiana*).—The strawberry grows wild throughout a large part of Europe, Asia, and North America, and in Chile, South America. The commercial cultivation of this berry extends throughout the states east of

Nebraska and Kansas, two of the most important centers being southern New Jersey and Delaware. Bees are required for the proper pollination of the strawberry, as a part of the flowers are perfect or hermaphrodite, and part pistillate. As the latter are more productive they are given the preference under cultivation. More failures have occurred among strawberry-growers from ignorance of the sex of strawberries than from any other cause. The blossoms yield nectar sparingly, and there is no record of a surplus of strawberry honey. Honeybees do not visit the bloom as frequently as is desirable, but it attracts many other insects. The perfect flowers are of value for pollen.

SULLA (*Hedysarum coronarium*).—A perennial plant closely related to sainfoin, and, like the latter, it is adapted to a deep limestone soil. The pink flowers appear in May and June and are reported to yield an excellent honey. It is not hardy and does not succeed above North Carolina and Arkansas. In southern Europe sulla clover is cultivated for hay and as a soil-improving crop. It requires the same treatment as alfalfa.

SUMAC (*Rhus*).—This genus is represented in the United States by about 15 species. Most of them are shrubs, but a few are small trees and one is a shrubby vine. The large handsome leaves are trifoliate, or odd-pinnate presenting a fern-like appearance. The pinnate-leaved species are highly ornamental as foliage plants, and in autumn display the most brilliant red and scarlet colors. The small flowers are borne in dense clusters, or panicles, at the ends of the branches or in the axils of the leaves. The stamens and pistils are usually in different flowers, one tree or shrub bearing only staminate flowers and another only pistillate. In the common staghorn sumac the staminate flowers are in large white clusters, while the pistillate are in dense green clusters, which stand well above the foliage. The white flowers yield both nectar and pollen and attract many more insects than do the green, which offer only nectar; but honeybees visit both kinds. The sumacs may be divided into two groups: the non-poisonous sumacs and the poisonous sumacs.

THE NON-POISONOUS SUMACS

Staghorn Sumac (*Rhus typhina*). This species reaches a height of 10 to 25 feet, and has orange-colored wood and crooked branches, covered with soft velvety hairs. The clusters of fruit are clothed with acid crimson hairs. The staghorn sumac grows in dry soil from Nova Scotia westward to Missouri. The flowers are visited by honeybees in large numbers, and, as the nectar is unprotected, by a great company of other insects. The flowers appear in June and July. (Fig. 108 and Fig. 109.)

Smooth Sumac (*Rhus glabra*). Upland sumac. Scarlet sumac. This species is an irregularly branched shrub, seldom more than 10 feet tall. It has a very wide distribution, extending from Nova Scotia to Florida and westward to Mississippi and Minnesota. In Connecticut, where much of the surface is covered with glacial moraines, it is very common in hillside pastures and along stone walls. The blooming period lasts for about three weeks, from July 8 to the beginning of August. The flowers secrete nectar very freely on hot clear days, but in cloudy, foggy, or cool weather the flow ceases almost entirely. If there are "hot waves" in July strong colonies will bring in 20 pounds of honey during an ideal day, and will store from 40 to 100 pounds each. But if there is much cool or rainy weather there may not be an average of 20 pounds to the colony. At its height the flow is very rapid and heavy. While the bees are busy on the bloom there is a very strong odor in the apiary, and the new honey is more or less bitter to the taste. Fortunately, the bitterness is only transient, and by winter the honey is edible. When pure the honey has a golden color. If properly ripened it has no noticeable odor, but is very heavy, and, like apple-blossom honey, waxes instead of candying. It is safe to say that much of Connecticut would be worthless to the beekeeper but for this plant.

The bloom also yields a large amount of pollen, great loads of which the bees bring in during a slow flow. Even during the height of the honey flow the bees

Fig. 108.—Staghorn Sumac (Rhus typhina). Staminate flowers. Photographed by Lovell.

Fig. 109.—Staghorn Sumac (Rhus typhina). Pistillate flowers. Photographed by Lovell.

gather pollen during the morning hours, before the sun has stimulated the nectaries. Later in the day little pollen is brought in.

In Georgia there are several species of sumac which are valuable, but the most important is *Rhus copallina,* common names of which are dwarf sumac and mountain sumac. This species extends from Maine to Florida and Texas and westward to Minnesota. In a few localities in North Georgia it is the main source of marketable honey. In that state it blooms in August.

In central Texas in the hilly sections of Coryell, Lampasas, Burnet, and other counties, several species of sumac are common. *Rhus copallina* and *R. glabra* are abundant, blooming the last of July and early in August. A surplus of 100 pounds per colony has been obtained, but the average is often not more than 25 pounds. About 50 colonies can be supported in one location. The honey is amber-colored, of heavy body, and has a fine flavor. In California poison oak (*Rhus diversiloba*) is abundant in the foothills of the mountains, and yields a light-colored honey of good flavor, which is not poisonous. From laurel sumac (*R. laurina*) and sugar bush (*R. ovata*) many beekeepers obtain an extraction. Two other sumacs are also valuable in this state.

THE POISONOUS SUMACS

There are about 17 species of poisonous sumacs, found in North America and Asia. Many persons are severely poisoned by coming in contact with these plants, while others are able to handle or even chew the leaves with impunity. Occasionally the exhalation given off by the foliage is poisonous to very susceptible individuals. A very distressing inflammation of the skin is produced, which in rare cases has been fatal. A few hours after exposure there is intense itching, followed by blisters, swelling, and fever. The malady commonly is at its height on the fourth or fifth day. Sometimes the eruption is confined to the part which has come in contact with the plant, in other cases it is more general. Relief may be obtained by the application of cold or ice water, which should be long continued. A solution of acetate of lead (a poison if taken internally) used externally is also beneficial, but in very severe cases a physician should be consulted. Poisoning is most frequently caused by poison ivy, a woody vine growing in both dry and damp locations, and often climbing trees, and by poison or swamp dogwood, a shrub or small tree often found in swamps.

Poison ivy (*Rhus Toxicodendron*), also called poison oak, extends from Maine to British Columbia, southward to Utah in the West and Florida in the East. A woody vine, climbing by rootlets, but often a low shrub. It may be readily recognized by the smooth shining compound leaves *consisting of 3 leaflets with entire edges.* Poison dogwood (*Rhus Vernix*), also called poison elder, is a swamp shrub, sometimes becoming a small tree, and has also a very wide distribution throughout the eastern states, but it does not extend west of Minnesota. The compound leaves are pinnate with 6 to 12 leaflets *with entire edges.* The leafstalks are often red-colored. This species is more virulently poisonous than the preceding. Both plants have small yellowish green flowers in clusters which yield much nectar, and the well-ripened honey is apparently harmless. A surplus from this source in the East seems to be seldom obtained.

Rhus Metopium, coral sumac, poisonwood, doctor gum, is a tree 40 to 50 feet tall, growing commonly on the extreme southern part of Florida peninsula and on the Keys, but found as far north as Palm Beach. Its common names are very numerous and misleading. It is often (but erroneously) termed "manchineel," from confusion with the tree of that name, which it closely resembles in many particulars; but it is more common than the real manchineel, which is far more poisonous. Both have an acrid sap which heightens the danger of confusion between them. The sap of *R. Metopium* is poisonous to a great many people, but it resembles poison ivy in being harmful only to certain skins.

Coral sumac, or poisonwood, yields honey of a high order and in great quantities. In some seasons the bees are said to go fairly wild with excitement over it.

While the honey is of good quality, its exact color, body, and flavor can probably never be determined with exactness, for it blossoms simultaneously with several other honey-bearing plants or trees (notably with the dogwood and also the pigeon plum), so that only a blend of several honeys is possible. The blend of the three is thick and appetizing, of good color, and usually brings good prices in the northern markets. It constituted most of the surplus of O. O. Poppleton, of Stewart, Florida, whose crop in 1909 was 28,000 pounds.

SUMMER FAREWELL (*Kuhnistera pinnata*).—Perennial herbs with glandular stems and compound leaves composed of 5 to 15 thread-like leaflets. The flowers are in sub-globular clusters subtended by small bracts, and resemble heads of the thistle family. The lobes of the calyx are bristlelike. The plant grows two feet tall, and in Florida blooms from the latter part of September until about the middle of November, when it is killed by frost. From its blooming so late in the season it is called "summer farewell." The flowers are white and are produced in such profusion that at a distance they appear like a solid mass, and small patches of them have been compared to drifts of snow — hence it is called "snow weed." It grows in sandy soil from North Carolina to Florida and Mississippi. Since 1906 it has spread rapidly in Columbia and Alachua counties, Florida, and in the fall many square miles of land are covered by its abundant growth. (Fig. 110.)

The honey, according to Wilder, is almost as clear as water and has an excellent flavor and a good body. It never ferments, but granulates at the beginning of cold weather. During the honey flow there is great activity among the bees, and there may be numerous swarms. Wilder reports that the average surplus ranges from 30 to 50 pounds per colony, and that he has obtained as high as 150 pounds from a single colony from this source. Summer farewell makes a better growth in dry than in wet seasons.

SUNFLOWER (*Helianthus annuus*).—The honey is amber-colored with a characteristic flavor. At Ventura, California, a carload of sunflower honey has been extracted, but this was an exceptionally large surplus.

Helianthus is an extensive American genus, embracing sixty or more species. The common sunflower grows wild throughout the West, especially from Minnesota to Texas, on the prairies and waste lands lying between the Rocky Mountains and the Mississippi River. In Nebraska it becomes "a veritable herbaceous tree," and completely takes possession of large waste areas, 10 to 25 or more acres in extent. The tall plants also grow along the roadsides and about the cities and towns. The stems yield a textile fiber, the seeds oil, and the flowers a yellow dye.

Many other species of sunflower are exceedingly common both in the West and South. The Jerusalem artichoke (*H. tuberosus*) is a good honey plant. In Contra Costa County, California, there are many acres of this plant growing wild. The tubers are used as a vegetable.

SWEET BAY (*Magnolia virginiana*).—Sweet bay or laurel bay is very abundant in the swamps at Valdosta, Lowndes County, in southern Georgia, and during two weeks yields from 15 to 20 pounds of amber-colored honey. The large white flowers are very fragrant.

SWEET CLOVER (*Melilotus.* Greek word from meli, honey, and lotos; the Greek lotos was a kind of clover, perhaps white sweet clover).—There are about 20 species of sweet clover, natives of Asia, Africa, and Europe. Nine species are found in France. Four have been introduced into North America. The sweet clovers were known to the ancient Greeks more than 2000 years ago, and in the Mediterranean region were valued as honey plants, as well as for forage and green manure. They are now distributed over the entire civilized world, usually growing on waste land; but in Australia, South Africa, and the United States they have been cultivated with remarkable success. Many species of sweet clover have been more or less utilized in the Old World, but only three are commonly cultivated in the United States: The white biennial sweet clover (*Melilotus alba*), the large yellow biennial (*M. officinalis*), and the small yellow clover (*M. indica*). (Fig. 111.)

Fig. 110.—Summer Farewell (Kuhnistera pinnata). Photographed by J. J. Wilder.

The large biennial yellow sweet clover (*M. officinalis*) is also called yellow melilot, balsam-flowers, heart's-clover, and king's-clover. It was introduced into this country from Europe, and grows well in waste places both in the northern and southern states. Yellow sweet clover is a biennial plant, storing in its roots the first season a reserve food supply, which is utilized in producing seed the second season. It closely resembles white sweet clover; but it has finer stems and does not grow as tall, seldom attaining a height of more than 3 to 5 feet. The flowers are yellow, in long racemes, as in the white, and bloom about two weeks earlier than those of the white species. It is thus valuable to the beekeeper, since it affords good bee pasturage earlier in the season. Nectar is secreted freely, and the honey does not differ essentially from that of white sweet clover. It is less easily exterminated, as the mower frequently passes over the decumbent stems, which thus remain to reseed the land. It is much less generally cultivated than the white species.

The small yellow annual sweet clover (*M. indica*), or bitter clover, grows wild in southern California and Arizona. It may easily be distinguished from the preceding species by its much smaller yellow flowers. The annual does not succeed well in other sections of the United States, and the biennial white sweet clover should be given the preference. In Ohio the annual was planted in a field which had been limed and inoculated with the proper bacteria. The seed germinated fairly well, but the plants grew so poorly that the crop was a complete failure. *Melilotus indica* is the only one of the sweet clovers which will make a satisfactory winter growth in southern California. It is suitable for a green-manuring crop and has been used in the citrus groves of both California and Arizona.

White sweet clover (*M. alba*) is also known as Bokhara clover, white melilot, bee clover, honey clover, tree clover, and honey lotus. White sweet clover was introduced into the United States by European colonists as early as 1738, but its value was not recognized to an appreciable extent until within the last 30 years. More than half of the states passed laws classing sweet clover as a noxious weed. Supervisors of roads were required in Ohio to mow it as well as Canada thistle, burdock, teasel, and other pernicious plants. Many farmers devoted a large amount of their time to endeavors to eradicate it, and there yet remain a few uninformed persons who regard it as a dangerous weed. But gradually its value became recognized, and to-day there is not an agricultural experiment station in this country that does not recognize its worth and approve of its cultivation.

White sweet clover is a biennial herb with smooth branching stems and compound leaves composed of three oblong leaflets. The first season it grows 18 to 20 inches tall and stores up in a very large tap-root reserve food, which enables the plant to make a rapid and vigorous growth the following season. The second year it grows 3 to 12 feet tall, blooms profusely, and dies after maturing its seed. The small white flowers resemble those of white clover, but are in long slender racemes instead of heads. The pod is egg-shaped, wrinkled, and contains a solitary seed. Young plants resemble alfalfa, both species belonging to the legume family; but it may readily be distinguished by the color of its bloom.

The plant has a strong odor, and the leaves a bitter taste due to cumarin. Cumarin is a vegetable substance usually obtained from the Tonka bean, but it also occurs in sweet clover and some other plants. It is well known to physicians, and has long been used as a corrective, tonic, and antiseptic in intestinal disorders. It imparts a characteristic flavor to certain kinds of Swiss cheese. Cumarin is believed to lessen the danger of bloating in cattle, which sweet clover causes much less frequently than the true clovers and alfalfa. It is much less bitter in early spring than later in the season. Cattle may at first refuse to eat the hay, but by sprinkling it with brine this difficulty may be readily overcome.

WIDE DISTRIBUTION AND ADAPTABILITY

White sweet clover is adaptable to great extremes of climate and soil, and will produce a valuable crop of forage in sections where alfalfa and red clover will not succeed. It is as vigorous in the severe climate of Quebec, Canada, as in middle

Alabama. It thrives in the semi-arid regions of Utah and Colorado as well as in western Washington, where there is a rainfall of over 100 inches. White sweet clover will grow on rocky limestone knolls and hills, almost bare of vegetation, and in a few years furnish excellent pasturage. If planted on poor sterile land it will deepen and enrich the impoverished soil and again render it productive; but it requires a compact seed bed and should not be planted on loose sand. It is more tolerant of poor drainage, overflow, and seepage than either alfalfa or the clovers.

Fig. 111.—White Sweet Clover (Melilotus alba). Photographed by Lovell.

It will grow well on wet lands near large reservoirs, which are useless for ordinary crops; but it can not be planted on the rice lands of the South. Provided they contain lime and the required bacteria, both clay and sandy loams will yield a good crop of this legume. It will endure more acid than clover and more alkali than alfalfa. While it grows most luxuriantly in fertile ground it can be seeded for pasture on the poorest fields. Eroded and gullied lands with thin and much depleted soils can be reclaimed and built up by its use. It is often abundant by the roadsides, on canal banks, and in waste places. It is destined to be the most valuable and most extensively cultivated leguminous crop in North America.

SWEET CLOVER HONEY

White sweet clover honey is white or nearly white; but, like alfalfa honey, under certain conditions of the soil it is light amber, or the darker color may be due to a honey from another plant. In portions of the East it is reported to have a greenish tint. The flavor is suggestive of vanilla, and by many it is regarded as a little too strong. When the nectar is secreted very freely the characteristic flavor is less pronounced. The body is medium. Sweet clover honey is now marketed by the carload, and the quality is generally admitted to be excellent, whether extracted or in the comb. It is an excellent honey for mixing with other honeys, and is often blended with that of alfalfa. The period of nectar secretion is long and the yield is heavy.

Occasionally there have occurred seasons when in certain localities it yielded little or no nectar. During a very dry year at Kenney, Illinois, bees were starving with 160 acres of sweet clover in full bloom. In 1914 sweet clover was reported to be a failure in Pendleton County, Kentucky. But usually the yield is very reliable, as is shown by numberless reports from many states. Hot, sultry weather, with sufficient moisture in the ground, is required to obtain the best results. It is to-day the most important honey plant in the United States, and the area under cultivation is yearly increasing. It will undoubtedly render beekeeping profitable in many sections where it is now only moderately successful.

The most important sweet clover areas, where it reaches its maximum development and its greatest value to bee culture, are in the north central states of Ohio, Indiana, Illinois, Iowa, and Wisconsin; among the limestone hills of Kentucky; in the limestone belt of Alabama and Mississippi; on the Great Plains in Kansas, Nebraska, and the Dakotas; and in the Rocky Mountain states of Colorado, Utah, Idaho, Wyoming, and Montana.

IN THE NORTH CENTRAL STATES

In Ohio, Indiana, Illinois, Iowa, Wisconsin, and Michigan, although sweet clover must compete with alsike, red clover, and alfalfa, its cultivation is steadily increasing. It grows spontaneously along tramped roadsides, abandoned roads, and in compact land everywhere. At one time an Ohio statute compelled its cutting as a noxious weed, like thistle and burdock, but it is now grown by hundreds of farmers under proper tillage. A few years ago the average farmer in Illinois ridiculed the claims of both alfalfa and sweet clover, but to-day on the banks of the Chicago Drainage Canal there are hundreds of acres of sweet clover. At Milledgeville it begins blooming early in July, and is in full flower at the time white clover and alfalfa have ceased to blossom. Where very abundant it has been known to yield nectar for two months. When pastured or mowed, it will bloom a second time and continue to bloom until after hard frosts. Bees have been seen on it in October when nearly all other plants were out of bloom. While it stands a drought well, it yields better when there are frequent rains. It is not unusual for the bees to store 50 pounds of honey per colony.

In Iowa the acreage of sweet clover has steadily increased until it is now found in every county in the state. There are two areas in which this valuable forage and honey plant is most abundant; one is in the extreme east, especially in Jackson County; the other is in the western part of the state along the Missouri River, from Sioux City for more than sixty miles southward. In Jackson County sweet clover is extensively cultivated and yields a surplus of honey every year, the average at Delmar for 37 years being 25 pounds of extracted honey. In northwestern Iowa, bordering the Missouri River, there are numerous hills of fertile loess soil, with sides so steep that they are liable to be badly eroded by heavy rains. When sweet clover is sown upon these hillsides it grows readily, and not only prevents the soil from washing away, but provides a supply of humus rich in nitrogen. In the region of Sioux City there are thousands of acres of sweet clover which support many colonies of bees, a single company operating 1200 colonies. It has been estimated that an acre of sweet clover is worth from $3 to $5 for bees alone.

ON THE LIMESTONE HILLS OF KENTUCKY

Sweet clover has had a wonderful development on the limestone hills of northern Kentucky, and a large area in the three counties of Pendleton, Bracken, and Robertson is devoted to its culture. Fifty years ago much of the land was planted with tobacco; in Pendleton County this was the chief agricultural industry. In this hilly country the fertile though shallow surface soil was gradually washed away by heavy rains, and the eroded and often gullied fields became bare and unproductive. Farm after farm was abandoned, and in many instances was sold for taxes. More than one-third of the population of Pendleton County moved away. Then sweet clover was introduced, apparently by beekeepers, and on the many limestone knobs and hills it found a most congenial home and multiplied apace, spreading in every direction. At first it was destroyed as a noxious weed likely to render the land even less valuable, but it outran the farmer and overran the fields. Gradually the soil became renovated and again became productive. Little by little the farmers returned, or new settlers bought up the abandoned farms. Dairy farming and the sale of sweet clover seed brought great prosperity and comfort. In Pendleton County alone there were at one time 50,000 acres of white and yellow sweet clover, that produced about half a million pounds of seed and a great amount of dairy products. Beginning about the first of June there is a continuous supply of bloom and nectar until late in the fall.

THE SWEET-CLOVER BELT IN ALABAMA AND MISSISSIPPI

In this section the white biennial, the annual white, and the biennial yellow sweet clover are found chiefly on the limestone hills and knolls of central and western Alabama and northeastern Mississippi. No other crop succeeds so well on this limestone soil, which in three years the sweet clovers deepen and improve so much that other crops may be profitably produced. In addition to renovating the fields it prevents the washing of hilly land and is excellent for fodder. In the black soil of the prairie section alfalfa is also grown. While sweet clover grows spontaneously in the limestone section, it has not extended to any great extent to the clay soil immediately adjoining; and so sharp is the line of demarcation that the abundance of sweet clover on the limestone soil and its absence on the clay soil a few feet away has often been remarked. In these two states it covers thousands of acres, and blooms in June, July, and the larger part of August. The larger apiaries range from 100 to 150 colonies, and not infrequently there are 200 or even 400 colonies in a single yard. Pellett reports one apparently well-authenticated record of 100 pounds per colony for 10 years, and states that along the line between Mississippi and Alabama for 100 miles north of Meriden there are many good locations. The farms are highly improved and there are numerous evidences of general prosperity.

Sweet clover grows along the rocky portion of the east coast of Florida, but in the interior there is not sufficient lime in the soil. In Texas the seasons are so dry that none of the clovers grow well except sweet clover. It is abundant in the northeastern part of the state in various places, and there are great areas of poor and waste land that can be planted with it to advantage. In many locations, where there are intervals without any bloom, it would be most helpful. In the western and southwestern part of the state it is too dry except along the streams. In Louisiana, along the rivers and bayous and around Shreveport all the clovers flourish.

IN THE WEST

Sweet clover has long been recognized as a valuable forage plant in the West, but in arid land it will not grow without irrigation. In many sections where irrigation is practiced and the water carries the seed into the alfalfa fields it is regarded as a weed; but in the Great Plains region in the states of Iowa, Oklahoma, Kansas, Nebraska, South and North Dakota extensive areas have become seeded with sweet clover. In Nebraska there are many scattered fields. In the eastern part of the sandhill district it has greatly improved the quality of the soil and increased the yield of hay. Since its introduction there has been an increase both in the quantity and quality of the honey. Sweet clover is adapted practically to all sections of Kansas, and in 1923 there were 48,000 acres under cultivation.

In South and North Dakota the future of sweet clover is believed to be of great promise. The sweet clover section extends more than 200 miles north of Sioux City, Iowa, and is rapidly spreading northward and westward. It is estimated that it will soon cover thousands of farms in the Dakotas, Wyoming, and Montana, and that this region will support thousands of colonies of bees and produce honey by the million pounds. The opportunities for beekeeping in South and North Dakota deserve careful consideration. In Colorado, on the western slope of the mountains, sweet clover grows vigorously and thrives in soils where alfalfa formerly died out on account of the alkali. It was extensively planted by the farmers, and large areas were reclaimed and rendered suitable for growing alfalfa. It was the history of sweet clover on the western slope that led to its introduction throughout the state. In the irrigated section surrounding Ferron, Utah, there are many farmers who are enthusiastic in its praise, and a large part of the land is devoted to this crop. It not only improves the poor soil but is considered in feeding value nearly equal to alfalfa. Throughout the Rocky Mountain region the onward march of afalfa and sweet clover offer much new bee territory, which will not be overstocked for a long time.

Annual White Sweet Clover (*Melilotus alba, variety*). Hubam clover. A few years ago the Iowa Agricultural College secured some 500 different lots of the seed of the common white sweet clover for trial, which were planted in greenhouses in January, 1916. All of the seed was supposed to belong to the common biennial (or two-year) white species. But about the first of March a few very large plants were observed by H. D. Hughes, in charge of the Farm Crops Section, which came from one special lot of seed. They were far superior in appearance to the other plants, and were nearly ready to bloom in less than three months from the time of seeding. By the middle of March they were 3 to 4½ feet tall and most of them were in full bloom, while the common biennial sweet clover was less than one foot high. There were 22 plants in this original lot, and they yielded enough seed in the greenhouse to grow a short row for each plant. They were not exactly alike, but varied in height and time of maturity. This seed was planted in the field the same year (1916), about the middle of June, and a thin seeding of oats was made with it. Other clovers were also planted at the same time for comparison. The oats were cut when the "heads were in the milk" without injuring the clover. The annual sweet clover plants were then about six inches tall. They now grew rapidly; and, like the 22 plants in the greenhouse, the seedlings differed greatly in height and time of blooming. A part bloomed 2½ months after seeding, while others required 3½ months. At 3½ months the best strain had reached a height of 4½ feet. During the same time the biennial, or common sweet clover, had grown only 12 to 14 inches, and the yellow sweet clover only 8 to 10 inches. Medium red clover, planted at the same time as the annual sweet clover, made a growth of only 3 to 5 inches, while the annual sweet clover grew 3 to 4½ feet. (Fig. 112.)

As soon as the plants had matured they were pulled and carefully examined. The root growth was found to be large and vigorous, but entirely different from that of the biennial sweet clover. The biennial at the close of the first season has a strong, large, succulent tap-root, much like that of the parsnip. At the top of this root about an inch below the surface of the ground there is a crown with 5 to 50 buds, which are ready to burst forth in early spring. But the root of the new clover was entirely different — there was no succulent tap-root or crown of buds formed for the renewal of growth the second season. The plant had made its full growth, bloomed, ripened its seed, and died — both stems and roots — clearly establishing the fact that this clover is an annual.

The annual sweet clover is believed to have originated on wild land in Alabama, and to be a sport or mutant of the biennial species. It probably did not first occur in a cultivated field, as in such a situation it would have been lost. It evidently grew on wild land for several years without anyone noticing that it was an annual. Recently it has been definitely established that it is still found on waste land in this state. Considerable sweet clover seed is harvested in certain sections of Alabama

1 2 3 4

Fig. 112.—Nos. 1 and 2, Hubam and annual yellow sweet clover planted the same day, cultivated and raised in the same row and under identical conditions. Nos. 3 and 4, Hubam and the old biennial sweet clover planted the same day, cultivated and raised in the same row and under identical conditions. (Courtesy Alabama Hubam Clover Association.)

by negroes, who either strip the seed from the standing plants or cut them down and thresh the seed out by hand. Thus the seed of the two clovers might have become easily mixed. Its subsequent discovery in an Iowa greenhouse was a fortunate event, which bids fair to be of great benefit to both farmers and beekeepers. (Fig. 113.)

Favorable reports from nearly all parts of the United States show that annual sweet clover can be successfully grown over a very wide area, and under very varied climatic conditions. It has succeeded from Alberta, Canada, to Texas, and from Pennsylvania to Oregon, and also in many foreign countries. Probably no other forage plant known will grow successfully under wider variations of soil and climate. At Ames, Iowa, the plants have often averaged a growth of over 1½ inches per day, with a maximum under greenhouse conditions of 2½ inches in 15 hours. Spikes of seed which measured 20 inches in length have been produced. Planted in rows 3 feet apart, the yield of seed was from 5 to 8 bushels per acre, while with closely spaced rows the crop of seed was over 10 bushels per acre. When seeded broadcast on a weedy and poorly prepared seed-bed the last of May, it overcame the weeds, made a growth of 5½ to 7 feet, and matured a seed crop. Individual plants in this state have attained a height of 9 to 10 feet.

The new annual white sweet clover fills a place which no other legume occupies. No other leguminous plant will so quickly furnish the farmer with hay and pasturage and the beekeeper with honey. A crop of hay, which the biennial sweet clover would require 15 months to produce, the annual variety will supply in four or five months. In much less time it will offer a rapid, rank-growing pasture to cattle and other stock. In localities where the natural grasses do not thrive well and the farmer or stockman has been dependent on the biennial form, a crop of forage can be harvested now in a single season. Hubam clover can also be used to advantage as a green manure crop. No cultivated legume will add more nitrogen to the soil. It has been estimated that 6.4 tons of dry sweet clover will furnish as much humus-forming material and as much nitrogen as 25 tons of the average farm manure. If the annual sweet clover is seeded with small grain in the spring, as wheat, rye, barley, or oats, after the grain has been cut it will make a good growth in the fall and can be ploughed under, putting the land in fine condition for corn in the following spring. As a cover crop in orchards it can be used to better advantage than many of the legumes now used for this purpose. On the other hand, it has been claimed that if the land is to remain permanently in sweet clover the biennial variety has the advantage of blooming longer and yielding a better crop of hay. It is less difficult to get a stand of it. The annual must be planted as early as the first of April in order to mature the seed, and this is not always possible. The relative merits of each clover are not yet fully established.

Fig. 113.—These Medina fields of biennial clover above, and of Hubam clover at the right, were both planted June 17, and photographed October 18. Note the difference in growth.

Finally, annual sweet clover secretes nectar as fully and freely as do the biennial white and yellow sweet clovers. It blooms three months after seeding, and the blooming period lasts for five or more weeks. Plants from seed sown early in March at Bay City, Texas, were beginning to bloom the last of May, and by the 15th of June had reached an average height of 4 feet and were a solid mass of flowers. Honeybees worked on the bloom from sunrise until evening, but were most numerous and seemed to secure the most nectar after 10 o'clock in the morning. A beekeeper at Joliet, Illinois, extracted six times at all yards where Hubam clover was growing. Many colonies produced 400 pounds or more of honey each. The honey, like that of the biennial variety, is light amber and has a pleasant aromatic flavor. A southern beekeeper reports that the honey does not differ in color, flavor, or body from that of biennial sweet clover. It granulates readily. Blended with another honey, as that of tupelo, the taste is improved. But a northern beekeeper describes its color as white to light amber without the greenish tinge observed in the honey from the biennial form, and the flavor as resembling that of white clover honey.

Before planting, the seed should be scarified. In an experimental test before scarifying only about 34 per cent. of the seed germinated; but after scarifying, 91 per cent. Nearly every seed company in the United States and many individual seed growers and farmers are now using the Ames hulling and scarifying machine, perfected and given to the world by the Iowa Agricultural Experiment Station. The soil should also contain an abundance of lime and the proper sweet clover bacteria. If the soil is acid and the bacteria are absent, the plants will probably not grow more than a foot tall. It is of the greatest importance, therefore, in fields where such conditions prevail, that lime should be worked into the surface, and the soil inoculated with soil gathered from an alfalfa or sweet clover field.

SWEET FENNEL (*Foeniculum vulgare*).—Cultivated in Europe for its sweet aromatic foliage. Escaped from cultivation, it grows wild in waste places from Pennsylvania to Louisiana. The large umbels of yellow flowers have been reported to yield a light amber honey in California. They appear, however, to be more attractive to wasps than to honeybees.

SWEET PEPPERBUSH.—See Clethra.

TARWEED (*Hemizonia*).—A genus of annuals, with viscid, unpleasantly scented foliage and yellow or white flowers, belonging to the Compositae. Yellow tarweed (*H. virgata*) yields in August for about 20 days a light amber honey of good flavor. It is common on the plains of the Sacramento and San Joaquin valleys. Another species (*H. fasciculata*) is abundant on the mesas of southern California. The honey is dark amber and ill-scented like the plant, granulating in two or three months after extracting. It is used chiefly for manufacturing purposes. Hay-field tarweed (*H. luzulaefolia*) is abundant in the hay-fields of the Sacramento and San Joaquin valleys. The lemon-yellow flowers appear in April and May. Coast tarweed (*H. corymbosa*) is common in the fields of the coast counties of Santa Cruz and Monterey. The yellow flowers bloom in June and July.

TEASEL (*Dipsacus Fullonum*).—A tall herb with prickly stems and small lilac-colored flowers in oblong heads. The linear bracts or leaves underneath the flower-cluster are spiny-pointed and are used for carding milled woolen cloth. In central New York it was formerly cultivated for this purpose, and, in 1910, 52 farms reported 110 acres of teasel. In 1919, 78 acres were harvested. The flowers are adapted to pollination by bumblebees, the corolla-tubes being more than 12 mm. long, or twice the length of the tongue of the honeybee; but as the tube flares at the mouth a part of the nectar can be reached by honeybees.

In the American Bee Journal (Aug. 18, 1886) the honey gathered from teasel is described by G. M. Doolittle, of Borodino, N. Y., as follows:

"Bees work on teasel all hours of the day, and, no matter how well basswood may yield honey, a few bees will be found on teasel. The honey from teasel is very thin and white—in fact, the whitest honey I ever saw; but it is not of as good flavor as either clover or basswood. This thinness of the nectar, and its coming just when basswood does, is the great drawback to it. Coming as it does with basswood makes it of no great advantage except that it usually lasts six to ten days after basswood is gone.

"As to what proportion of my honey has come from teasel the past fifteen years, I should say about one-tenth, some years more and some years not a single pound. In 1877 I got the largest crop, while from 1878 to 1882 little if any was obtained."

Teasel is so rare as to be of only minor importance; and in view of the length of the corolla-tube it seems rather probable that its value has been much overrated. Fuller's teasel. Card thistle. Clothiers' brush.

TENEZA (*Leucaena pulverulenta*).—A large Mexican tree growing in the rich bottom-lands in Hidalgo and Cameron counties, Texas, along the Rio Grande. It was formerly placed in the genus *Acacia*. The leaves are pinnate, and the white flowers, which are produced in great abundance, are in globose heads. It is reported as important at Brownsville, in the extreme south of Texas, and to yield nectar from April to September. Also called Tennazza.

TEXAN EBONY (*Siderocarpus flexicaulis*).—A small evergreen tree or shrub with pinnate leaves and fragrant yellow flowers, growing on bluffs in southern Texas and northern Mexico. At Brownsville it blooms in June, but, like many other Texas plants, it blossoms several times if there is much rain. It is reported to yield a small surplus of honey of good quality. This species resembles an Acacia in both foliage and flowers, in which genus it was formerly placed.

THOROUGHWORT.—See Boneset.

THISTLE.—See Canada Thistle, Blue Thistle, and Napa Thistle.

THYME (*Thymus Serpyllum*).—A creeping perennial plant, with sweet aromatic foliage, cultivated as a culinary herb, and growing wild from Nova Scotia to Pennsylvania. The purplish flowers, which are in dense whorls, bloom from July to August and are very attractive to bees. The nectar is very abundant and has an aromatic flavor.

TIE-VINE (*Ipomoea trifida*).—At Victoria, Victoria County, Texas, from 60 to 120 pounds of honey are gathered in spring and summer from horsemint and cotton and in the fall from tie-vine and aster. Tie-vine grows in the bottom-lands and black lands, and blooms in late summer and fall. The flowers are pink tinged with blue. The honey is mild-flavored, heavy, and amber-colored.

TITI.—The titi family, or Cyrillaceae, contains but 6 species, which are found only in America. They are shrubs or small trees growing in wet land, or swamps, and along rivers. There are 3 species in the southern states, which are valuable as honey plants. The honey has been frequently described as dark colored and poor flavored compared to northern white honey.

Black titi (*Cliftonia monophylla*). Buckwheat-tree. Ironweed. Spring titi. A smooth evergreen shrub, or small tree, common in swamps in Georgia and Florida and westward to Louisiana. The white fragrant flowers are in long racemes, which are drooping when young but finally become erect. In southwestern Georgia the blooming period extends from the last of February to the first of April. It is a very reliable honey plant and yields a large surplus. When the weather is fair the bees bring in the nectar very rapidly, working early and late on the bloom. The honey is light amber in color, with a rather thin body, and after extraction granulates quickly. Its flavor is a little strong with a slight bitter after-twang. As a comb honey it sells readily in the southern market. (Fig. 114.)

Small-leaved or red titi (*Cyrilla parvifolia*) is an evergreen shrub, 6 to 10 feet tall, growing in swamps and along streams from Florida to Louisiana. The leaves are oblong, leathery, shining green above but paler below. The numerous small white flowers are in racemes and appear in February and March. It yields an amber-colored honey, which is strong-flavored, but suitable for baking purposes.

White titi or ivory bush (*Cyrilla racemifolia*) is also a swamp shrub, or small tree, but it is more widely distributed than the preceding species, extending from Virginia to Florida and westward to Texas. The large much branched bushes, from 5 to 10 feet tall, are during the last half of May covered with innumerable small white blossoms. The flowers are in narrow, dense racemes, which are clustered at the ends of the twigs. It is the last of the spring honey plants to bloom and the

Fig. 114.—Black Titi, or Spring Titi (Cliftonia monophylla). Photographed by J. J. Wilder.

bees work on the bloom very diligently, but it yields nectar only sparingly. Although the bloom lasts for thirty days, it is seldom that more than two supers, or 50 pounds of honey, are secured. A sample of the honey from Mt. Pleasant, Alabama, is a dark reddish-amber color, with a pleasant characteristic fragrance. It has a good body and a mild flavor, which was more pronounced when it was first gathered. It is considered a good table honey. Summer titi.

TOBACCO (*Nicotiana Tabacum*).—There are about 50 species of this genus, all of which, with the exception of two species, are natives of America. The source of the larger part of the tobacco of commerce is *N. Tabacum*, a coarse, rank-growing annual, with a single unbranched stalk, 6 feet or more tall, bearing large, oblong leaves covered with long soft hairs which exude a viscid juice. The stem terminates in a cluster of rose-colored flowers, which have a funnel-formed corolla 2 inches long. Nectar is secreted at the bottom of this tube on the lower side of the ovary. The flower is adapted to pollination by butterflies and moths; and in Jamaica 100 moths, belonging to several species, were observed to visit the bloom during three evenings. Gould saw humming-birds visit the blossoms in Mexico. But the throat of the corolla-tube is somewhat inflated so that honeybees are able to creep within and gather nectar. If the flowers hang downward the nectar flows toward the entrance and is in consequence more easily gathered. In a neglected field of tobacco in Porto Rico, full of grass and weeds, in which both the leaves and flowers inclined downward, honeybees were observed gathering nectar abundantly; while in adjoining fields, which were well cultivated, the flowers stood erect and not a bee was seen on the bloom.

At White Plains, N. C., the farmers, in 1919, for the first time let their tobacco bloom late in the season. A large quantity of nectar was gathered. The unripened honey had the flavor of green tobacco. Hundreds of acres of tobacco with myriads of flowers are offered to honeybees, and they store honey as rapidly as during the earlier flow.

In Connecticut the honey flow from tobacco comes between the flows from buckwheat and fall flowers. Thus beekeepers need not fear the storing of tobacco honey, as the sections of early honey have been removed before it begins to yield, and it is largely used for winter, insuring a good fall flow, which in this state has been the exception rather than the rule. As a winter food for bees it is open to no objection. A cool dry summer checks the secretion of nectar to a great extent. One hundred pounds per colony in sections have been reported from this source. The honey has a dark brownish color and compares not unfavorably with buckwheat honey. No disagreeable results follow its free use in the family as an article of food. Sections of tobacco honey are reported to sell as well as any of the darker grades of honey.

TOCALOTE.—See Napa Thistle.

TOOTHACHE-TREE.—See Prickly Ash.

TORNILLO (*Strombocarpa odorata*).—Screw bean. A common large shrub in the river valleys in the southern portion of New Mexico, with pinnate leaves and yellow flowers in crowded spikes. A minor honey plant in New Mexico. The pods contain a large amount of sugar and are very sweet.

TOYON.—See Christmas Berry.

TREE OF HEAVEN (*Ailanthus glandulosa*).—Chinese sumac. Varnish tree. Introduced from Asia, and widely cultivated for ornament, but growing wild in the eastern United States. In Texas and California the greenish white flowers, which open in June, yield an abundance of poorly flavored nectar. The staminate flowers are ill-scented. The pinnate leaves also bear nectaries.

TULE MINT.—See Mint.

TULIP TREE (*Liriodendron Tulipifera*).—Other vernacular names are whitewood and yellow poplar from the varying colors of the wood, canoe-wood from the use made of it by the Indians, and saddle-tree from the arrangement of the leaves in the bud. This magnificent tree belongs to the same family as the Magnolia, and among American deciduous-leaved trees is surpassed in size only by the plane or buttonwood, to which it is superior in symmetry and in the attractiveness of its foliage and flowers. Its height is usually from 60 to 90 feet, but in favorable localities it may grow 140 to 180 feet tall, with a diameter of 4 to 12 feet. Michaux measured a tree near Louisville, Ky., which at five feet from the ground was 22½ feet in circumference and exceeded 120 feet in height. The tulip tree is one of the handsomest of American ornamental trees, growing in a conical form, offering an extensive shade, and putting forth in May or June an immense number of large greenish-yellow flowers. The peculiar-shaped leaves easily distinguish it from all other forest trees. They are four to six inches long, four-lobed, with the end abruptly truncated, or broadly notched, and have a smooth bright green surface. The bark, which is broken into large flat ridges, has a very bitter taste and was used by the Indians as a remedy for intermittent fevers. (Fig. 115.)

The slightly fragrant bell-shaped flowers are two inches long, solitary, and terminal. The calyx is composed of three oval concave sepals of a pale greenish color, which finally become reflexed. There are six large yellowish-green petals, each of which is marked at the base with an irregular crescent-shaped, bright orange-yellow spot. The stamens are numerous with short filaments. In the center there is a cone-like mass of pistils (carpels). The seeds are winged and form a dry cone 3 inches long, which falls apart in autumn. The flowers are very frequently visited by bees and also by humming-birds.

The tulip tree is found in rich woods from Massachusetts and Michigan southward to Florida and Mississippi and westward to Arkansas and Louisiana. It succeeds best in a fertile loamy soil, such as occurs in river-bottoms and on the borders of swamps. As a source of honey it is important in southern Virginia, West Virginia, Kentucky, Tennessee, Maryland, North Carolina, South Carolina, and northern Georgia. In southern Virginia on the Piedmont Plateau tulip tree and sourwood are the only plants which yield a large surplus. In the rugged wooded region of southwest Virginia tulip tree, sourwood, black locust, and basswood furnish a large amount of honey. Along the Ohio River in West Virginia tulip tree is likewise abundant. On a tract of land 625 acres in extent near the Ohio River between the Great Kanawha and Big Sandy Rivers there were counted 16,987 trees, of which 858 were tulip trees. This characteristic tree was at one time very common in Kentucky; and, although merchantable trees have been largely cut for lumber, it still furnishes a part of the surplus in the less thickly settled sections of the state. Young trees are rapidly springing up and beginning to bloom. On the ridges and tablelands of the eastern and central regions of Tennessee tulip tree and sourwood are the most important sources of honey. The former blooms about the first of May and yields heavily for about two weeks.

In Maryland above the "fall line" on the Piedmont Plateau tulip tree is sufficiently abundant to yield a honey crop regularly. On the Coastal Plain it never furnishes a surplus. Formerly in central Maryland it was one of the main surplus-honey plants, and it is still important in Montgomery County, where it is associated with chestnut, walnut, and maple. But it has been so largely cut for pulp wood that there has been a great decrease in the quantity of honey obtained. No other honey plant in North Carolina has so wide a distribution as tulip tree. It is found in all parts of the state except in the eastern lowlands. It blooms from May 10 to 30, the date varying somewhat in different localities. Tulip tree is likewise widely distributed in South Carolina, but is most common in the Piedmont region. It extends over northern Georgia, where it is usually a reliable source of honey, and it is also found in the mountainous section of northeastern Alabama.

The nectar may be seen in both large and small drops on the orange-yellow portions of the petals, on the inner side, which thus serve as both nectaries and nectar guides. The time of blooming varies with the conditions of the weather from the last of April to the first of June. When the blossoms are late in opening and the

Fig. 115.—Tulip Tree (Liriondendron Tulipifera). Photographed by H. Garman.

weather is warm and dry, the honey flow is very much heavier than when the bloom is early. Under such conditions there are few if any better honey plants than the tulip tree, and each flower will yield not far from a spoonful of nectar. When the flowers appear early in the season the flow is often interrupted by cold rains. A large quantity of honey is stored even when the trees are scarce, and one or two supers are often filled from this source alone. Where the trees are abundant there is little danger of overstocking, and it has been estimated that 200 colonies could not take all of the nectar within their range. Unfortunately, there are to-day few such locations, and they are in regions difficult of access.

The honey obtained from tulip tree is bright amber when new, but it becomes darker with age, and very thick, so that it closely resembles molasses. In quality it is fair, somewhat strong, but with a rather pleasant flavor. It is in good demand locally throughout the South, but it does not sell well in the general market. As it is gathered early and does not command the highest prices, it can be used to advantage in brood-rearing and increasing the strength of the colonies for gathering the lighter-colored honeys which come later. The tulip tree is a host for an abundance of plant lice in late summer, which furnish considerable honey-dew.

The seed should be sown as soon as ripe in moderately dry fertile soil, and should be protected during the first winter. The wood is soft and fine-grained and is easily worked; it is usually nearly white, but in some localities is yellowish. It shrinks badly in drying and consequently is not adapted to exposure to the weather.

When dry it resists decay and is rarely attacked by insects. It may be used for sections and brood-frames, but is very unsatisfactory for hives. It is suitable for door panels and wainscotting and for the manufacture of carriages, furniture, and various small articles. As the wood is light and strong, the Indians used it in building great canoes, capable of carrying 20 persons or more.

TUPELO (*Nyssa*).—*Nyssa* is a small genus containing only 7 species, of which five occur in North America and two in southern Asia. Four of the American species are trees, and one is a shrub. The leaves are alternate, thick and leathery, almost entire, oblong, or obovate, and are brilliantly colored in autumn. The flowers are small, greenish, and appear with the leaves. The stamens and pistils are usually in different flowers on different trees, the staminate are clustered, and the pistillate solitary or two to three together. The name of the water nymph *Nyssa* was given to this genus on account of the aquatic habit of the species.

White Tupelo (*Nyssa aquatica*). White gum. Cotton gum. Water tupelo. Tupelo gum. Swamp tupelo. In river swamps in the coast region from southern Virginia to northern Florida, westward to the Nueces River, Texas; northward through Arkansas, west Tennessee and west Kentucky, and southern Missouri to the lower Wabash River, Illinois. A large water-loving tree, attaining a height of 100 feet and a diameter of 4 feet. The bark is dark brown in color, ridged and broken into small scales. The leaves are thick, oval, pointed at the apex, dark green and shining above, paler and pubescent below. The small greenish flowers open in April and May; the staminate are in dense round heads; the pistillate or fertile are solitary on slender stalks. The blue-purple fruit ripens in September. The wood is soft, but can be used for crates and packing-boxes. In the older floras the Latin name of this species is given as *Nyssa uniflora*.

The honey of white tupelo has a very mild exquisite flavor and a thick body, and is very light in color with a pale lemon hue, which renders it very attractive in glass containers. The bulk of this honey is produced in the extracted form and shipped northward in 30-gallon barrels. It is in great demand among northern dealers in honey, who prefer it because it does not granulate. The nectar is secreted very copiously and a great amount is collected by the bees in a few weeks; but they are not numerous enough to harvest more than a small part of it. In pine-barren ponds the white tupelo is often a small tree, which may be readily mistaken for a distinct species.

Black Tupelo (*Nyssa biflora*). Black gum. Water gum. Water tupelo. This species has a much more restricted range than white tupelo, extending only from Montgomery County, Maryland, to Florida and central Alabama. A large tree attaining a maximum height of over 100 feet, with a rough, ridged, dark-brown bark. The oval leaves are smaller than those of the white tupelo, smooth on both sides, blunt-pointed, with entire margins. The blossoms appear in April and May, and the dark-blue, plum-shaped fruit in early fall; the staminate flowers are clustered, the pistillate are two together, instead of solitary as in the white tupelo. Black tupelo throughout its range is usually associated with white tupelo, but it often extends to higher land. It is abundant along lake margins and on the bottomlands of small streams. The fruit, which falls into the water, sometimes accumulates in large heaps in sheltered coves.

The beginning of the honey flow is determined by the length of time the lowlands are covered by water. If there has been no overflow in early spring the trees in northern Florida will bloom in March, and the honey flow will last for three weeks. But if there has been much rain and the rivers have flooded the bottomlands, the blooming time will be much delayed. The honey when first gathered is thick, light in color, and very mild in flavor; but with age it grows darker colored and stronger flavored.

Sour Gum (*Nyssa sylvatica*). Pepperidge. Black gum. This large forest tree, 100 to 150 feet tall, is the giant of the tupelos, and has a much wider distribution than any other species. It extends from the Kennebec River, Maine, Ontario, and southern Michigan to the Kissimmee River, Florida, southeastern Missouri, and the Brazos River, Texas. The seventh edition of Gray's Manual ranks the black tupelo (*N. biflora*) as a variety of this species, and undoubtedly beekeepers very often confuse the two trees. While pepperidge grows on high land it requires moist soil. In North Carolina the black tupelo (*N. biflora*) is common in the southeastern swamps, while pepperidge (*N. sylvatica*) extends westward to the center of the state. Both species are called black gum in this state, and the beekeepers do not distinguish carefully between them.

The bark and leaves of the sour gum or pepperidge are very similar to those of the black tupelo. The flowers open from April to June, according to the locality; the staminate are in dense clusters, the pistillate 3 together. While the bloom yields nectar, it is apparently of much less value to the beekeeper than either the white or black tupelo.

Fig. 116.—Black Gum (Nyssa biflora). Photographed by J. J. Wilder.

Ogeche Plum (*N. Ogeche*). Ogeche lime. Wild lime tree. Gopher plum. A small tree, reaching a maximum height of 60 feet, but seldom more than 30 to 40 feet tall. Common in the river swamps of South Carolina, Georgia, and Florida. The greenish-yellow flowers appear from January to May, the staminate in round heads, the pistillate solitary. The fruit is red and very acid. In Florida it has been reported to bloom just before the white tupelo and to yield a white, thin honey.

Bush Tupelo (*N. acuminata*). This species is a mere shrub, growing 6 to 10 feet tall, and peculiar to the pineland swamps near the coast of Georgia. The bark is smooth and the branches and twigs red. The honey is thick and white, closely resembling that of white tupelo, but has a greenish tinge.

The white and black tupelos (*N. aquatica* and *N. biflora*) are commonly found in the same swamplands from Virginia to Florida. In the swamps near Norfolk,

Virginia, they are important sources of nectar, and only rarely does the beekeeper fail to obtain a surplus. In the southeastern swamps of the Costal Plain of North Carolina the tupelos are very abundant, while gallberry covers much of the higher land. A large number of colonies of bees are successfully operated in this section, but much of the region is still unoccupied by beekeepers. In the river swamps of South Carolina both white and black tupelo are again common and yield a large portion of the crop of honey. All five species of tupelo native to the United States occur in southwestern Georgia. It is in this section of the state that the largest and most advanced beekeepers are located, who operate thousands of colonies of bees.

The most famous section of Florida for beekeeping is the northwestern part of the state along the Apalachicola and Ocklocknee rivers, where white tupelo, black tupelo, and spring titi are abundant. From this section comes about one-third of the honey crop of the entire state. About 50 miles from the point where the Apalachicola River enters the Gulf of Mexico the river has low banks, the main channel breaking up into small streams which wind through the marshland. This strip of bottom-land is about 10 miles wide and is covered by a luxuriant growth of tupelo trees. As it is overflowed in the rainy season, it is necessary to place the hives on platforms, 6 to 10 feet high. The season opens with the blooming in March of black titi, and a little later in April and May the flowers of the tupelos open. So copious is the flow that an average of 70 pounds per colony is extracted each year, and in some seasons 100 or 150 pounds; 250 barrels of extracted honey have been secured in 26 days. Vast quantities of nectar go to waste, and it is doubtful if this region will ever be fully stocked with bees. A full crop is obtained three years in five, and there is never less than a quarter of a crop. But many do not care to live in a desolate, unwholesome swamp, where malaria and mosquitoes are prevalent, roads are absent, and the only signs of civilization are sawmills. Other objections are the shortage of pollen and the absence of late-blooming plants to maintain the strength of the colonies in the fall.

In the southeastern corner of Alabama the surplus honey plants in spring are titi, the tupelos, and gallberry. The yards, which are usually small in size, in a favorable season average 60 pounds of surplus per colony. Although there are a large number of colonies in this region, it is considered better adapted to queen-rearing than honey production. At Mount Pleasant on the Alabama River the main crop comes from the tupelos, holly, blackberry, and velvet bean. This is one of the best locations in the pine barrens. Beyond Alabama the black tupelo (*N. biflora*) ceases to be an important source of honey, although it has been reported as far west as Louisiana; but the white tupelo (*M. aquatica*) and the sour gum or pepperidge (*N. sylvatica*) have a much wider distribution. In the Yazoo Delta, Mississippi, the white tupelo does not appear to be one of the surplus-making plants, since none of the beekeepers secure a surplus before June, and the white tupelo blooms much earlier. In the flood plains of Louisiana white tupelo yields a veritable flood of nectar, and the bees are busy on the bloom from early dawn until sunset; but the flow is of short duration. The river swamps of eastern Texas mark the western limit of the white tupelo. The banks of the Nueces River are lined with white tupelo and sour gum (*N. sylvatica*), the belt of timber being from one to two miles wide. In southern Arkansas white tupelo is valuable, and sour gum has also a wide distribution. White tupelo is also listed as a honey plant in western Tennessee and western Kentucky.

TURPENTINE WEED.—See Blue Curls.

TURKEY MULLEIN (*Eremocarpus setigerus*).—Also called Yerba del Pescado in California, because the Indians in that state used the strong-scented foliage to stupefy fish in small streams, in order that they might catch them by hand. In Orange County, California, it is known as "woolly white drought weed," since it is covered with a grayish-white pubescence. It springs up in dry grain fields after the grain has been harvested and gives a silvery appearance to this stubble land. According to Jepson it is very abundant in the interior of California and in the plains of the Sacramento and San Joaquin valleys. It blooms in late summer, yielding an amber-colored honey.

VANILLA-PLANT (*Trilisa odoratissima*).—Deer's tongue. A perennial herb, 2 to 3 feet tall, belonging to the Compositae. The lower leaves are spatulate or tongue-shaped, the upper, oval and sessile; the whitish or purple flowers are in terminal clusters. In southwestern Florida this plant has been reported to cover scores of acres and to yield nectar plentifully.

VARNISH TREE.—See Tree of Heaven.

VERVENIA.—See Phacelia.

VERBENA.—See Vervain.

VERVAIN, PURPLE (*Verbena hastata*).—Purple vervain and hoary vervain (*V. stricta*) occasionally yield a surplus at Center Point, Iowa, and vicinity. They grow best on low moist pasture land, and in an average season, while in some fields quite thick, they are for the most part very scattering. Two or three seasons of excessive moisture and rather cool weather are required to render the plants sufficiently abundant to yield a surplus, and such conditions have occurred only twice in the past 24 years. Then nearly every pasture was blue with them, and they remained in bloom for four or five weeks. The low rich pasture lands looked like a great blue sea. The flowers began to open in July and the blooming period lasted through August, until the colonies had filled the sections in two or three supers. The honey is water-white, or fully as white as clover honey, and were it not for a greater or less admixture of heartsease honey, which is always abundant at this season of the year, it would be a mild-flavored honey. The comb has a bluish tinge due to a little of the blue pollen being mixed with the cappings, giving it a pretty appearance, but which would prevent its passing for clover honey. It is not quite so thick as clover honey, and does not granulate as quickly; in fact, comb honey from this source did not candy at all.

The purple vervain grows in damp fields and pastures from Nova Scotia to Oregon, and southward to Florida and Texas. It is absent from the southwestern states, although found in California on the lower islands of the Sacramento River. It is a perennial herb, growing from 3 to 6 feet tall. The stems are square, and the leaves oblong, lance-shaped, and roughish. The purple flowers are small, sessile, and in long erect spikes, as shown in the photograph. The corolla-tube is a little shorter than the tongue of the honeybee, so that it can easily gather the nectar. (Fig. 117.)

In New England it is so rare as to be of no value as a honey plant. At Oak Ridge, Passaic County, New Jersey, honeybees are reported as working on the flowers, and it is listed among the honey plants of this state, but it is of little value. It also finds a place in the honey flora of Nebraska. Besides honeybees, bumblebees, solitary bees, wasps, flies, and butterflies frequently visit the bloom.

In Texas blue vervain (*V. xutha*), which grows in sandy soils from Louisiana to southern California, is the source of a small amount of honey. It blooms from April to August.

In the dry open hilly country of western California the spreading vervain (*V. prostrata*) is not uncommon. It is reported to be a good bee plant. The European vervain (*V. officinalis*), according to M. P. Fabreques' *Bee Flora of Spain*, produces a bitter astringent honey. This certainly seems improbable.

VELVET BEAN (*Mucuna utilis*).—In 1909 only 12,560 acres were reported; but in 1918 there were 4,600,000 acres under cultivation. It is grown chiefly in North Carolina, South Carolina, Georgia, Florida, Mississippi, and Alabama, where the bees gather nectar from it. The vines are too coarse and densely matted to be used for a hay crop, but ground together the bean and pod produce a palatable and nutritious feed. If the bean is thrashed out a more concentrated feed is obtained. The nectar is very thin when brought into the hive, but it thickens as the honey ripens. The comb honey is white, and the extracted honey has been reported as

Fig. 117.—Purple Vervain (Verbena hastata). Photographed by Lovell.

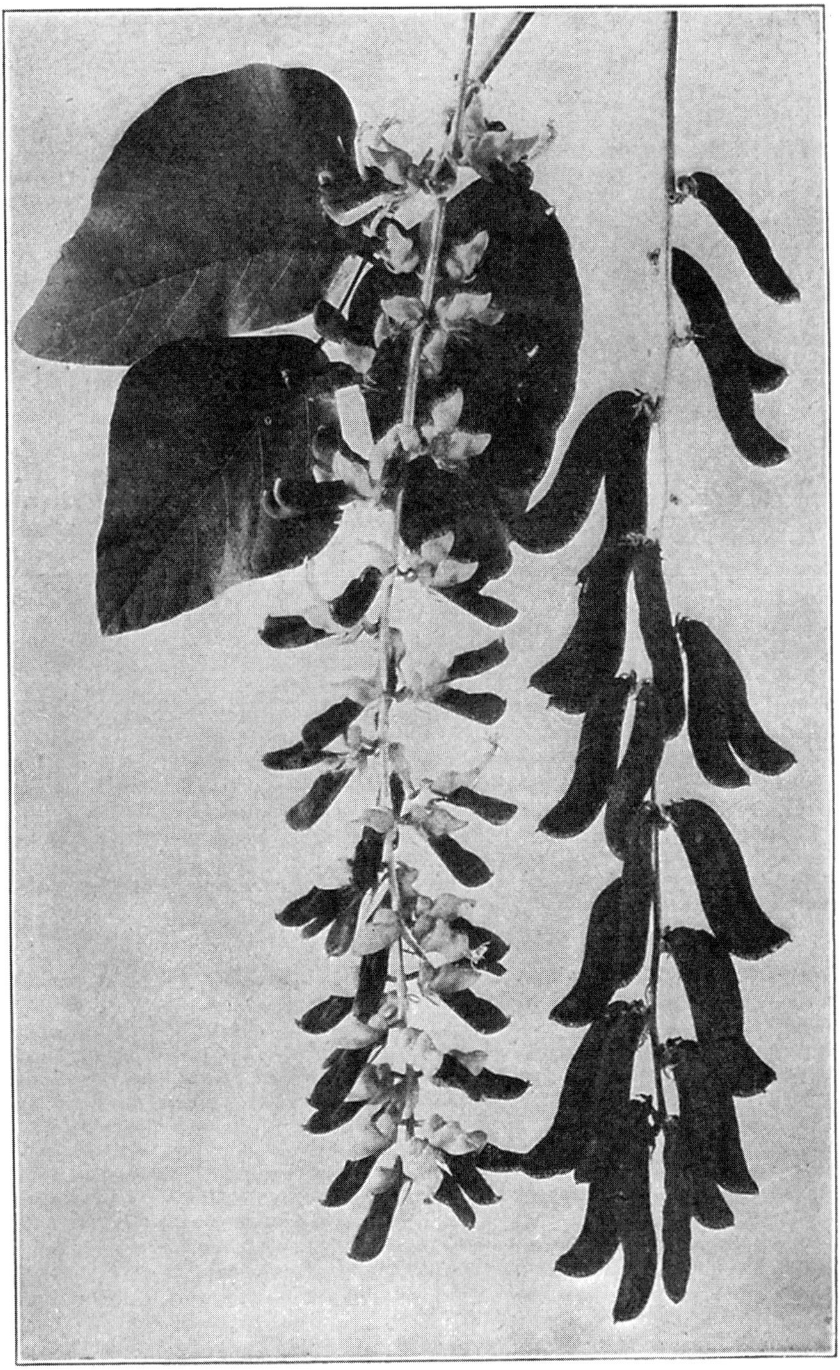

Fig. 118.—Velvet Bean, one-third natural size. Photographed by Bureau of Plant Industry, United States Department of Agriculture.

wine-colored; but a sample from Georgia is light brown, with the odor of vinegar and a mild acid flavor. It granulates quickly with the approach of cold weather. (Fig. 118.)

VETCH (*Vicia villosa*).—Sand or winter vetch is an excellent soiling crop, which has been grown successfully in the wet coastal regions of Oregon and Washington, the dry prairies of South Dakota, and the rich loams along the Gulf of Mexico. The corolla tube is 12 millimeters long and the flowers in the experimental garden were visited only by bumblebees. But there are extra-floral nectaries on the leaves, which honeybees have been seen to work faithfully in western Washington. The honey is white, and mild in flavor. The common vetch (*V. sativa*) is at times a source of a heavy white honey gathered from the leaf nectaries. The flowers of the common vetch (*V. Cracca*) are visited by honeybees. (Fig. 119.)

VIBURNUM.—See Black Haw.

VINE.—See Grape.

VINEGAR WEED.—See Blue Curls.

VIRGINIA CREEPER.—See Clematis.

WATERMELON (*Citrullus vulgaris*).—The flowers are monoecious, and dependent on bees for pollination. Watermelons are grown chiefly in the South and in New Jersey, Delaware, and southern Illinois. See Cucumber.

WATTLE—See Acacia.

WHITE BRUSH (*Aloysia ligustrina*).—Mexican heliotrope. A small shrub with sweet aromatic foliage, which is eaten by sheep and goats, and white flowers tinged with violet. In southern Texas, west of the Colorado River, it forms impenetrable thickets, often many acres in extent. About 5 days after every rainfall of more than an inch, according to H. B. Parks, it blooms from 5 to 7 days. In 1923 near San Antonio it bloomed seven times, the first time in May and the last time in November. In 1924 it bloomed only in June and November. During the short blooming period bees store from 3 to 6 pounds of honey. But if there are numerous rains about a week apart white brush may remain continuously in bloom, as in 1919, when well-filled supers of honey were obtained. The honey is light amber-colored with a mild flavor. (*Lippia ligustrina.*)

On the Edwards Escarpment westward to southern Arizona much larger yields of a similar honey are obtained from *Aloysia Wrightii.*

WHITE SWEET CLOVER.—See Sweet Clover.

WILD ALFALFA (*Lotus glaber*).—Deer-weed. Wild broom. A shrubby plant, 2 to 5 feet tall, belonging to the pulse family, or Leguminosae. In the western states there are some 40 species belonging to the genus *Lotus.* In California wild alfalfa is common throughout the Coast Ranges and in southern California, blooming from June to September. As the plant dies out every two or three years, it is not a reliable honey-producer every season. Says Richter: "Some years in some sections yielding twice as much as the sages; this is true for either the coast or the valley side of the Coast Ranges, yet a good wild alfalfa honey flow on the east side does not necessarily mean that such is the case on the west side. Beekeepers report wild alfalfa honey as being white, light amber, amber at times with a characteristic green tinge. It is one of the main honey plants of the Coalinga district." It comes up in ground that has been burned over, and is of little value as a forage plant, as the stalks, though fine, are tough and woody. Where it has water it grows throughout the year; but if it is dependent on rainfall and not on irrigation, it dries up and drops its leaves about the first of July, after maturing its seed crop. Then the stalks change in color from green to a reddish tint. It is this plant which gives its hue to wide areas of the plains and hills at this season of the year.

WILD BUCKWHEAT (*Eriogonum fasciculatum*).—This bushy shrub, which is two to four feet tall, is an important honey plant in California. The small white

Fig. 119—Purple Vetch (Vicia Cracca). Photographed by Lovell.

flowers are in umbellate clusters, which bloom from spring until fall. In southern California the honey flow comes in July and August, usually later than that of the sages; but when it comes at the same time, the sage honey is colored amber. At Acton, in Antelope Valley, Mendleson has obtained crops of pure wild buckwheat honey, but in Ventura County it is always mixed with honey from other plants. The honey is dark amber but has a fine flavor. A chemical analysis shows that it is very similar in composition to that of cultivated buckwheat, but it contains a slightly larger per cent. of sucrose; the color is much lighter and the flavor much better than that of the cultivated buckwheat. This shrub is very common on the plains and mountain slopes of southern California. (Fig. 120.)

WILD CUCUMBER (*Echinocystis lobata*).—A tall climbing vine, cultivated for arbors and porches. The small white flowers are very numerous. It has escaped from cultivation, and grows wild along the rivers from New England to Texas. It is often very abundant in damp woodlands, where, as in the valley of a stream at Humboldt in the southeast section of Nebraska, it festoons the trees on all sides. It is reported to yield a light amber honey of good quality in localities in the bottom-lands of the Missouri and Mississippi rivers. (Fig. 121.)

Near Kansas City there is a Missouri River bluff about 3 miles long and 150 feet high. This bluff is completely covered with wild cucumber vine. It blooms for 4 or 5 weeks and yields large quantities of nectar. Four stands of bees averaged 60 pounds each from this source, there being little else for them to work on. The light amber-colored honey has an excellent flavor, and is very thick.

WILD CARROT.—See Carrot.

WILLOW (*Salix*).—At Borodino, New York, bees have often made a gain of from six to ten pounds of honey from willow bloom, and one season they stored fifteen pounds. The honey was light in color and had a pleasant aromatic flavor. It was gathered from the white willow (*Salix alba*), the weeping willow (*S. babylonica*), and a golden willow, probably a variety of the white species with yellow twigs. In warm clear weather the bees work eagerly on the bloom, and the nectar is so abundant that it can be seen glistening in the sunlight. But as it is often cold, windy, cloudy, or rainy, honey and pollen from the willows is not at all certain.

In Georgia the black willow (*Salix nigra*) grows along streams throughout the state. It blooms in March, and in a few localities yields a surplus of honey of medium quality. In Louisiana there are immense areas of willows, and a great amount of willow honey is produced. A strong colony may gather 100 pounds from this source. As white clover and the willows bloom at the same time, it is probable that some of the honey credited to white clover comes from the willows.

The black willow is also common in Texas, where it is valued for both nectar and pollen. In California, in Sacramento, San Joaquin, Fresno, and Tulare coun-

Fig. 120.—Wild Buckwheat (Eriogonum fasciculatum). Left, a single plant in bloom. Right, a large mass of bloom. Photographed by G. M. Huntington.

Fig. 121.—Wild Cucumber (Echinocystis lobata). Photographed by Lovell.

ties a surplus of dark honey with a strong flavor is stored from the willows. Late in the season a large amount of honey-dew is gathered from the foliage. "In western Washington the willows are all valuable. Abundant everywhere, bloom-

Fig. 122.—Pussy Willow (Salix discolor). Left, staminate catkins; right, pistillate catkins. Photographed by Lovell.

ing from mid-February to the end of March, and, if the weather is fair, causing the bees to breed up rapidly."

The earliest willow to blossom in New England is the glaucous or pussy willow (*Salix discolor*). (Fig. 122.) On a calm warm day in April the sweet odor may be noticed several rods away, and the air is filled with insects hovering about the bloom. Besides honeybees there are female bumblebees, the only caste of bumblebees in spring on the wing, and a great number of solitary bees, belonging to the genus *Andrena*, gathering pollen for brood-rearing. Ants often climb the stems, and moths resort to the flowers in the evening. It is soon followed by the riverbank willow (*S. longifolia*), then there come a great variety of species, as the silky willow (*S. sericea*), the beaked willow (*S. rostrata*), the heart-leaved willow (*S. cordata*), the white willow (*S. alba*) (Fig. 123), and the shining willow (*S. lucida*).

The very small flowers are destitute of both petals and sepals, and are crowded together on an elongated stem or axis forming a cluster called an ament or catkin. The stamens and pistils in all species are in separate flowers, which are borne on different plants, some producing only staminate, others only pistillate flowers. In a staminate catkin of the pussy willow (*S. discolor*) there are about 270 flowers and in a pistillate catkin 140 flowers. The multitude of bright yellow anthers renders the staminate blossoms very conspicuous. All of our species furnish both pollen and nectar, but it would, of course, be useless to look for pollen on pistillate shrubs or trees. The nectar is freely secreted in both kinds of flowers on the tips of minute glands, which in the pistillate flowers may be found at the base of the ovary. (Fig. 124.)

WILLOW-HERB (*Epilobium angustifolium*).—Fireweed. Indian pink. Rose bay. A perennial herb, 2 to 8 feet tall, with long lance-shaped leaves, and handsome red-purple flowers in long spikelike racemes. After forest and brush fires it springs up in great abundance, and flourishes for about three years, when other plants crowd it out. Wild raspberry, another good honey plant, is one of the first plants to replace it, and goldenrods, asters, Canada thistle, and various shrubs also soon spring up and occupy the land. But the length of time fireweed offers a good location for beekeeping varies greatly in different parts of the continent. Near Maniwaki in the Gatineau Valley, about 100 miles north of Ottawa, the location had become practically worthless for honey production six years after a fire had swept over the land; but 200 miles north of Ottawa, half-way between the city of Quebec and Lake St. John, there was still a large amount of fireweed in bloom 15 years after a forest fire. While in the upper part of the Lower Peninsula and also in the Upper Peninsula of Michigan, fireweed was until recently a reliable honey plant, yet in Tuscola County, farther southward, according to Hutchinson, although very common it never yielded a pound of honey. On the Canadian Pacific Railway in British Columbia there are localities in the Rocky Mountains where fireweed blooms year after year and shows no signs of diminishing.

Fireweed is adapted to a greater variety of soils than either alsike clover or white clover. Moist ground and a cool temperature are favorable to its growth; but drainage is necessary; and, if the soil is swampy, both growth and secretion are poor. While fireweed thrives best in clay soils and particularly in soils rich in humus, as in the decaying remains of fallen trees, it will grow well northward for a time in rather sandy soils or on rocky ground after a fire. (Fig. 125.)

DISTRIBUTION OF WILLOW-HERB

Willow-herb is widely distributed in the northern part of Europe, Asia, and North America. In eastern North America it extends from Labrador southward along the Appalachian Chain to North Carolina. It is abundant in New England, and in northern Michigan, Wisconsin, and Minnesota. A few years ago there were thousands of acres of this plant in northern Michigan without bees to gather its sweetness. A large part of southern Michigan (the Lower Peninsula) was formerly covered with white and red pine, which has now been largely cut for lumber. During the first dry season after the cutting, fire burns over this stump land, and two or three years later the growth of willow-herb comes to maturity. A few years ago it produced large quantities of honey, but as the pine has been largely lumbered the prospect is that willow-herb in the Southern Peninsula of Michigan has had its day. In the Upper Peninsula it is at present a most valuable source of surplus. Blooming at midsummer, it prolongs the honey flow until the middle of August. It is easily eradicated by cultivation, but it will be many years before the bee-

Fig. 123.—White Willow (Salix alba). Staminate catkins. Photographed by Lovell.

keeper will not be able to profit from the bloom dotting the cut-over lands. (Fig. 126.)

Willow-herb is common in the maritime provinces of Canada; in Quebec; in Ontario, particularly in the Rainy River district, and on the clay lands; also around Lake Temiskaming; in Manitoba, especially around Lake Winnipeg and in the low

Fig. 124.—Willow (Salix Nutallii). Photographed by Richter.

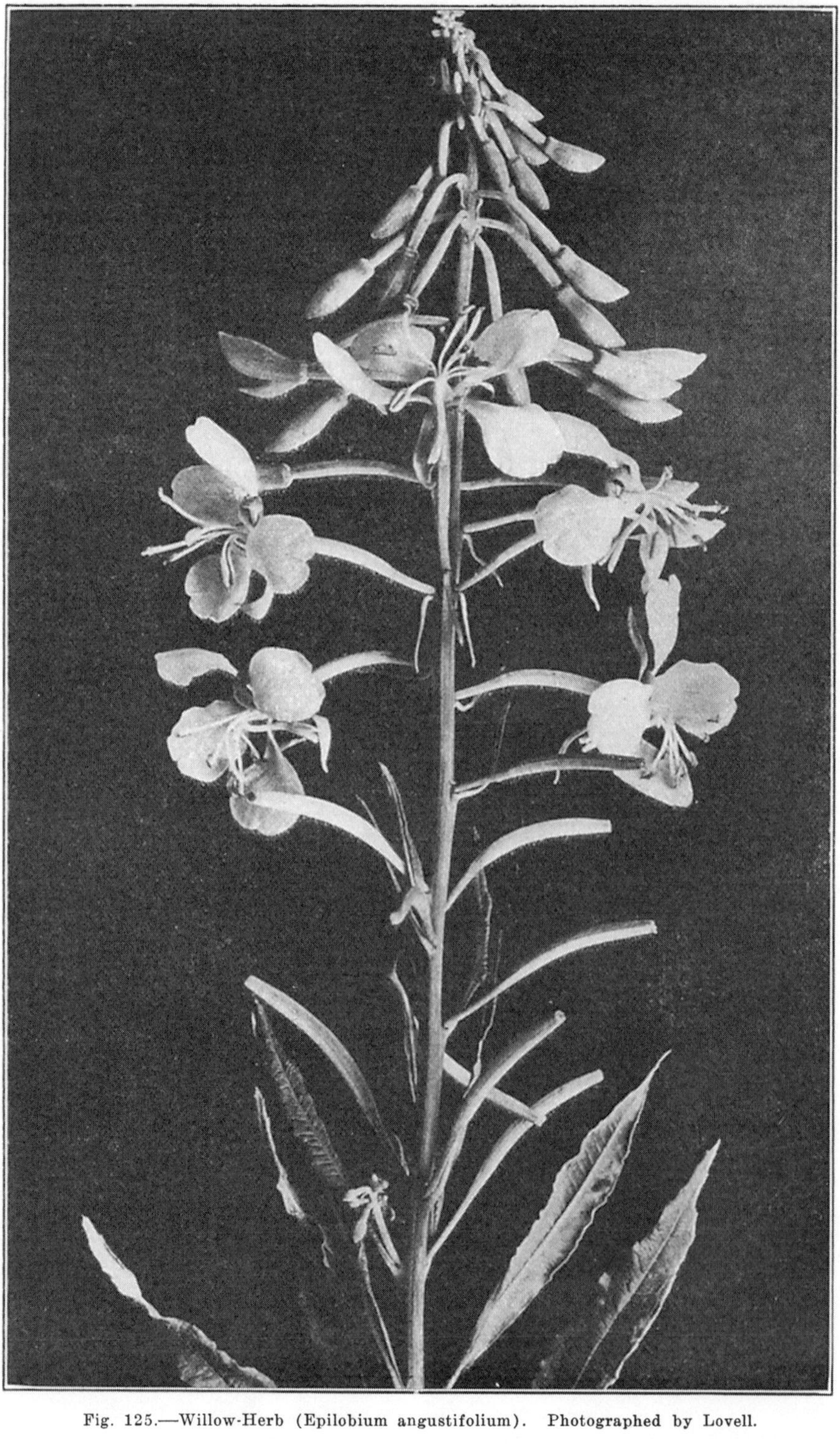

Fig. 125.—Willow-Herb (Epilobium angustifolium). Photographed by Lovell.

Fig. 126.—Willow-herb in northern Michigan. No. 1, the blossom; No. 2. the plant; Nos. 3 and 4, its habitat. Photographed by E. R. Root.

moist lands of eastern Manitoba; in northern Saskatchewan; and in central and northern Alberta. But it is most abundant in British Columbia both in the mountains and on the coast. It reaches its highest development both in the height of the plant and in the size of the flower-cluster in the lower Fraser Valley. At Hector. B. C., at an altitude of 5200 feet, and at Glacier at an elevation of 4000 feet in the Rocky Mountains and Selkirk Range on the Canadian Pacific Railway there are large patches of fireweed in bloom year after year. Between Lacombe and Edmonton in central Alberta fireweed springs up and blooms in wheat fields in places where the grain has failed to grow. It is also fairly common on scrubland.

In the warmer valleys of the southern part of British Columbia the plants begin to grow so early that the blooming period may close before the end of the summer; but in the North they continue to flower until killed by about five degrees of frost. In northern Ontario a killing frost may come as early as the last of August. Travelers to the Yukon and other parts of the far north of Canada have observed that fireweed is prevalent as far as the forest extends, even to the delta of the Mackenzie River.

In the rain belt of eastern Washington and Oregon, in the lumbered regions, there are immense areas of fireweed, which perhaps offer as promising a bee pasturage as is to be found in the United States. In Washington its acreage is probably equal to that of any two other honey plants. At present it does not support as many colonies of bees as alfalfa, partly because of the absence of good roads, and partly because beekeepers do not realize its possibilities as a honey plant. As in other states, it is confined largely to the burned-over areas in the sections of coniferous forests, but there is probably not a county in Washington in which it does not occur. It ranges in altitude from sea level to the upper timber line. In the northeastern timbered section it is very common, and in some localities it is the leading honey plant; but it is also becoming more abundant along the irrigating ditches in the Yakima Valley and in other irrigated valleys, although here it is only a minor honey plant. It is also very important in northern Idaho. In eastern Oregon it is equally abundant, and areas of 100 acres or more thickly covered with fireweed occur. It remains at its best for four or five years, depending upon rainfall and soil conditions. Gradually other vegetation crowds it out. A second fire will temporarily increase its abundance, but the second period of growth is usually short, since the roots of many other perennial plants survive in the ground. After forest fires it appears in abundance in the Sierra Nevada of California. *Willow-herb has a more northern range than any other honey plant of the first rank.*

HONEY FLOW

Willow-herb blooms in July and August, but the period of blooming is influenced by altitude, latitude, and rainfall. The flowers are usually red-purple in color, but at Monteith, Ontario, Sladen observed solitary stalks of a white-flowered variety. The nectar is secreted by the green fleshy top of the ovary, where it is protected from rain, and yet is easily accessible to insects. On the outer side the nectar is enclosed by the dilated bases of the stamens and above by a ring of hairs around the style. The flowers are visited not only by honeybees and bumblebees, but likewise by many solitary bees, flies, and butterflies. Bumblebees are common, and one was observed to make 37 visits in a minute. The pollen is pale greenish purple and is bound together by fine viscid threads. The anthers mature before the stigma, and cross-pollination regularly takes place. The flowers are odorless.

Cool nights and warm days, as in the case of many other honey plants, cause the secretion of the largest amount of nectar. The honey flow lasts longer than that of clover. In the Gatineau Valley north of Ottawa, it begins one or two weeks later than clover, or about July 10, and lasts until Sept. 5. It thus covers the larger part of the summer, or the months when the colonies are strongest. A colony on scales in a large apiary at Montcerf, Quebec, 100 miles north of Ottawa, gained 20 pounds per day for several days during August; and the average yield for six years was 144 pounds per colony, of which probably 100 pounds was from fireweed.

In northern Michigan over 250 pounds of honey per colony have been stored from fireweed; and 100 and 125 pounds of surplus year after year have been reported from this source. In this region until recently no plant furnished more honey than willow-herb, and if the pasturage were permanent a beekeeper would find in such a location a bonanza. It yields nectar, says Hutchinson, during weather that would stop all storing from basswood or clover, and bees have been seen bringing in honey at a fair rate with a cold wind blowing from the North. "To my knowledge it has failed only once in a dozen years." Sometimes a drop of nectar can be seen at the base of each petal. At times several pounds of honey may be brought into the hives in a few hours.

For four consecutive years a good crop of honey has been obtained at Melford in northern Saskatchewan. A beekeeper near New Westminster, British Columbia, writes: "Last year my two best colonies gave 550 pounds each. I am satisfied that most of it, if not all, came from fireweed, which grows here in great profusion." Eighteen miles southeast of Tacoma, Washington, an average of 120 pounds per colony has been secured entirely from fireweed. It has been reported that willow-herb is occasionally unreliable in western Washington, and that hundreds of acres in full bloom may not yield a pound of honey. If there is very little rain during May and June the crop will be light. The largest average crops are secured within 50 miles of the ocean. Heavy fogs followed by warm clear days give the best yields. In the vicinity of St. Maries, Benewah County, Idaho, a large area of land, which has recently been cleared of forest, is covered with a luxuriant growth of fireweed, which yields nectar until killed by severe frosts. The total number of colonies of bees in this locality probably does not exceed 100, but it is estimated that there is ample room for at least 2000 colonies. Unfortunately the best areas for fireweed honey production are difficult to reach and are, consequently, seldom utilized by beekeepers. The loss of the apiary from forest fires must also be guarded against in many locations.

WILLOW-HERB HONEY

Hutchinson, whose knowledge of willow-herb honey was based on an experience covering many years, described it as follows: "Willow-herb furnishes the whitest and sweetest honey I have ever tasted. The flavor is not very pronounced, but there is a suggestion of spiciness." According to Sladen: "Fireweed honey is almost water-white, has a good density, and a very mild flavor. It granulates soon after extraction." In some instances the honey has been described as being as clear as water. The comb is also very white and tender.

WILD SENNA (*Cassia marilandica*).—The yellow flowers are nectarless; and, as in the case of partridge-pea, are visited only by bumblebees, upon which they are dependent for pollination. Nectar is secreted on the upper side of the leaf stalks near the base by club-shaped nectaries. Reported to be a good honey plant in Louisiana. See Partridge-pea.

WITCH HAZEL (*Hamamelis virginiana*).—A tall shrub with yellow flowers, blooming in late autumn, and maturing its seeds the next season. It is very common in damp woods from Nova Scotia to Florida and west to Nebraska. As it blooms so late and is abundant, it is helpful in preparing the bees for winter.

WOOLLY WHITE DROUGHT WEED.—See Turkey Mullein.

WOODBINE (*Psedera quinquefolia*).—A woody climbing vine with small green flowers in clusters. Leaves digitate, of five leaflets. Common in thickets and often cultivated for ornament. Nectar is secreted in minute drops at the base of the ovary. Honeybees resort to the flowers in great numbers.

YELLOW CROCID.—See *Crocidium multicaule*.

YELLOW JESSAMINE (*Gelsemium sempervirens*).—Evening trumpet-flower. A shrubby twining vine with large, yellow funnelform flowers. In Florida it blooms from February to March, and, although bees visit the blossoms, a surplus seems never to be obtained. It is useful for spring stimulation. The honey has been reported to be poisonous, but this has been denied. Some severe cases of vomiting with the usual symptoms of poison have been reported. (Fig. 127.)

YELLOW STAR THISTLE (*Centaurea solstitialis*).—Barnaby's thistle. Introduced into California about 25 years ago, it has spread over Sonoma, Napa, Solano, Sutter, and a large part of Butte counties. Regarded as a pest by the farmers, it has become an important source of honey to the beekeepers. About one-half of the honey crop of Butte County, which averages not far from sixty tons, is gathered from this plant. The bright yellow flowers bloom from July to the middle of October. Star thistle yields a heavy nearly white honey, which is very sweet and has a greenish yellow tinge like olive oil. It is considered equal in quality to any white honey in California, and commands a higher price than the light amber honeys. It may be destroyed by repeated plowings, but will still continue to thrive by the roadsides and in hedgerows and neglected fields. On overflowed land it often grows to the height of 7 or 8 feet, forming impenetrable thickets.

Fig. 127.—Yellow Jessamine (Gelsemium sempervirens). Photographed by E. R. Root.

YELLOW SWEET CLOVER.—See Sweet Clover.

YELLOW POPLAR.—See Tulip Tree.

YELLOW TARWEED.—See Tarweed.

YELLOW TOPS.—See Tarweed.

YELLOW-WOOD (*Cladastris lutea*).—An ornamental tree, growing 50 feet tall, found in rich woods in Tennessee and Kentucky. In Tennessee, where abundant, it yields a small surplus. The white flowers, which resemble those of the locust, are in large drooping clusters and open in May. The wood is yellow.

YERBA BUENA (*Micromeria chamissonis*).—A trailing herb, with small white flowers, common in woods near the coast from Humboldt County southward to southern California. Listed as a fair honey plant in California. Blooms in June.

YERBA DULCE (*Baccharis angustifolia*).—The odor of the flowers can be perceived for a long distance. A branched shrub, 3 to 6 feet tall, with leathery, resinous leaves and numerous heads of yellowish flowers, blooming in the fall. Abundant in brackish marshes in the southern states, especially in Texas, where there are thousands of acres. Honeybees visit the flowers in great numbers, and in the coast prairies gather a surplus of mild amber-colored honey. In Arizona, water motor, or bottom willow (*B. glutinosa*), is common in the river valleys, growing chiefly on land that is sometimes inundated. It also covers large areas along the banks of the Rio Grande. Near Phoenix, Arizona, a large surplus of excellent light-colored honey is obtained from this species. The Spanish name gautemote has been corrupted into water motor. Desert bloom (*B. sarathroides*) is a fall source of nectar near Tucson.

YERBA SANTA (*Eriodictyon californicum*).—Mountain balm. A low shrub common over extensive areas in the Coast Ranges of California, often associated with chamise. Blooms in June and July, and yields an amber-colored honey of good quality.

YUCCA.—Bear-grass. Spanish dagger. Adam's needle. Spanish bayonet. The yuccas, of which there are more than a dozen species in this country, are very abundant on the semi-arid lands of the southwestern states, occurring in such large numbers in some localities that they form "straggling forests." Some species have fibrous stems 20 feet high, while others are almost stemless. The flowers are large, white, bell-shaped and pendulous, and borne in great branched clusters. There are three oblong nectaries in a flower, enclosed by the partitions that separate the three cells of the ovary, which open externally at the base of the flower, where the nectar escapes through a capillary pore. Very little nectar is secreted, as it is of no use in the pollination of the species. The flowers are pollinated by small moths, which fly in the evening, and take no food in the adult stage, as the alimentary canal is functionless. A few flies and beetles also visit the flowers, but they do not bring about pollination. In Florida the common Spanish bayonet (*Yucca filamentosa*) is never visited by bees, and it is probably nowhere important to beekeeping. *Yucca Whipplii,* common in the chaparral belt of the mountains of California, is the noblest of the yuccas. From a crown of leaves near the ground there arises a stalk 12 feet tall, bearing an immense cluster or panicle of white flowers. Coquillet, who investigated its manner of pollination, says: "I did not observe a single butterfly or wild bee of any kind visit the flowers, although all of these insects were quite abundant in the vicinity." Richter reports it as eagerly visited by honeybees, and yielding a surplus; but, as it secretes nectar sparingly, it should probably be classed as a minor honey plant.

PART IV.

A SURVEY OF BEEKEEPING IN THE UNITED STATES

More nectar goes to waste than is gathered. Fifty to eighty per cent. of it is lost simply because there are no bees to gather it.—E. R. Root.

It is essential that the beekeeper should study thoroughly any given location to determine its possibilities in honey production before he concludes to settle down permanently.—W. D. Wright.

PART IV

INTRODUCTION

Very few beekeepers are acquainted with the conditions controlling bee culture outside of the localities in which they have engaged in this industry. This statement was recently confirmed by one of the most prominent apiarists in the United States, who added that he himself did not know in what section of the state in which he had lived for fifty years the largest surplus of honey was produced. There are large areas in this country, the adaptation of which to beekeeping is wholly unknown to the general public, and in regard to which hardly a line has ever been printed. A few states have published descriptions of the honey floras found within their borders. Occasionally a practical beekeeper has prepared a list of the more important species of plants supposed to be valuable as sources of honey. But there is very little literature in existence to-day to guide the beekeeper in his choice of a location, or to inform him as to the results he may expect to obtain in the different states. Yet there is no phase of bee culture in which the majority of beekeepers express a greater interest, nor one which more closely concerns their welfare.

From a commercial point of view it is desirable to review the conditions of beekeeping in the United States by states rather than by natural or physical units. But the state boundary lines are artificial lines, and seldom coincide with the lines dividing the physiographic regions. On the contrary, they often divide into two parts areas which are uniform in climate, soil, and honey flora. But it is by states that statistics relating to the number of beekeepers and colonies of bees, and the amount of honey and wax produced, are gathered. Each state has its own associations, conventions, inspectors, and laws for the promotion and protection of beekeeping. Reports on bee culture, honey plants, prices, sales, weather, and general crops, are issued by states. Bulletins and articles treating of bee culture commonly confine their attention to state areas. The limits and positions of the states can easily be determined by reference to any school geography, while the boundary lines of regions based on their physical features are given only in special works. From a strictly practical point of view, therefore, it will be more convenient and helpful to gather and arrange the data on beekeeping by states.

While the beekeeper seeking a good location for the production of honey will give his attention chiefly to such artificial divisions as states and counties, the distribution of the honey plants is clearly determined, not by political boundaries, but by altitude, rainfall, soil, and temperature. The eastern third of the United States, for example, is covered largely with a hardwood forest due to its ample rainfall and humid summers; while the great forests of the western states consist mainly of conifers adapted to dry summers. The Sierra Nevada and Cascade Mountains, by checking the moisture-laden winds from the Pacific Ocean, reduce the region east of these ranges to a desert. In order to understand the distribution of the honey plants within the United States a knowledge of the physiographic regions into which it is divided is necessary.

Manifestly it is impossible to divide the United States into regions based on the honey plants alone. Such an attempt would result in a few large, ill-defined regions and a great many small ones, which would often conflict with each other, or would leave large intermediate areas which would not fall under any of these divisions. The honey plants form a wholly artificial group, and are not sharply divided from many other plants. They include herbs, shrubs, and trees, and belong to a great variety of plant families. There is no soil, no extreme of rainfall or temperature, to which some species is not adapted. Their distribution is extremely erratic. The only characteristics which the honey plants have in common are that they secrete nectar freely, and are very abundant in one or more localities.

The United States comprises eight natural regions which are recognized by physiographers, geographers, and geologists, and which are used by botanists in studying the distribution of the endemic flora or native vegetation. They differ to a great extent in temperature, rainfall, soils, agriculture, and geological history. Their honey floras are in the main distinct, although the same honey plant is often found in two or more of them. Their limits are usually well defined, and we shall meet them again in every state in considering the conditions of beekeeping which they greatly influence.

The following eight physiographic provinces, it is believed, will serve more satisfactorily as honey-plant regions than any others that can be proposed:

1. The northeastern, or New England region, which includes New Brunswick, New England, and the Adirondack Mountains of northern New York.
2. The Appalachian Highlands. This region includes the mountain ranges extending from the Catskills in New York to northern Georgia and Alabama; the Piedmont Plateau on the eastern side of the mountains, which extends from northern New Jersey to eastern Alabama; and the Alleghany Plateau on the western side of the mountains, which extends from southern New York to northern Alabama.
3. The Coastal Plain of the Atlantic Ocean and the Gulf of Mexico.
4. The Central Lowlands north of the Gulf Coastal Plain, in which are included the prairies.
5. The Great Plains, or the semi-arid belt east of the Rocky Mountains.
6. The Rocky Mountains.
7. The desert region west of the Rocky Mountains and east of the Sierra Nevada and Cascade Ranges, including the sagebrush plains of eastern Washington, the lava beds of eastern Oregon, the Snake River Desert of southern Idaho, the sagebrush deserts of Utah, western Colorado, and northern Nevada, and the cactus deserts of southern Nevada, southern California, Arizona, New Mexico, Texas west of the Pecos River, and northern Mexico.
8. The Pacific Coast, or the region west of the Sierra Nevada and the Cascade Ranges.

A brief description of each of these regions follows, and additional details in regard to their characteristics and boundary lines will be given in the descriptions of the different states, which are divided into sections according as the territory of each is occupied by one or more physiographic regions.

1. The New England, or northeastern region, which includes New Brunswick, New England, and northern New York, was entirely covered by the great ice sheet, and its vegetation wholly destroyed. After the retreat of the ice the land was occupied by a forest consisting of a few kinds of trees, white pine, spruce, hemlock, rock maple, birches, and oaks predominating. The mountains and highlands were swept bare, or were covered by a thin layer of coarse drift. In the valleys and on the plains the glacial soils are generally sandy or clay loams; but limestone soils occur in many localities, as in the Champlain Valley, Vermont, the Berkshire Valley, Massachusetts, and in St. Lawrence County, New York, where white clover and alsike clover are usually dependable. But over great areas the soils are neutral or

acid, and white clover, even if present, yields little nectar, as in eastern Massachusetts. Raspberry and fireweed are the common honey plants on the neutral or acid soils of the Adirondack and White Mountains and on many other rugged areas. The dry barrens of New Brunswick, the blueberry barrens of Maine, and the outwash plains of Massachusetts and Rhode Island have acid soils and produce an abundance of blueberries and huckleberries. The soils of the many swamps and bogs are likewise acid. Goldenrod and sumac are common over this region.

2. The Appalachian Highlands, as the result of the abundant and uniform rainfall, are covered with a magnificent forest of hardwood trees, unrivaled elsewhere in the United States. The pine trees of the New England region are almost wholly absent; and instead of a few species of trees there are several hundred. Within an area of one square mile 75 different kinds have been counted. In western North Carolina, where this forest reaches its highest development, the size and variety of the trees are surpassed only in tropical woodlands. Trees, as would be expected, are the principal sources of honey, as basswood, tulip tree, sourwood, sumac, locust, redbud, magnolia, maples, honey locust, holly, willows, sour gum, and persimmon. There is also a great variety of shrubs. The flora of this region is a very ancient one, and contains many types of plants not found elsewhere in North America. The Appalachian Mountains were elevated above the sea at the close of the coal period, and this region is thus much older than the Coastal Plain.

3. The Atlantic and Gulf Coastal Plain was the last portion of the United States to rise above the waters of the ocean. The Atlantic Plain is separated from the Appalachian region by a well-defined escarpment known as the "fall line," which is the eastern limit of the hardwood forest. The soils are largely sandy, and a great pine forest extends from Virginia to eastern Texas, known as the pine barrens. Thousands of acres of marine and river swamps occur in this region on which gallberry, black titi, black tupelo, and white tupelo flourish. Only plants adapted to an acid soil will grow in these swamps, and the clovers are consequently absent. Another characteristic of this region is the annual cultivation of millions of acres of cotton, a plant which requires 200 frostless days. Florida was the last portion of North America to rise above the ocean. The Coastal Plain, which is easily distinguished from every other region in the United States, includes southern New Jersey, Delaware, eastern Maryland, eastern North Carolina, southern Georgia, Florida, southern Alabama, Mississippi, Louisiana, southern Arkansas, and eastern Texas.

4. The Central Lowlands or Upper Mississippi Valley is largely a grassland known as the prairies. Much of the valley during the Upper Cretaceous was an inland sea; and after the water receded large areas were occupied by temporary lakes. At the close of the Glacial Period thousands of square miles in Michigan, Ohio, Indiana, Illinois, Iowa, Missouri, Kentucky, and Tennessee were covered by the wind with a fine fertile soil known as loess. The prairies owe their origin to the fineness of this soil, in which the grasses have formed a dense turf on which it is difficult for trees to encroach. Throughout this region white clover is a very important honey plant, and in favorable seasons is the source of an enormous surplus. Sweet clover also succeeds well over much of this area. In the river valleys the flowers of many hardy Compositae display great sheets of brilliant colors, as Spanish needles, sunflowers, asters, goldenrod, crownbeard, Rudbeckia, and gum-plant. Near the center of the Mississippi Valley is the upland region called the Ozark Plateau.

5. The Great Plains is a well-known and well-defined belt of land with an average width of about 300 miles, lying between the Mississippi Lowlands and the Rocky Mountains, and extending from Canada to the Rio Grande River. During a

large part of the Upper Cretaceous Period this region was the bottom of an inland sea which extended from the Gulf of Mexico to the Arctic Ocean, and divided eastern North America from the western portion. During the Tertiary Period the land was raised above the waters, and for a time possessed a much milder climate than at present. The Great Plains to-day form a vast barren tableland 4000 to 5000 feet in altitude, which is broken by numerous canyons, and dotted with buttes and isolated mountains. It is subject to great extremes of heat and cold, and to high winds. It is a treeless area, not because, like the prairies, it has a fine soil and a dense turf, but because of the aridity of the climate. Trees can not survive where the rainfall is less than 26 inches. Sagebrush, greasewood, and bunch grasses are the common forms of vegetation. Commercial beekeeping throughout the Great Plains is dependent on irrigated alfalfa and sweet clover. In the Dakotas, Nebraska, Kansas, and Oklahoma, alfalfa, when grown without irrigation, is only partially reliable; but the acreage of sweet clover in these states is reliable and very large, and is increasing every year. Probably 50 per cent. of the honey crop comes from this source in the Great Plains, which is destined to become a great sweet-clover belt. But in eastern Montana, Wyoming, Colorado, and New Mexico irrigated alfalfa seldom fails to yield a surplus of honey. The areas planted with alfalfa are described in detail under the different states.

6. The Rocky Mountain region comprises western Montana, eastern Idaho, western Wyoming, northeastern Utah, central Colorado, and north-central New Mexico. The summits of the mountains are bare of vegetation, but on the flanks are coniferous forests composed of pine, hemlock, and spruce. The aspen (*Populus tremuloides*) is a very common tree, and willows are abundant in the mountain meadows. In the northern Rocky Mountains, in the fertile valleys of western Montana and Wyoming, there are large areas under irrigation from the mountain streams, and alfalfa and sweet clover yield immense crops of honey. In the southern Rocky Mountains there is a much smaller number of colonies of bees. In the mountains the winters are long and severe and the snowfall heavy. Colonies dwindle greatly in size before spring, and a large part of the short summer has passed before they are strong enough to store a surplus. The native honey plants are not important, and large areas are covered with softwood forests. The Rocky Mountains were elevated to the height of 11,000 feet in comparatively recent times.

7. West of the Rocky Mountains and east of the towering Sierra Nevada Range is a vast desert region extending from northern Washington southward across the United States into Mexico. The northern portion of this arid area comprises the sagebrush plains of eastern Washington, the lava beds of eastern Oregon, the Snake River Desert of southern Idaho, western Colorado, Utah, and northern Nevada. There is no sod except along the streams, and probably nine-tenths of the vegetation is sagebrush. In the southern portion of this region, which includes southern Nevada, southeastern California, Arizona, New Mexico, and western Texas, as well as north-central Mexico, the heat becomes intense, and the annual rainfall decreases to a few inches, and sagebrush is replaced by cactus, yucca, and agave. The few streams vanish in the sands, or in "sinks," or terminate in small saline lakes. There are areas of alkaline flats which are so highly charged with salts that they are bare of vegetation, or produce only greasewood and a few other plants. Except for the coniferous trees on the mountains and the mesquite forests in the south, the region is treeless. It is a desolate, forbidding country "where iron will not rust, tin tarnish, nor flesh mortify." But millions of acres of this desert land are being reclaimed by the United States Reclamation Service and by state and private enterprises. On the great fields of irrigated alfalfa commercial beekeeping is a very reliable industry, and thousands of colonies of bees are operated by individuals or

by companies. In the mesquite districts good results are often obtained, but they bear no comparison to those obtained in the irrigated areas.

8. The Pacific Coast Region comprises the sections of Washington, Oregon, and California west of the Cascades and Sierra Nevada Ranges, and the Coast Ranges in southern California. The rainfall in the northern part of this region is very heavy, exceeding 100 inches annually; but on the mountains of southern California, the home of the bush sages, it is hardly sufficient for growing crops, due to the higher temperature and the diminishing winds from the ocean. The Coast, Cascade, and Sierra Nevada Mountains are covered with a magnificent coniferous forest composed of giant trees, but hardwood trees are comparatively rare. This strip of land rose above the waters of the Pacific Ocean near the middle of the Tertiary Period, completing the western border of North America. The land is thus the youngest, geologically, in western North America. This region is clearly separated from every other by its age, geological history, climate, and flora. The honey plants are described in detail in the subsequent pages.

The advantage of dividing the United States into eight regions characterized by different physical features, climates, floras, and geological history is apparent. A comparison of these regions answers many questions in regard to the distribution of the honey plants, which a consideration of the honey floras by states alone would not make clear. There is no difficulty in understanding why white clover reaches its maximum development in the prairies, or why trees are so valuable as sources of honey in the Appalachian region; or why gallberry, gum-trees, and cotton are confined to the Coastal Plain; or why mesquite and agave are restricted to the southwestern states; or why California has so many honey plants. Everywhere rainfall, temperature, soil, and altitude affect the conditions of beekeeping. In the following descriptions of the states each state will be divided into sections according to the natural regions found within its borders; for example, western Oregon belongs to the Pacific Coast region, and is wholly unlike eastern Oregon, which belongs to the desert region; while in North Carolina conditions differ widely in the Coastal Plain, the Piedmont Plateau, and the mountainous section.

Many letters have been received inquiring as to the average size of the apiaries and the number of pounds of surplus which can be secured in the different regions. Others desire information in regard to the soil, climate, and honey plants. Farmers who have pursued beekeeping as a sideline often desire to move to a state where they can produce a larger amount of honey, or make beekeeping their chief occupation. A very general desire is expressed for information as to the best locations for beekeeping. So many factors must receive consideration that a satisfactory reply is difficult to give. "Beekeeping pays in almost every state in the Union," writes E. R. Root, "but each section has its own special difficulties. In the North there may be great losses from the long and severe winters. In the South the winters are so mild that bees may fly during nearly every month, with the result that the stores are rapidly consumed and many colonies are likely to perish from starvation. For example, in southern California the queens, from continual brood-rearing, become exhausted, and in the spring are unable to build up strong colonies in time for the honey flow. It is a great mistake for anyone to move from one section of the country to another without first making careful investigation of the locality in which he proposes to make his new home. Before engaging in beekeeping in the South a northern beekeeper would do well to familiarize himself with the unknown conditions he must meet by working a year in a southern apiary."

But the value of a locality depends on the personal character and methods of the beekeeper as well as on the honey flora and climate. "The poor quality reported for many regions for beekeeping," says E. F. Phillips, "is probably due to poor beekeeping. In many sections adequate trials of commercial beekeeping have not been made. The prevalence of foul brood is not a serious drawback to a genuine beekeeper, although it spoils the fun of the amateur. Other things being equal, it may be advisable for a thorough-going beekeeper to chose a location where foul brood has cleaned out competition."

While it is important that the beekeeper should be familiar with climate and soil and should be acquainted with the best methods of bee culture, it is vitally important that he should be familiar with the honey plants. "If he is not well informed in regard to the honey flora," says Wilder, "he will certainly do as thousands of others have done — locate bees in sections where little if any surplus can be secured. This lack of knowledge is putting many out of the bee business. I have seen hundreds of apiaries in localities where there was only a meager sustenance for the bees and a surplus could be obtained only occasionally. But if he has a thorough knowledge of the honey plants, and can single them out from the thousands of other plants, know when they are in bloom, and the approximate amount of honey each will yield under normal conditions, he can form, as he goes to and fro, north, east, south and west, over the country, a correct estimate as to how much honey can be produced in every locality he visits."

The principal sources of table honey in the United States are white clover, alsike clover, sweet clover, alfalfa, orange bloom, and the mountain sages. Commercial beekeeping is very largely dependent on these plants, and without them very little surplus honey would reach the big markets. Their distribution is of very great interest to all who seek a living from the production of honey.

The best white clover territory is found in the north-central states, especially in the states bordering on the Great Lakes, where the soils formed during the Glacial Age are rich in lime. This region, sometimes called the "white clover belt," includes the Champlain Valley of Vermont, northern and central New York, northwestern Ohio, northern and eastern Indiana, northern Illinois, a belt running east and west through the Lower Peninsula and the southern portion of the Upper Peninsula of Michigan, portions of Wisconsin and Minnesota, and eastern Iowa. White clover is also a valuable honey plant in Washington and Oregon west of the Cascade Range, and over a large area in Canada. The surplus secured in successive seasons varies greatly in different localities, as has been described under WHITE CLOVER in PART III. Alsike clover is very generally grown in the states north of Kentucky and east of the east boundary line of the Dakotas. There are small areas in southwestern Idaho, northern California, and in the coast regions of Washington and Oregon. In Ontario, Canada, hundreds of acres are grown exclusively for seed. It is almost unknown in the southern states, but is grown to a limited extent in Virginia, Tennessee, Missouri, and Louisiana, especially around Baton Rouge.

The largest amount of sweet clover honey is produced in the north-central states (Ohio, Indiana, Illinois, Michigan, Wisconsin, and Iowa), northern Kentucky, the Black Belt of Alabama and Mississippi, the Great Plains region, and in the Western Highlands (Montana, Wyoming, Colorado, and Idaho). While sweet clover honey is produced in nearly all portions of the white clover territory, the amount is small compared with that obtained from the clovers. In northern Kentucky and in the Black Belt of Alabama and Mississippi sweet clover is the main reliance of the beekeeper. In northwestern Iowa and in North Dakota and South Dakota the acreage is very rapidly increasing and excellent new locations for bees are constantly becoming available. In Kansas it is grown in more than 90 counties. Under irrigation in the Rocky Mountain states very large yields are obtained, especially in Wyoming.

Alfalfa honey is produced commercially chiefly in the states west of the Missouri River in semi-arid regions where irrigation is practiced. Immense crops are secured in Montana, Wyoming, Colorado, New Mexico, Idaho, Utah, Arizona, and in the Great Central and Imperial valleys, California. In the southwestern section the honey is darker in color and slightly stronger in flavor than in the intermountain region. The alfalfa territory is nearly all occupied at present, but with the extension of the area under irrigation there will soon be a larger area of this valuable honey plant.

Practically all of the orange and mountain sage honey comes from southern California within a radius of one hundred and fifty miles of Los Angeles. Florida produces a small amount of orange honey, but not much of it finds its way into the markets. Although the yield in California is limited, irrigation of the groves renders it fairly reliable. But in order to obtain a surplus the colonies must contain a great number of bees at the beginning of the orange flow, and this is difficult to secure since the colonies dwindle from excessive activity during winter, and the queens wear out from laying eggs and are incapable of producing a great amount of brood in the spring. This difficulty may be remedied to a large extent by the introduction of new queens and by having a super of stores above the colony.

The mountain-sage region of California is restricted to the Coast Ranges, extending from the foothills in San Benito and Monterey counties to San Diego County in the southwestern part of the state. The larger portion of the sage honey comes from Ventura, San Diego, and Los Angeles counties, and from western Riverside and San Bernardino counties. The crop is dependent on sufficient rainfall.

While most of the white honey of the highest quality which comes into the commercial market is produced in the colder parts of the country, many light-colored honeys of good flavor are gathered in the southern states, as those obtained from gallberry, white tupelo, black tupelo, sourwood, palmetto, cotton, huajilla, and catsclaw.

Sourwood is chiefly valuable as a source of honey in the mountainous regions of North Carolina, South Carolina, and Tennessee. This fine honey is nearly all consumed in the localities in which it is produced. Undoubtedly the best section for the production of gallberry and tupelo honey is southeastern Georgia. "The commercial beekeeper," writes E. R. Root, "will find that Georgia has better opportunities to offer than any other southern state. Alabama, Mississippi, and Louisiana are better adapted to raising bees in early spring than to the production of honey. More bees in package form are shipped from Alabama than from any other state. Neither in quantity nor quality is the honey produced in these three states equal to that of Georgia. Owing to the wide distribution of bitterweed much of it is dark and bitter, and for this reason many apiaries are run mainly for the sale of bees in packages. Texas and California alone offer advantages to the beekeeper comparable with those of Georgia, but in some years both suffer so severely from dry weather that no surplus is stored, or there may be an insufficient amount of honey gathered for wintering."

Cabbage palmetto and scrub palmetto cover large areas in central and southern Florida, and yield a large amount of pale yellow or nearly white honey. Scrub palmetto honey is considered one of the finest honeys of Florida. Huajilla and catsclaw are the sources of a large surplus of white, mild honey on the Rio Grande Plain in southeastern Texas. The cotton belt comprises eastern North Carolina, eastern South Carolina, all of the Gulf states, Arkansas, and Oklahoma. Cotton is a very important honey plant on the Black Prairie of Texas, where it yields a mild light-colored honey in great amount. It is also the chief source of surplus in the bottom-lands of Arkansas.

A brief preview of the best locations for beekeeping in the different states, and an enumeration of the more important honey plants, will prove helpful as an introduction to the more extended descriptions which follow.

New England to-day offers very moderate inducements to the specialist; but if alsike clover were more commonly planted instead of red clover the bee pasturage of this section would be greatly improved. Beekeeping is successfully pursued, however, in Aroostook County in Maine, the Champlain Valley in Vermont, and in the Berkshire Valley in Massachusetts. In no part of this area have better results been obtained than in the Champlain Valley, where the surplus comes from white clover and alsike clover, and the soils are of limestone origin.

Of the eastern states, New York stands foremost in the number of beekeepers and in the production of honey. According to the location, large crops are secured from white clover or buckwheat. On the glacial till soils of St. Lawrence and Jefferson Counties white clover yields an immense amount of white honey. A second white-clover belt extends from Buffalo to the Hudson River, in which are located a great number of apiaries among the Finger Lakes and around Syracuse. The southern portion of the state, especially the southwest corner, is the great buckwheat country, where this plant is usually a reliable source of honey. Another important center for beekeeping is in the eastern part of the state in Schenectady County, where the clovers, buckwheat, and basswood are abundant.

Southeastern Pennsylvania is in a very high state of cultivation and few follow beekeeping as a vocation. Along the north-central border and in the northwest corner thousands of acres of buckwheat are grown. Within the mountains there are many fertile valleys with limestone floors, where the clovers flourish and yield well. Southwestern Pennsylvania, the region of the great steel mills and oil fields, affords little pasturage for bees. New Jersey, Delaware, and Maryland are not considered beekeeping states from a commercial point of view and the average surplus per colony is low.

"In all the southern states," writes E. F. Phillips, "the box hive is far more abundant than the movable-frame hive. It would probably be a fair and conservative estimate to state that 75 per cent. of all the colonies are in 'gums.' The number of colonies per square mile is higher here by far than anywhere else in the country, which indicates good beekeeping conditions. With the continuation of the educational work now being done in North Carolina, South Carolina, Mississippi, and Louisiana, the number of 'gums' is decreasing. In North Carolina probably 50 per cent. of the bees are now in 'patent gums.' At present the South is relatively free from brood diseases — due, probably, to the small amount of movement of bees under the backward methods employed. However, both European and American foul brood are making their appearance, and beekeepers should be ready to combat them. The solution of the problem in the South lies in the development of relatively few extensive beekeepers who will practice migratory beekeeping. If the difficulties of transportation can be overcome, the South can produce enormous crops of honey."

In the southeastern states the crop of honey comes chiefly from the swamps or from the mountains; and in both locations the honey plants are mostly shrubs and trees. Much of the soil in the southern states is acid, and white clover is found only to a limited extent. The higher average annual temperature, the greater rainfall, and several consecutive honey flows make it necessary to modify the methods of beekeeping in use in the North.

Around Norfolk, Virginia, near the swamp land, a surplus is gathered nearly every year from gallberry, aster, and gum trees. The best part of the Piedmont Plateau is directly east of the Blue Ridge. West of this mountain range is the

Great Limestone Valley of Virginia, the most fertile section of the state. In the valley of the Shenandoah, where blue thistle is abundant, good success with bees has been obtained. The smaller limestone valleys in the extreme southwest corner of the state are most promising, and the area of sweet clover and white clover is yearly become greater. There is little commercial beekeeping in West Virginia. In the eastern mountainous section old-time equipment is still common, and the average yield per colony is small. The largest number of colonies of bees is found in the Ohio River Valley in Kanawha and Roane counties. Beekeeping is in an undeveloped condition in West Virginia.

In eastern Kentucky, in the mountains and coal-fields, as in eastern West Virginia, the country is thinly populated, and the methods of beekeeping are backward. Beekeeping is prosperous in Kentucky in the famous Blue Grass region, or Lexington Plain. It is a limestone area, and in Pendleton and Bracken counties there is a great acreage of sweet clover. In the lowlands west of the Tennessee River there are many colonies of bees near Paducah. In Tennessee the eastern Tennessee River Valley is one of the best locations for beekeeping in the state. The fertile limestone soils support an abundant honey flora. Another excellent region is the Central Valley around Nashville, where most of the commercial apiaries are located. The Mississippi Slope seems to possess great possibilities.

The Coastal Plain of North Carolina, with its vast area of swampland covered with gallberry, gum trees, huckleberry, and blackberry, offers great opportunities to the specialist. In the future, gallberry honey, according to E. R. Root, may be sold by the carload. The cotton belt is the poorest part of the state for beekeeping. In the western portion of the Piedmont Plateau and in the mountains a large surplus is secured from sourwood and tulip tree. There are in South Carolina 10,000,000 acres of pine barrens which are largely destitute of honey plants. Beekeeping here is in a very undeveloped condition. Good opportunities may be found along the coast (Horry County), and in Pickens and Oconee Counties in the northwestern part of the state.

In southeastern Georgia, where the gallberry, white tupelo, and black tupelo are abundant, are located the largest apiaries in the state. The crop is usually reliable, and an average surplus of 100 pounds per colony is often obtained. As has been stated above, southeastern Georgia, in the opinion of E. R. Root, offers better locations for beekeeping at the present time than any other region in the South. Throughout northern Georgia there is little commercial beekeeping, but many farmers have a few colonies.

Of all the states in the Union, Florida has the mildest winter climate. Winter losses, when they occur, are due to worn-out queens and starvation. Almost constant breeding wears out the queen and the bees, and uses up the stores. Bees are found in all portions of the state, but because so much honey is used in brood-rearing during almost every month, a very large surplus is seldom obtained. While there are specialists who successfully make beekeeping their sole occupation, as a rule it is not advisable to depend entirely on it for a livelihood. Florida is located too far south for the shipment of package bees to the North. The Apalachicola River region in northwestern Florida is an excellent region for beekeeping in some seasons, when black titi, white tupelo, and black tupelo yield well; but there is a shortage of pollen and the country is a desolate swamp. In Manatee County, on the west coast, there are probably as many colonies of bees as the flora will support. In middle Florida the flow from orange bloom and the palmettos is often unreliable, so that this section is only a fair locality. The mangrove was formerly the source of a large amount of honey in southern Florida, but the great freeze cut it down and it has only partially recovered.

In Alabama and Mississippi the best region for beekeeping is the Black Belt, or sweet clover belt, a tract of land extending from Union Springs, Alabama, to Noxubee County, Mississippi, whence it follows the state line northward to Tennessee. Thousands of acres of sweet clover flourish in this fertile soil, and the apiaries range from 50 to 200 colonies. In northern Alabama beekeeping is much neglected, but with better methods would probably be fairly profitable. The most fertile soils in Mississippi are found in the Yazoo Delta, where there is a dense acreage of cotton; but none of the honey plants are of great value except holly, without which, asserts a beekeeper, it would not pay to keep bees. Moderately good opportunities may be found in all parts of the state near swamps and in river valleys.

In Louisiana the alluvial lands along the Red and Mississippi rivers and in the Atchafalaya River Basin are well adapted to honey production. Much of the honey gathered in this state is dark in color and inferior in flavor.

The southeastern half of Arkansas belongs to the Coastal Plain, and in the river valleys there are fertile alluvial soils. The honey flora of the lowlands is dependable and there are good opportunities for engaging in beekeeping on a commercial scale. On the bottom-lands cotton secretes nectar very freely. In the southwestern counties most of the surplus comes from holly and black gum.

In southeastern Texas a score or more of thorny trees and shrubs, as huajilla, catsclaw, agarita, and coma, yield nectar so copiously that in a favorable season it is almost impossible to overstock this region with bees. On the Black Prairie, which extends from San Antonio to the north border line, cotton is a reliable honey plant, and seldom fails to yield a large surplus. West of the Pecos River agriculture is dependent upon irrigation, and alfalfa is the main dependence of several large honey-producers.

The growth of white clover on the Miami soils of northwestern Ohio, eastern Indiana, and central Michigan is unsurpassed in any other portion of the United States, except perhaps in parts of Wisconsin and Minnesota, noted elsewhere. In periods of depression beekeeping will persist here when it disappears elsewhere. In southern Ohio white clover is less dependable. In southwestern Indiana large crops of honey are secured from blue vine, or climbing milkweed, and in the northwestern corner Spanish needles and boneset in the Kankakee swamps are the sources of a large surplus. In Upper Michigan alsike clover is very abundant and hundreds of acres of unoccupied territory invite the beekeeper.

In northwestern Illinois, Stephenson County leads in the production of honey, the surplus coming from white clover, sweet clover, and heartsease. In the central portion of the state the best locations are found along the Illinois River and the smaller streams. A large acreage of sweet clover has been planted in this section. On the Mississippi River, where there is a wide valley, according to Pellett, there is an abundance of Spanish needles, heartsease, and boneset, which assure a fall flow. There is an immense acreage of apple trees in southern Illinois, but beekeepers report only medium crops of honey.

The northern portion of Wisconsin is a most promising section for honey production, and very large crops of honey are secured. Good crops are also produced in the southern and western parts of the state. The center is the poorest area for bee culture. In the southeastern corner of Minnesota white clover is a very reliable honey plant, and very rarely fails to yield a surplus. With the more general cultivation of sweet clover and alsike clover the western prairie section of the state should yield a large amount of honey. In the cut-over region of northern Minnesota the beekeeper will find large crops and little competition.

In Iowa there are two sweet-clover regions, one in the eastern part of the state, and the other in the western part extending northward from Sioux City. There is also much white clover in eastern Iowa. Fair results may be obtained in almost every county in this state. The majority of successful beekeepers in Missouri are located near the Missouri and Mississippi rivers or their tributaries. When white clover fails on the uplands, a crop may usually be obtained from the fall flowers on the river-bottom lands. The southern part of the state is a much poorer section for the production of honey than the northern.

In the 17 arid and semi-arid western states, North Dakota, South Dakota, Nebraska, Kansas, Oklahoma, Texas, Montana, Wyoming, Colorado, New Mexico, Idaho, Utah, Nevada, Arizona, Washington, Oregon, and California, there were, in 1920, about 19,000,000 acres under irrigation, and there were included in enterprises about 16,000,000 additional acres. In that portion of the United States lying east of these states the normal annual precipitation exceeds 26 inches, and is so distributed through the year as to provide sufficient water for the growing of general farm crops. In the Great Plains the normal annual precipitation ranges from 25 inches in the eastern portion to 15 inches in the western portion. In the Rocky Mountains there is a heavy snowfall on the summits, but in the valleys the precipitation is light and irrigation is necessary for growing crops. In the arid or desert region west of the Rocky Mountains and east of the Sierra Nevada Range the precipitation ranges from 8 inches in southern Idaho to 2 inches in southwestern Arizona. West of the Cascade Range and the Sierra Nevada there is a great variation in the rainfall. In western Washington and Oregon it is the heaviest of any part of the United States, but in late summer there is a dry period when irrigation is desirable. In California there is a well-defined dry period in summer and an equally well-defined wet period in winter. In the northern portion of the state the climate is nearly similar to that of Oregon, but in the southern portion irrigation is often necessary.

The streams in the western states have the common characteristics that they are subject to heavy floods in the spring and become very low in late summer. It is thus necessary to store the flood waters for use later in the season. The low water flow of the streams is exhausted, but there is a very large supply of flood water available for storage. The future extension of irrigation depends on whether the value of the crops will justify the storing of the flood waters, or the use of ground water by pumping. In 1919, water for about 85 per cent. of the acreage irrigated was supplied by streams and for 7 per cent. by wells. The other sources were chiefly lakes and springs.

In eastern North Dakota and eastern South Dakota there has been a very rapid increase of the area of sweet clover. The climate stimulates a very heavy flow of nectar. At present there are not many commercial beekeepers in this region, but there are hundreds of small ones. This region gives promise of being one of the best for bees in the whole of the United States. In the irrigated areas of the Black Hills there are thousands of acres of irrigated alfalfa. Excellent results are obtained in the Belle Fourche Valley.

In Nebraska, Kansas, and Oklahoma the future of beekeeping will depend largely upon the sweet clover acreage, which is steadily increasing. There are more than two million acres of alfalfa in these three states, but in the eastern portions it yields very little nectar, while in the central and western portions, where irrigation is practiced, it is a good honey plant. Beekeeping in Nebraska is most successful in the eastern part of the state along the Missouri River, where the honey plants are white clover, alsike clover, and sweet clover, and in the valley of the Platte River. Honey production in Kansas is most successful along the Arkansas River in the alfalfa and sweet clover belt. Good locations for beekeeping in Oklahoma are found on the bottom-lands of the larger streams, as the Canadian and Washita rivers.

In Montana the irrigated lands along the Yellowstone River and its southern tributaries offer the best locations for beekeeping. The surplus comes largely from sweet clover, as alfalfa is often cut so early that it furnishes only a little nectar. Ravalli County, in the heart of the Rocky Mountains, where there is a large area under irrigation, is a reliable and profitable location for beekeeping. There are more bees, and more honey is produced, in the Big Horn Basin of Wyoming than

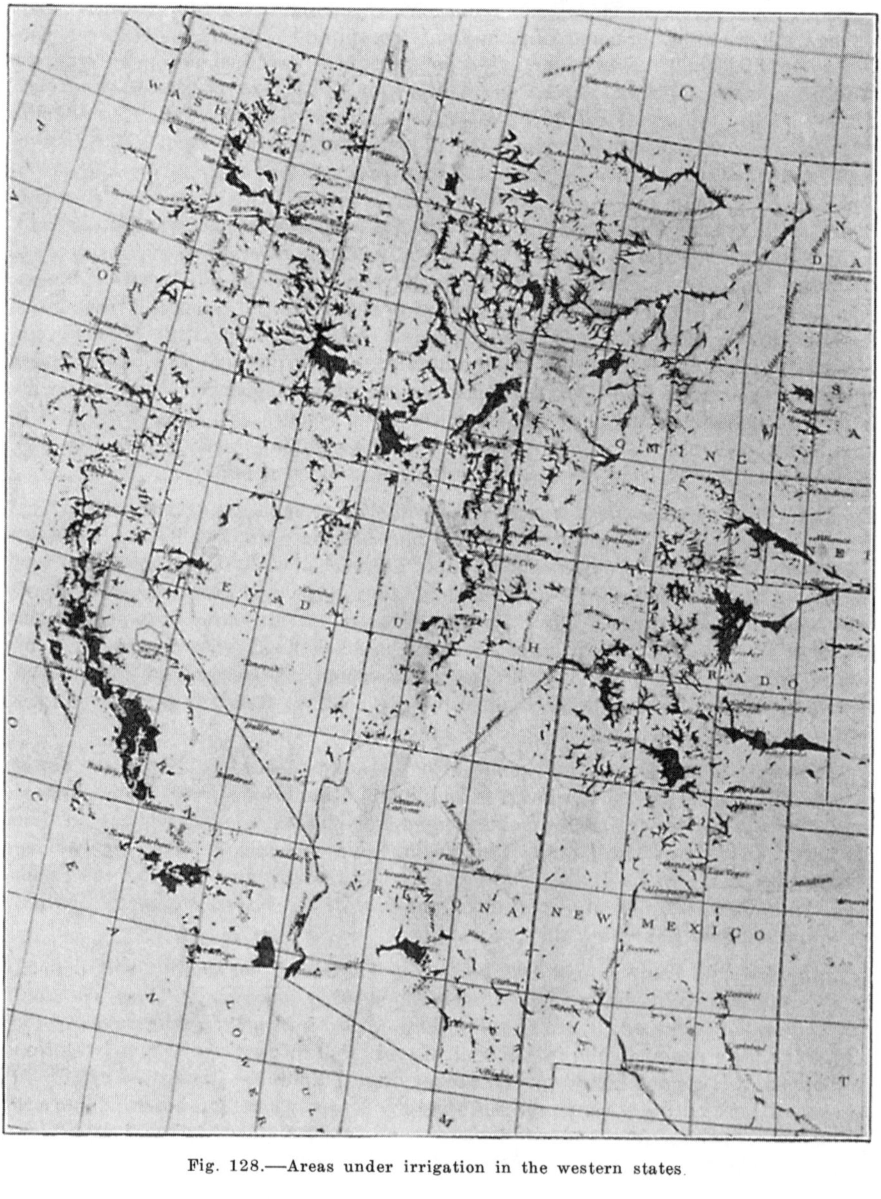

Fig. 128.—Areas under irrigation in the western states.

in any other section of the state. In Laramie County, in the southeastern corner, a large amount of honey is secured from successive crops of irrigated alfalfa. Good results are also obtained in Fremont County.

The largest irrigated areas in Colorado are in the valleys of the South Platte and Arkansas rivers; and it is only in the river valleys that beekeeping is an important industry in the Great Plains region of the state. There are few bees in the Rocky Mountain parks, but immense crops of honey are gathered on the western slope. Eastern Colorado is well stocked with bees; but in the western portion there is room for many more apiaries. Agriculture and beekeeping in New Mexico are mainly restricted to the valleys of the Rio Grande, Pecos, and San Juan rivers.

Among the mountains of northern Idaho, as at Sandpoint, good crops are gathered from white clover, alsike clover, buckwheat, and fireweed. In the southern portions of the state where alfalfa is extensively grown under irrigation, and also a large acreage of alsike clover, commercial beekeeping is successful in the Boise Valley in the southwest, at Twin Falls in the south, and in the vicinity of Idaho Falls in the southeast.

Most of the beekeepers of Utah are located in the mountainous agricultural belt extending through the center of the state, in the Uintah Basin, south of the Uintah Mountains, and in Emery County. The larger portion of the irrigated area and consequently of the alfalfa acreage lies in the central tract, where water is brought down from the higher levels. There are many commercial beekeepers in this region. Alfalfa is grown extensively for seed in the Uintah Basin and affords an immense pasturage for bees; but, in the absence of a railroad, transportation is expensive. In Emery County a total failure of the honey crop has never occurred. Nevada is largely a desert, but great crops of honey are produced in the western counties from irrigated alfalfa.

Maricopa County and the Salt River Valley stand easily in advance of any other county in Arizona both in agriculture and bee culture. One of the most noted irrigation systems in the world, the Salt River project, is capable of irrigating more than 200,000 acres of land. Alfalfa and cotton cover a large acreage. In the Yuma country, in southwestern Arizona, where a great area is irrigated from the Colorado River, and alfalfa, cotton, and mesquite are abundant, there is room for many more colonies of bees. The Arizona deserts, near Phoenix, according to E. R. Root, can furnish thousands of pounds of bees in packages in early spring for the honey flows of California and many other western states.

The average annual rainfall in Washington and Oregon west of the Cascade Range exceeds one hundred inches, coming between November and May, but the summers are almost arid. In the lumbered section of the Coast Range fireweed, or willow-herb, offers wonderful possibilities; but colonies of bees require special management and the flow is not always reliable. East of the Cascade Range there are many large honey producers in the Yakima Valley, a rugged area of land in the eastern foothills of the mountains. Irrigated alfalfa is the chief honey plant. In eastern Oregon commercial beekeeping is probably more successful in Umatilla County than in any other county in the state. Good crops of honey are also secured in Malheur County.

California is surpassed in area only by Texas, but in climate and physical features it offers greater contrasts than any other state. The conditions of beekeeping are exceedingly diversified. In Los Angeles County and the adjacent territory is found the most dense area of bees in the United States. On the other hand, at no great distance away in the San Bernardino desert there may be no rain for an entire year and bees would quickly perish in this arid region. The climate of northwestern California resembles that of western Oregon, but the dry season is more pronounced.

In the Great Central Valley irrigated alfalfa is the chief source of surplus, but there are many other honey plants which yield a large amount of nectar. In the southwestern counties of Santa Barbara, Ventura, Los Angeles, and San Diego, and in the western portions of San Bernardino and Riverside counties, beekeeping reaches its highest development in California. The principal honey plants are the orange, mountain sages, Lima bean, and wild buckwheat. In 1900, the Imperial Valley was a hot sandy desert, but twenty years later there were nearly 200,000 acres of irrigated alfalfa and cotton in this valley a hundred feet below sea level. The honey flow is very reliable and there are few unoccupied bee ranges.

During the past 20 years the number of colonies of bees on farms has rapidly grown smaller, largely as the result of brood diseases. There are many localities in which practically all of the small yards have been destroyed. Commercial beekeepers have welcomed the disappearance of the farmer beekeeper and bought up many small yards. The farmer beekeeper almost uniformly neglects his bees, and his few colonies become centers for widely spreading American and European foul brood. Their destruction has been a great benefit to the honey producer; but it has been the cause of an almost incalculable loss to agriculture. For the proper pollination of fruits of all kinds, of many vegetables, flowers, buckwheat, and many fodder plants, there ought to be at least a few colonies of bees on every farm. A few large apiaries can not and do not perform the work of pollination equally well. For example, during the period of fruit bloom there is often much rainy and cloudy weather; and unless there are colonies of bees actually in the orchards much of the bloom will not be pollinated. Reese has estimated that in West Virginia alone the loss from this cause amounts to a million dollars annually. The dependence of the farmer and fruit-grower on a large number of small apiaries is very generally recognized; and it is very desirable that there should be an increase in the number of colonies of bees on farms under a system of inspection which shall assure that they receive proper care.

The natural or physical units into which the United States and the several states have been divided in Part IV are based largely on an article entitled "Physiographic Divisions of the United States," by Nevin M. Fenneman, Annals of Association of American Geographers, Vol. VI, pp. 19-98, 1917. Following the request of members of the United States Geological Survey, the Bureau of Forestry, the division of Pomology, and many geographers, the Association of American Geographers appointed a committee to devise a systematic division of the United States. This committee consisted of five eminent geographers and geologists, who were assisted by a number of geologists selected because of special familiarity with various parts of the United States. The results of this committee's work are incorporated in the article mentioned above. There is little reason to doubt that the divisions here given will continue in the main to be recognized as valid. The boundaries of most of the regions coincide with geological lines; and the regions differ to a great extent, as has already been pointed out, in elevation, topography, vegetation, and geological history.

While the United States census figures for 1900, 1910, and 1920, relating to bees and honey, have been carefully considered in the preparation of the following descriptions of the different states, they have been quoted only occasionally. The census reports on beekeeping are restricted to farms of three acres or more, producing $250 worth of agricultural products. But a great number of commercial beekeepers and many small honey producers live in cities and towns and own comparatively little land. It is estimated that, in the eastern portions of Massachusetts, New York, and Pennsylvania, where there are many large cities, not more than one-third of the beekeepers are included in the census figures. In California, again, which has doubtless a larger number of beekeepers and produces a greater amount of honey than any other state, probably three-fourths of the beekeepers do not live on farms and consequently have not been included in the returns of the census enumerator. In the more strictly agricultural states, such as Maine, New Hampshire, Vermont, Virginia, North Dakota, South Dakota, Washington, and Oregon, the figures are more nearly accurate; and in thinly settled rural counties they are approximately correct. But in general the number of beekeepers in the several states is twice or three times as large as the figures given in the census reports. For this reason, except in special instances, they are seldom used in Part IV.

The descriptions of the conditions of bee culture in the forty-eight states are based on replies to a great number of questionnaires which were sent out in 1919 and 1920; and the statements, strictly speaking, apply to those years and to prior years. While the size of the apiaries and the surplus obtained vary more or less from year to year, conditions in most localities have remained essentially unchanged. An apiary is an experiment in bee culture, which shows the results which may be expected in the production of honey in a locality during a term of years. This experience has a permanent value and continues to be instructive even after the apiary has disappeared. All of the following records, therefore, remain valuable, although in some instances the apiaries may no longer be in existence.

In the preparation of the descriptions of the various states the author desires to acknowledge his indebtedness to the following distinguished apiarists for much information and many helpful suggestions. Responsibility for the facts as presented, however, should rest with the author alone.

Alabama, T. Atchinson, J. M. Cutts, and Mell Pritchard; Arizona, Mrs. M. G. Loveitt; Arkansas, J. V. Ormond; California, L. L. Andrews, M. H. Mendleson, and G. H. Vansell; Colorado, Prof. C. P. Gillette and F. A Rauchfuss; Connecticut, A. Latham; Florida, Frank Sterling and J. C. Goodwin; Georgia, S. V. Brown and A. R. Irish; Idaho, Roy Rabbit; Illinois, A. L. Kildow and A. G. Gill; Indiana, C. O. Yost; Iowa, Frank C. Pellett and F. B. Paddock; Kansas, Dr. J. H. Merrill; Kentucky, Prof. H. Garman; Louisiana, Prof. E. C. Davis; Maryland, E N. Cory; Massachusetts, Dr. Burton N. Gates; Michigan, Prof. R. H. Kelty, B. F. Kindig, and Ira D. Bartlett; Minnesota, Prof. Francis Jager; Mississippi, Prof. R. B. Willson; Missouri, Prof. Leonard Haseman; Montana, W. A. Petzoldt and B. J. Kleinesselink; Nebraska, H. C. Cook and J. H. Wagner; Nevada, G. C. Schweis; New York, Charles Stewart and Prof. R. B. Willson; New Hampshire, Wm. H. Wolff; New Jersey, E. G. Carr; North Carolina, C. L. Sams, B. E. Eckert, and Frank H. Lathrop; North Dakota, Prof. R. L. Webster; Ohio, Chas. A. Reese and E. R. Root; Oklahoma, C. F. Stiles; Oregon, Prof. H. A. Scullen; Pennsylvania, George H. Rea and C. N. Green; Rhode Island, A. C. Miller; South Carolina, E. S. Prevost; South Dakota, L. A. Syverud; Tennessee, J. M. Buchanan and Prof. G. M. Bentley; Texas, Dr. F. L. Thomas, B. I. Solomon, and H. B. Parks; Utah, F. B. Terriberry and M. A. Gill; Vermont, J. E. Crane; Virginia, Dr. W. J. Shoene and B. Anderson; Washington, Prof. H. A. Scullen; West Virginia, Charles A. Reese; Wisconsin. Dr. S. B. Fracker, Prof. H. F. Wilson, and C. D. Adams; Wyoming, C L. Corkins, Charles Ranney, and O. Hamm.

THE HONEY PLANTS AND CONDITIONS OF BEEKEEPING IN THE DIFFERENT STATES

MAINE

Total area, 38,040 square miles. About 140 miles from the coast a watershed crosses the state from west to east, dividing its surface into two drainage slopes. The northern slope, which has an extreme width of 80 miles, is so poorly drained in its middle and western portions that there are many swamps and lakes and it remains an uninhabited wilderness. The southern slope in its western portion is hilly or mountainous, but eastward becomes more level in Hancock and Washington counties. The coast, which is noted for its scenic beauty, is fringed with a series of narrow rocky ridges running far out into the sea, between which are deep fiord valleys forming bays and river mouths. Granite is the prevailing rock in the western portion of the state and is abundant in every county; but there are many beds of limestone. All of Maine was glaciated, and as a result there are large areas of barren rocks, and the hills are covered with a thin sterile soil mixed with gravel and boulders. The more fertile valley soils are underlaid by a deposit of clay and boulders known as hardpan. Of the 19,000,000 acres within its borders only five and one-half million acres are in farms. Much of the land is unfit for agriculture and can never be cultivated.

There are no good locations along the coast, and the counties bordering on the ocean have but few bees. The honey plants are willows, maples, dandelion, fruit bloom, white clover, alsike clover, raspberry, sumac, willow-herb, and goldenrod, but none of them are sources of a large amount of honey. There is usually a fair flow in June and July from the clovers, while in the fall the goldenrods are the main dependence of the beekeeper for winter stores. Near Bangor, on the Penobscot River, the locations are equally poor, and more honey is stored from wild raspberry than from any other plant. In Hancock County, near Ellsworth, there are about a dozen apiaries of 6 colonies each. As high as 48 pounds of surplus have been obtained, but the average is much less. In Washington County, in the southeast corner, clover is of little value, and goldenrod is the chief dependence. There is one yard which contains 32 colonies. At Addison an apiary of 28 colonies reports good results from clover, willow-herb, and goldenrod. At Machias the blueberry barrens are important.

In the central part of the state, at Dover, conditions are better; bees are fairly profitable, and a failure is almost unknown; but there is no one who makes beekeeping his sole occupation. The apiaries average from 15 to 20 colonies, but one numbers 100, and another 63 colonies; a surplus of 40 to 60 pounds is obtained. The honey plants are willows, maples, dandelion, raspberry, clover, and goldenrod, which yield a nearly continuous flow from May 15 to Sept. 15. Basswood has not blossomed for several years, and is of little value. In one year there was a good flow from orange hawkweed — never known to occur before in central Maine.

The western part of the state is very hilly, and reaches an altitude of 2000 feet above sea level. Apiaries range from 5 to 20 colonies, with perhaps an average of 19. Lewiston, Woodstock, and Bethel are fair locations for 25 to 50 colonies. In this section of the state this industry is increasing. The honey plants are similar to those already enumerated, but buttonbush is of value near the swamps.

The northern slope, except along the eastern border, has so gentle a descent, and is so poorly drained that it is largely covered with swamps and lakes, and is of little importance to bee culture. The eastern part of Aroostook County, in which there is the largest undivided area of arable land in New England, is far the best section of Maine for beekeeping. Yards of 30 colonies are common, and there are a few of one hundred or more. In the vicinity of Presque Isle there are from five to six hundred colonies. The surplus is gathered chiefly from white clover and alsike clover, which bloom from June 15 to the middle of August. There is no dependable flow in the fall; but raspberry, goldenrod, and willow-herb are important in localities. The nights are so cold when goldenrod blooms that little nectar is usually secreted; but a fair flow may be expected about once in four years. The long winters are, however, a serious drawback, as the bees are confined to the hives from November 20 to May 1, and the springs are late and variable.

There is a profusion of wild currants, gooseberries, blackberries, raspberries, and strawberries throughout much of the state. The pussy willow, riverbank willow and white willow, red maple, rock or sugar maple, choke cherry, wild red cherry and black cherry, haw, sumac, black locust, cornels, Viburnum, and apple trees are abundant. White clover and alsike clover succeed inland better than on the coast. The bushy goldenrod and tall hairy goldenrod are the most valuable species of *Solidago*. Asters are of little importance; but *A. paniculata* attracts many bees. Basswood, boneset, Canada thistle, heartsease, milkweed, and sweet clover are rare or devoid of nectar. Dandelion in southern Maine is valuable chiefly for pollen.

NEW HAMPSHIRE

Total area, 9341 square miles. Physiographically the state may be divided into three sections — the region of the White Mountains, or Grafton and Carroll counties; northern New Hampshire, or Coos County; and the region south of the White Mountains, which includes seven counties. The total number of colonies of bees is very small, and there are no apiaries which number more than 50 colonies. Beekeeping is pursued only as a side line. The larger apiaries are located in the vicinity of the Connecticut River; but there are many small yards in the valley of the Merrimac and in the southern portions of the state. Along the coast there is a belt about three miles wide which is unsuitable for beekeeping.

The north-central portion of the state, a rudely circular area of 1400 square miles, is occupied by the White Mountains, a group of round-topped mountains which were formerly densely forested with spruce and pine. Where the land has been cleared and burned over, wild raspberry, willow-herb, and low and high blueberries furnish fairly good bee pasturage.

Northern New Hampshire, or Coos County, is crossed by numerous mountainous ridges which are separated by wide rolling valleys. At Colebrook, within a radius of five miles, there are a few bees. The average yield per hive is very poor, but a surplus of fifty pounds is occasionally obtained. It is the general opinion that there is not sufficient bee pasturage to permit extensive beekeeping. At Stewartstown, five miles northward, there are seven apiaries which range in size from 1 to 20 colonies. The average surplus is twelve pounds per colony. The only honey plants of importance are wild raspberry, white clover, willow-herb, goldenrod, willows, and wild cherry. There is at present greater freedom from bee diseases in this region than in the southern part of the state. Both in Coos County and in the White Mountains the winters are very severe, and the bees are usually wintered in cellars. Northern New Hampshire is largely a rugged unreclaimed wilderness which is thinly populated.

In the counties south of the White Mountains there are many small apiaries, and a few which contain from 20 to 50 colonies; but beekeeping is always combined with some other vocation. The honey flora is wholly inadequate to support a large number of colonies. West of the Merrimac River the country is broken by hills and low mountains; but east of this river much of the land is less than 500 feet above the sea, and there are numerous picturesque lakes. The best locations are found in the valleys of the Connecticut and Merrimac rivers and in the southeast portion of the state, where the soil is fairly fertile. At Lebanon there is an apiary of 50 colonies; and in the southwest corner of Cheshire County there are reported to be fair locations for a few beekeepers. The honey plants are locust, raspberry, basswood, the clovers, sumac, goldenrod, and aster. At Nashua, on the Merrimac River, there are not far from 40 beekeepers within a radius of five miles. At Greenville, near the south border, there are half a dozen apiaries which contain from 6 to 20 colonies.

At Durham, in the southeast corner of the state, the surplus is obtained from raspberry, white clover, and alsike clover. Other species of value are willows, dandelion, fruit bloom, sumac, goldenrod, and aster. Basswood is rare, but maples are very common. Of minor importance are milkweed, mustard, willow-herb, and buckwheat. At Exeter the honey plants are the same as those at Durham, except that wild raspberry is less abundant. A surplus of more than 50 pounds per colony is seldom obtained.

The apiaries in New Hampshire rarely exceed 5 to 10 colonies, and the average is about 5. With the elimination of brood disease, and the introduction of better methods, the status of beekeeping in the southern section should be much improved. For those who have some other vocation, beekeeping as a side line can often be pursued to advantage; but for the specialist this state can offer little inducement.

VERMONT

Total area, 10,212 square miles. In Vermont, as in Maine and New Hampshire, the great ice sheet planed down and grooved the rocks, gouged out the valleys, and covered the surface with glacial drift. The best arable land in the state is found in the Champlain Valley, where the underlying rock is limestone. The soils are well adapted to fruit-growing, and the clovers flourish throughout this section. About 1500 species of flowering plants occur in the state, of which 100 are shrubs and trees. Maples, willows, sumac, cherry, raspberry, cornel, and blueberry furnish both nectar and pollen. The pulse family is represented by 46, and the Compositae by 171 species — the two families which contain the largest number of honey plants. Vermont is divided into two nearly equal portions by the Green Mountains, which are densely covered with black spruce and other evergreen trees.

The whole section of the state west of the Green Mountains is generally good for beekeeping. The best flows are obtained in Addison and Rutland counties from white clover and alsike clover. From Rutland to Fairhaven up to the state line clover is abundant; but it does not yield as well as on the clays near Lake Champlain, as the soil is often sandy or gravelly, and lacking in lime, although limestone ledges are common and great marble beds of commercial importance occur in Rutland County. Apiaries run from 60 to 150 colonies. Ten miles farther southwest, nearer the lake, at Shoreham, the apiaries range from 25 to 120 colonies. During 35 years only one season has been reported in which there was no surplus. Although with proper management this district would probably maintain additional colonies, it is reported as fully stocked; but farther south, in Orwell, Benson, and Fairhaven there are good locations.

In Addison County the main dependence for surplus is white clover and alsike clover, which are reliable nearly every year. Basswood, which was formerly important, has been so largely cut for timber that it is now only a small factor in securing a crop. Its loss is made good by alsike clover, which thrives well on the heavy clay soils, and is common along the roads and in waste places. Sweet clover is coming in slowly, and may be of value later. Raspberry is not abundant here, but dandelion, fruit bloom, and goldenrod are valuable. Near Middlebury an experienced apiarist has 1000 colonies divided into yards containing over 100 colonies each, and located about three miles apart. From an apiary of 140 colonies, spring count, in 1919, 8000 pounds of extracted honey were obtained. By feeding sugar for winter it has been found possible to increase greatly the size of the apiaries.

In Grand Isle County, in Lake Champlain, 10,000 pounds of honey have been secured from a single apiary. At Alburg there are from 10 to 150 colonies to a yard. The good sites are all occupied in this county. The principal honey plants are white clover and buckwheat, while dandelion, apple, aster, and goldenrod are of minor importance; but there are years when the fields are white with clover bloom, and yet little surplus is obtained. Great humidity, with hot sultry days and occasional showers, are required to stimulate the best flow.

Bennington County, in the southwest corner of the state, is a fair location, but is not as good as Addison County. The honey plants are clover, basswood, buckwheat, and a few others already mentioned. Wild marjoram (*Origanum vulgare*) has become naturalized in this county, and affords excellent pasturage, one beekeeper securing 50 pounds of comb honey per hive from this source. It is called locally "horsemint."

Eastern Vermont, or the region east of the Green Mountains, is less suitable for beekeeping than the western section. About one-third of the colonies reported in the state are found in this region. Most of the apiaries are small and scattered, averaging about 15 colonies, leaving large areas in which there are no bees. The chief honey plants are clover, raspberry, and goldenrod. Basswood occurs to some extent near Montpelier, running southwest. In the extreme north, east of Lake Memphremagog, raspberry is abundant. With proper methods, in eastern Vermont good results on a small scale may be expected in this industry.

MASSACHUSETTS

Total area, 7800 square miles. As the result of glaciation the higher mountainous ridges have been swept bare of fine-grained soils, and a deep deposit of glacial till has been made on the flanks and crests of the lower hills. With the retreat of the ice, extensive sandy and gravelly outwash plains were formed in all of the larger valleys. The east and southeast portions of the state are level sandy barrens well watered by ponds; the center is hilly; and the west, rugged and mountainous. Massachusetts may, therefore, be divided into three principal physical regions: The western highlands, including the counties of Berkshire, Franklin, Hampshire, and Hampden; a central and northeastern hilly region, or the counties of Worcester, Middlesex, and Essex; and a relatively level sandy southeastern region comprising the counties of Norfolk, Bristol, Plymouth, and Barnstable. The Connecticut Valley, which has the most fertile soils in the state, is, from the standpoint of bee culture, often ranked as a distinct region. It differs materially from the rest of the state. South of Pittsfield, Berkshire County, says Gates, seems to prove as good a bee section as parts of eastern New York.

Massachusetts is only fairly well adapted to intensive bee culture, for there are no large areas of arable land covered with white clover, as in the Champlain Valley of Vermont or in Aroostook County in Maine. Beekeeping is pursued almost exclusively as a side line. From the honey producer's standpoint the Berkshire Valley, south of Pittsfield, in the west, Worcester County in the center, and the Clethra country in the southeast offer the most desirable locations. In 1919 the four western counties produced more than one-half, and Worcester County nearly one-sixth of the total surplus of honey reported from the state.

The Berkshire country is a rugged mountainous highland, traversed by long wooded ridges broken by deep valleys. It is a region of great natural beauty and productiveness; but, with its high altitude, the winters are severe and the springs late and cold, which renders it more exacting for beekeeping than the southeastern part of the state. In the Berkshire Valley the soils have been derived from the glaciation of the underlying limestone, and are among the best in the state. The more important honey plants are willows, maples, raspberry, dandelion, fruit trees, locust, clover, sumac, basswood, buckwheat, milkweed, goldenrod, and aster. Raspberry is abundant, and there are hundreds of acres of wild thyme (*Thymus Serpyllum*) in Berkshire County. This species also occurs in Hampshire and Franklin counties. At Pittsfield and southward good crops of clover honey are secured, and also some pure raspberry honey, and there is a nearly continuous flow from a succession of other flowers.

The central and northeastern section is generally hilly, becoming more level toward the coast; and, owing to the fact that it contains a number of large cities, many beekeepers are found here, though it offers few inducements to the specialist. The chief honey plants are willows, fruit bloom, clover, raspberry, basswood, and goldenrod; but none of them yield very abundantly. Dandelion, sumac, Clethra, locust, and sweet clover deserve mention. The winters are severe; disease occurs locally, and the flow of nectar is irregular. Throughout Worcester County blueberry and huckleberry are abundant. Three acres of Scottish heather (*Calluna vulgaris*) have been planted on the estate of Bayard Thayer, near Lancaster, which secretes nectar freely in its new home.

A peculiar phase of beekeeping in Massachusetts is the demand for bees for pollinating cucumbers in greenhouses (see Cucumber). In Worcester, Middlesex, Essex, Suffolk, Hampshire, Hampden, and Franklin counties, cucumber growers purchase more than 2000 colonies annually for this purpose, and many beekeepers devote their entire attention to raising bees for sale to them.

The southeast section is a sandy plain, generally level except for an elevated ridge south of Plymouth, diversified with numerous ponds and dry barrens, and partially covered with yellow pine. The ocean tempers the climate, which is milder here than elsewhere in the state. The sweet pepperbush or Clethra (*C. alnifolia*) region extends from Middleboro to Sandwich on the Cape and to New Bedford on Buzzard's Bay (see Clethra). Near North Rochester there is a swamp of 35 or 40 acres of this shrub with other large expanses of it elsewhere. It yields nectar abundantly, but only in periodic years. Three colonies are reported to have stored 900 pounds of this honey; but as a rule there is about one poor year in every three or four. On Cape Cod the huckleberry grows everywhere, and the cranberry covers 7000 acres but does not secrete nectar freely. The cranberry-growers use many bees for pollinating the cranberry blossoms. Without plenty of bees the crops are short. The salt-marsh goldenrod (*Solidago sempervirens*) thrives in beach sand, and is the source of a large amount of honey in the fall.

According to Gates, the honey production of Massachusetts is only one-tenth of what it might be. Allowing an average of 100 to 126 acres to support a colony of bees, based on the experience of large beekeepers who maintain a series of out-yards, and eliminating 500 square miles as probably unavailable for bee pasturage, there remain about five million acres for forage in Massachusetts, which would support approximately forty to fifty thousand colonies of bees. In two (or perhaps three) years out of five, honey-dew is to some extent injurious east of the Connecticut River.

RHODE ISLAND

Total area, 1248 square miles. The surface is a rolling plain with a mean elevation of only 200 feet above the sea. It is strewn with glacial drift, which furnishes a poor soil for vegetation; but, owing to the proximity of the Gulf Stream, the climate is milder and more equable than that of any other New England state. The mean annual temperature is 50 degrees F., and the mean annual rainfall 50 inches.

Rhode Island offers little opportunity for commercial beekeeping; but many small apiaries can be maintained to advantage for honey and the pollination of fruit trees, berry bushes, buckwheat, cucumbers, squashes, and many other plants. But it is generally agreed by all beekeepers that there is little inducement for any one in Rhode Island to engage in this industry for the purpose of making a living. Most of the northern part of the state, and nearly all the western half, except a narrow strip along the shore of the Bay, is covered largely with birch and oak. Around Chepachet in the north there are not more than 15 or 20 colonies in a radius of 20 miles. Bristol County is a little better region, and some 40 colonies are located around Warren. The islands in Narragansett Bay, Prudence Island and Conanicut Island, offer fair opportunity for a few apiaries. But the southwestern section along the bay and coast, where sumac abounds, and in the vicinity of the Great Swamp, north of Wordens Pond, is best adapted to beekeeping. Here there are several apiaries of 30 to 50 colonies.

The honey flora is varied, but none of the species are sufficiently abundant to afford large yields. Willows, maples, fruit trees, raspberries, blueberries, huckleberries, and choke cherry are the earliest sources of honey. There are large areas of locust, which grows 50 to 60 feet tall, and this tree is yearly becoming more abundant. The sumacs (*Rhus glabra* and *R. copallina*) and *Viburnum dentatum* are common in the southern section. Clematis is widely distributed, and yields an excellent honey, but it is not reliable every year. Clethra is abundant in all lowlands, but is very erratic and uncertain. Buttonbush is also common in swamps. In many villages and cities the European basswood has been extensively planted. Sweet clover thrives at Lime Rock, and white clover and alsike clover add to the surplus in favorable seasons. Vast areas of wind-swept dunes in the coast region are covered with goldenrod. Honey-dew in some seasons is very abundant; but, as it is often bitter and dark-colored, it spoils the honey with which it is mixed.

CONNECTICUT

Total area, 4956 square miles. The state is divided by the Connecticut River Valley, which has an area of 600 square miles, into two upland sections which rise from the low seashore to a height of over 1500 feet along the north border. The soil is not very productive, as the original soil was removed during the glacial age; but it is more fertile in the Connecticut Valley, where thousands of acres of tobacco are grown. The forests have been largely cut away.

It is the opinion of a veteran Connecticut beekeeper that this state can not be considered a good location for beekeeping, and should not be ranked more than third rate when compared with New York. The honey flora does not differ greatly in the different counties. In the eastern upland the ravages of disease have nearly exterminated the domestic bee. In Windham County, in the northeast corner of the state, practically all the bees have been killed off, and there is here an opening for one or two beekeepers who will be content with small crops. In New London County conditions are a little better; but only a small percentage of the bees survive. Nine-tenths of the farmer beekeepers have abandoned the business. The largest apiary is the home apiary of Allen Latham, of Norwichtown, which numbers 70 colonies. His out-apiaries seldom exceed 20 colonies, but they contain nearly 60 per cent. of all the bees within a radius of 10 miles. The more important honey plants are huckleberry, the clovers, sumac, and goldenrod, while fruit bloom, locust, maple, and dandelion are of value. Mustard in the oat fields has become a source of honey in June. Of the plants which yield a surplus none is reliable every year except sumac. The smooth sumac (*Rhus glabra*) flourishes on rocky hillsides and in poor sandy or gravelly soils. In the rocky glacial drift it is found in profusion, and much of Connecticut would be worthless without it. The clovers are often not dependable, and goldenrod yields only every other year, or even less often.

The southwestern part of the state is regarded as best adapted to bee culture. At Stamford the apiaries range from 1 to 20, and there is one of 40 colonies; but ordinarily there are not in this state more than 5 or 6 colonies to an apiary. The honey plants are clover, basswood, sumac, asters, and goldenrod, while willows, fruit bloom, maple, and locust are of secondary importance. Litchfield County, in the northwestern corner, which contains the highest land in the state, is much better adapted to bee culture than Windham County, and is the home of a number of successful beekeepers; but Hartford County, which lies in the Connecticut Valley, probably contains more beekeepers than any other county along the north border. A large area is devoted to the culture of buckwheat, and thousands of acres of tobacco are grown. Where the tobacco is "shade grown" and permitted to bloom, it yields a dark honey which is excellent for winter stores.

NEW YORK

Total area, 49,204 squares miles. The state is triangular in outline, extending 300 miles along the Hudson River from north to south, and 326 miles from west to east. In the northeastern part a roughly circular area, sparsely settled and too rugged for agriculture, is occupied by the Adirondack Mountains, between the short ranges of which are numerous glacial lakes. The southern and western portions of the state form an elevated plateau, sloping northward toward Lake Ontario, in which the great continental ice sheet plowed out the Finger Lakes and many other valleys, removed the original soil, and left a thin sterile covering on the uplands which to-day comprise the larger part of the buckwheat-honey country of New York. But south of Lake Ontario the land is more fertile, and a narrow strip along the lake is devoted to fruit-growing. The two regions are separated by the Mohawk Valley, the former bed of a glacial river, of which the Mohawk River is the successor. During the recession of the ice in the glacial era the waters of the Great Lakes were dammed up, and a great river flowed over the ledge at Little Falls, finding an outlet in the valley of the Hudson. On the retreat of the ice, morainic deposits compelled this ancient river to seek a new outlet through the St. Lawrence River, which is nearer sea-level.

The proposed Adirondack Park, which includes portions of St. Lawrence, Franklin, Essex, Hamilton, and Herkimer counties, is the poorest section of the state for beekeeping. The valleys are filled with stony and sandy loams, which are unsuitable for growing clover, while large surfaces of the crystalline rocks of the mountain flanks and summits were swept bare of their covering by the ice. According to the state census for 1917, Hamilton County had only 136 colonies of bees, 193 acres of buckwheat, and 544 acres of fruits. In northern Herkimer County the area in farms does not exceed 6 per cent. of the land, and an apiary of 75 colonies is too large for the best results. In Warren County, on Lake George, the apiaries usually contain less than 10 colonies, but there are a few which number over fifty. The honey plants are white clover, alsike clover, basswood, and sweet clover. Buckwheat is of little value here. The larger part of Lewis County also offers few attractions to the specialist, but at Lowville there is some good clover territory.

North of the Adirondack region are enormous deposits of almost pure sand. The soil is acid and the clovers do not thrive upon it; but raspberry is abundant and yields well. Raspberry honey is produced in commercial quantities year after year near Massena Springs. It is the only source of surplus, and about 50 pounds of extracted honey per colony is obtained. As little honey is gathered in the fall, it is often necessary to feed for winter stores. Even in this locality wild cherry and other small trees and shrubs are reducing the raspberry area. A portion of Franklin County should be included in the raspberry region, as from Malone southward. In the raspberry section black bees are the rule, and the methods of beekeeping are poor.

Westward of the Adirondacks, along the St. Lawrence River in St. Lawrence and Jefferson counties, there is a remarkable belt of white and alsike clover, producing an unsurpassed grade of white honey. There is no finer clover area in this country. In St. Lawrence County there is a large area of calcareous glacial till soil, and a small area of this soil is also found along the northern edge of Franklin County. It also occurs in the southern part of Jefferson County; but the northern part is covered by glacial terrace soils which are nearly neutral or tend to acidity. The secretion of nectar by the clovers is less reliable on the terrace soils, and "strangely enough," says E. F. Phillips, "some of the best beekeepers have established outyards in poor clover areas through lack of knowledge of soil differences." Other nectar-yielding plants of value are willows, soft maple, dandelion, wild raspberry, basswood, and 300 acres of fruits. There are 6000 acres of buckwheat; but it is of little value in this region. Commercial apiaries in St. Lawrence County average about 100 colonies, but there is one at Pope Mills in which there are over 200 colonies. The average surplus in a good year is 50 pounds, but occasionally 100 pounds of comb honey per hive are obtained. Once in about eight years there is an entire failure. Both American and European foul brood occur to some extent; and half the colonies in some yards are lost by outdoor wintering — a needless loss, however. A famous location for beekeeping is Black River, in Jefferson County. A crop is obtained nearly every year, the average surplus per hive being 75 pounds. All the honey plants are reliable each year, as a rule, except buckwheat. There are not many small yards in this section, which is fully stocked with bees.

In the eastern part of the state there is an important section for beekeeping south of the Mohawk River and west of the Hudson River, including Montgomery, Schenectady, Schoharie, and Albany counties, and the southern portions of Saratoga and Fulton counties. In a circular area 45 miles in diameter, with Schenectary County as a center, Charles Stewart estimated that there were, in 1920, over 90,000 colonies of bees. In the spring, bees build up rapidly on willows, maples, dandelion, and fruit bloom; and in summer a surplus is stored from white clover and alsike clover, sumac, basswood, and buckwheat. There are over 40,000 acres of buckwheat besides many large apple-orchards. In southern Fulton and northern Montgomery counties, according to the experience of Charles Stewart, extending over 35 years, 100 pounds of extracted, or 50 pounds of comb honey, is a fair average. In the vicinity of Mayfield, Fulton County, there are 12 commercial apiaries. The surplus per hive may be as low as 20 pounds, and as high as 100 per hive. Usually it ranges from 40 to 60 pounds. At Delanson the Alexander home yard numbers 500 colonies. There are few if any opportunities for establishing new apiaries in this section.

In the counties of Delaware, Otsego, Broome, Chenango, Madison, and Oneida the apiaries range from a few colonies to over 100. The surplus comes chiefly from the clovers, buckwheat, and occasionally partly from basswood and raspberry. The area of buckwheat comprises over 40,000 acres. None of these counties stand in the front rank in the production of honey.

New York is one of the leading states in fruit-growing, and the number of fruit trees under cultivation exceeds twenty-two million. In the southeastern counties of Albany, Columbia, Greene, Dutchess, Ulster, and Orange there are four million fruit trees, largely apple trees. Besides fruit bloom, buckwheat, white clover, alsike clover, sweet clover, basswood, sumac, locust, goldenrod, and aster are valuable sources of nectar. Sweet marjoram (*Origanum vulgare*) is reported to yield a white honey of fine flavor near Poughkeepsie and in the southern Lake George region.

Fair to good locations may be found in Washington and Rensselaer counties east of the Hudson River, but the soil here is not calcareous. Near Fort Edward, Washington County, there are six apiaries which range from 75 to 200 colonies. A crop is obtained usually eight years out of ten, and in a favorable season the average surplus is about 100 pounds. There are over 3500 colonies in this county. Rensselaer County contains fewer bees, but at Indian Fields there are several commercial apiaries, one of which numbers 200 colonies.

While all of the soils in southern New York, south of the Mohawk Valley, are almost wholly derived from glacial till, a part of them are calcareous, or contain lime, and a part are neutral or acid. A knowledge of these different soil types and the areas they cover will be found very helpful to the beekeeper in locating his apiaries. Lying between the southern boundary of the state and a line drawn from Buffalo to Troy there extends from Lake Erie to the Catskill Mountains and farther northward nearly to the Hudson River a belt of land about 60 miles in width, on which the soils are poor and non-calcareous, either neutral or acid — in some instances quite acid. Throughout this belt the clovers do not prosper, and the secretion of nectar by these plants is less dependable. But buckwheat is grown by the thousands of acres, and is a very reliable yielder of nectar, and the main dependence of the beekeeper.

North of this belt there is a second belt, covered with soils of glacial origin, but containing lime, which extends nearly across the state. In the west it is only about 20 miles wide, but it runs northward along the east end of Lake Erie and southward among the Finger Lakes, while farther east it sends out prolongations in both directions. The soil types of this belt are calcareous, and richer in humus than the more sterile soils of the buckwheat region, and are valuable for general farm crops. The clovers grow well throughout this area, and the secretion is dependable. South of Lake Erie there is a strip of glacial lake soil which is suitable for fruit-growing, although less rich in lime than the country south of it. New York well illustrates the importance of a knowledge of soil properties, not only by the farmer but by the beekeeper as well. The clovers succeed well on the calcareous glacial till soils, but grow poorly on the non-calcareous soils in the southern part of the state.

In the west-central part of the state, Onondaga County, with its five thousand colonies of bees, is widely recognized as an excellent region for beekeeping. Around Syracuse the apiaries range from 100 to 150 colonies, with a few small yards. At Marietta, at the north end of Otisco Lake, there are three commercial apiaries with a total of 500 colonies, which secure a fair surplus nearly every year. As there are only 4000 acres of buckwheat in the county, the beekeeper is forced to rely largely on white clover, alsike clover, sweet clover, basswood, fruit bloom, sumac, locust, goldenrod, and aster. The limestone soil of this region is especially well adapted to the growth of the clovers. There are 35,000 acres of alfalfa, and, surprising to relate, this plant, which throughout the eastern states is commonly nectarless, is reported to be often a good honey plant near Syracuse. For many years after its introduction a bee was seldom seen on the bloom; but during the past 10 or 12 years at East Syracuse it has yielded from 10 to 30 pounds of surplus on an average every other year. While bees work on the bloom more or less every season, unless the weather is very dry and hot (conditions akin to those in the arid states) it is of little importance. The first bloom, in June, secretes very little nectar; but the second bloom, in July and August, when the weather is dryer and hotter, furnishes the surplus. Except where it is grown extensively, the fields are mowed before it begins to blossom. The possibility of obtaining a strain of alfalfa which will yield nectar in the east is deserving of consideration.

Excellent results are also obtained in Cayuga County among the Finger Lakes, where there are 4500 colonies of bees and 10,000 acres of buckwheat, and the same number of acres of fruits. American foul brood has been exceedingly bad in the Finger Lake region for several years. European foul brood is also always ready in this region to wipe out weak colonies, of which there are many, due to poor methods of wintering. The southern half of Seneca County is supposed to contain more bees to the square mile than any other county in the state. Near Geneva, at the head of Seneca Lake, there are about 50 commercial apiaries with a total of 4000 colonies. The crop fails only about one year in ten, and an average surplus of 75 pounds is not unusual. One specialist has operated as high as 1200 colonies in 20 apiaries, and produced from 30,000 to 70,000 pounds of honey. Reports from Ontario County are much less favorable. Owing to foul brood and heavy winter losses only a very few beekeepers have been successful commercially. A beekeeper at Naples writes: "I have been trying to produce comb honey for 40 years, but I should have starved had I not established out-apiaries in adjoining counties. Almost any other county would be better."

In the counties along the southwest border, and in Erie County, on Lake Erie, there is a dense acreage of buckwheat and numerous colonies of bees. Fifty-five thousand acres of buckwheat are under cultivation in the counties of Tioga, Chemung, Steuben, Allegany, Cattaraugus, Chautauqua, and Erie.

In Tioga County 50 colonies are, as a rule, as many as should be placed in one apiary; but in 1913 a much larger number would have been possible. In Chautauqua County the apiaries range from 50 to 100 colonies, and the honey flow is very reliable. A beekeeper at Mayville reports that in 40 years there has been only one year in which he secured no surplus, and that feeding for winter has never been necessary. In addition to buckwheat and fruit bloom the honey plants are the maples, locust, white clover, alsike clover, sumac, basswood, and goldenrod. The yield from buckwheat varies. In some years it is heavy, in others very light.

On a belt of land south of Lake Ontario and east of Lake Erie is the most dense area of fruits in the United States, and more than nine million fruit trees are reported in the counties of Oswego, Wayne, Monroe, Orleans, Niagara, Genesee, Erie, and Chautauqua.

Apple trees are by far the most abundant, and comprise the largest acreage; Niagara has 23,000; Wayne, 35,000; Monroe, 17,000, and Orleans, 19,000 acres. In Chautauqua County there are 24,000 acres of vineyards which apparently do not yield nectar. Fruit bloom is a valuable source of nectar in Wayne, Monroe, Orleans, and Niagara counties. If the weather is fair a gain of 20 pounds per colony is sometimes obtained during apple-bloom; but in New York bees get little more than a living from this source four years out of five. At Macedon, Wayne County, near the swamps along the Erie Canal, the apiaries range from 20 to 40 colonies. Near the city of Rochester there are some 20 commercial apiaries which will average 100 colonies. In a good year a surplus of 120 pounds may be expected. In a radius of five miles there are 75 small yards of about a dozen colonies each. As high as 200 pounds of comb honey has been secured per hive from 25 colonies. In Niagara County commercial apiaries do not exceed 100 colonies. The honey plants are white clover, alsike clover, sweet clover, basswood, goldenrod, dandelion, boneset, catnip, Canada thistle, and buckwheat.

Long Island forms the northern extremity of the Coastal Plain. The flora growing on its sandy hills will not support more than 40 or 50 colonies of bees in a single apiary. The total number of colonies on the island does not much exceed five hundred.

The best locations for beekeeping in New York are very irregularly distributed, or, as a beekeeper aptly expresses it, "the good locations occur in pockets." It is usually better to restrict the size of the apiary to 75 or 100 colonies. In the more thickly populated sections of the state there are few openings for new yards; but in the more remote regions there still remain unoccupied sites. Of the outlook in New York, Charles Stewart writes: "Men of wide experience are having good success; but there are many failures."

PENNSYLVANIA

Total area, 45,126 square miles. According to its physical features Pennsylvania may be divided into three regions: A triangular open area southeast of the Appalachian Mountain system; the Appalachian Mountains, which traverse the central portion of the state from northeast to southwest; and the Alleghany Plateau, comprising the section west of the mountain ranges, or more than one-half of the state.

The southeast section is a picturesque country of rolling hills and well-watered fertile meadows, with a mild equable climate. A productive limestone soil covers the larger part of York, Lancaster, Berks, and Chester counties. Fruits and cereals, tobacco, potatoes, asparagus, cucumbers, muskmelons, and, near Philadelphia, flower-growing, occupy large tracts of the land, which is in the highest state of cultivation. Lancaster County has been called the "garden of the state." A dense forest once covered this whole area; and of the nectar-bearing trees there still remain in greater or less abundance the tulip tree, hackberry, sweet gum, persimmon, wild cherry, sugar maple, and red maple. Wind-pollinated trees, as oaks, beech, elm, hickory, and buttonwood are also common. Cornels, Viburnums, huckleberries, buttonbush, willows, Kalmia, and Rhododendron, with many annual and perennial herbaceous plants, unite to make this the most varied flora of the state.

But, although an ideal agricultural region, this section is only fairly well adapted to beekeeping, as much of the land is devoted to the cultivation of the cereals and truck crops. The large number of colonies of bees is to be attributed partly to the dense population, which is equal to that of the whole western half of the state. In Lehigh County the apiaries are small, a few of the larger ranging from 20 to 40 colonies. A fair surplus is obtained once in two or three years from the clovers, buckwheat, goldenrods, and asters. Most of the beekeepers are farmers who have only one or two colonies in the back yards. The apiaries in Montgomery County are also small, and are maintained for a home supply of honey. In the suburbs of Philadelphia, while small beeyards are very numerous, there are several apiaries which range from 60 to 100 colonies. In a good year 60 pounds of extracted honey per colony are often obtained, but the crop is uncertain. At Media, in Delaware County, there are three apiaries which together contain 240 colonies. The crop fails about once in four years. The honey plants are fruit bloom, black locust, white clover, and alsike clover. Near the Delaware River there is no spring or early summer honey flow, but there is a good fall flow from bur marigold, which yields an excellent honey. Bees are sometimes moved to this locality to take advantage of this late flow. Reports of the average surplus in this county vary from 40 to 125 pounds. The only other honey plants are the clovers, goldenrods, and asters.

In Lancaster County, "where almost every foot of land is under cultivation," dairy farming is one of the principal industries, and there is a large amount of alsike clover, which is the chief dependence of the beekeeper. The fall flowers are not sufficiently abundant to yield a surplus. Within a radius of ten miles of the city of Lancaster there are four apiaries which number about 150 colonies each, while the number of small yards probably exceeds one hundred. The profit in this county is hardly sufficient to warrant following beekeeping as a sole vocation; and the region, moreover, is fully stocked. At Wrightsville, York County, there is an apiary of over 50 colonies; but more than three-fourths of a crop is seldom obtained. Near Waynesboro, in Franklin County, there is an apiary of 20 and another of 40 colonies. In a good year nearly 50 pounds of section honey may be secured. There are many yards in which kegs and boxes are used for hives.

The Central, or Appalachian Province, has a uniform width of 50 miles, and consists of parallel mountainous ridges, between which lie deep and often fertile valleys. To the north of the Susquehanna River there are extensive anthracite coal-fields, large open tracts, forests, numerous lakes and bogs, and mountain ridges and slopes well wooded. There are a number of small apiaries scattered through the mountains, and a few of larger size. Wayne County, in the northeast corner, contains many bees. The honey plants are willows, maples, dandelion, wild raspberry, white clover, alsike clover, milkweed, basswood, goldenrod, and aster. Pike County is almost destitute of beekeepers, but a few small yards are kept by farmers. Lackawanna County, in the heart of the mountains, the seat of a great coal-mining industry, is also a poor location. Most of the surplus comes from alsike clover and buckwheat, and beekeeping is pursued chiefly as a side line. Schuylkill is another mountainous county which has few bees. In the strict meaning of the term, there are no commercial apiaries. At Ringtown the yards average about 15 colonies, and the beekeepers are satisfied with 20 pounds per colony. Bee pasturage is scant, and bees often do not store enough for their winter needs. The outlook would be very discouraging if alsike clover were not largely replacing red clover. Sweet clover is grown to some extent. A beekeeper writes: "I do not consider this a bee country, and I do not believe that much honey will ever be produced here."

But in the valley of the Susquehanna much better conditions prevail. In the vicinity of Harrisburg there are large apiaries which in a good year are very successful. At Liverpool, farther up the river, there are specialists who operate several hundred colonies. But in Montour County, on the west side of the mountains, beekeeping is generally neglected. A few farmers have from 1 to 6 colonies, and usually get only a small crop.

South of the Susquehanna the dominant trees are hardwood species, as the nectar-secreting tulip tree, basswood, sugar maple, and locust, and the nectarless oak, beech, hickory, and walnut. Deep-lying within the mountains, walled in by ridges 1000 feet high, are many fertile valleys with limestone floors, where the clovers and all leguminous plants flourish. But on the tops of the mountains at an elevation of 2000 feet the soil is poor, the winters are cold, and bees must be protected. There are seldom more than 40 colonies in one yard, and the yield is not good. The honey plants are wild red raspberry, white clover, locust, buckwheat, and huckleberries which secrete nectar well when it is warm and not too dry. In Blair, Huntingdon, and Cambria counties there are many bees, chiefly in small yards. A few apiaries contain 75 colonies. The winters are severe, and the bee pasturage is only moderately good. Somerset and Bedford counties on the south border have a large acreage of buckwheat and fruit trees.

It is in the third section, or bituminous-coal region, known as the Alleghany Plateau, which covers the western half of the state, that commercial beekeeping is most successful. Along the north-central border, in the northwest corner, and in in the western part of the state, thousands of acres of buckwheat are grown. Two-thirds of the entire acreage of this cereal in the United States are found in New York and Pennsylvania. The leading industry is dairying, and alsike clover is extensively planted for hay. Commercial apiaries obtain a surplus of 100 pounds per colony. In summer there is a good honey flow from the clovers, and in fall from buckwheat. Many colonies have been destroyed by European foul brood in this section.

Farther westward in McKean, Cameron, Elk, Forest, and Warren counties buckwheat and alsike clover are much less abundant, and there is a marked decrease in the number of bees. Native wild species of value for nectar are maples, dandelion, sumac, prickly ash, wild cherry, raspberry, swamp milkweed, goldenrod, and aster. Honey from prickly ash is extremely bitter and ruins the honey crop in some places. Forty colonies form as large a yard as is considered desirable. In Cameron, Elk, and Forest counties, less than 20 per cent. of the land is in farms.

There are many colonies of bees in the northwest corner of the state, in Erie, Crawford, Mercer, and Venango counties, where there is a large acreage of buckwheat. This section is also a center for dairy farming, and alsike clover is abundant. South of Lake Erie a large area of land is planted with apple, pear, and peach trees, small fruits, and truck crops.

In the central-western portion of the state (Butler, Clarion, Jefferson, Indiana, Armstrong, and Clearfield counties) the acreage of buckwheat exceeds 70,000 acres, and the number of colonies is large. Conditions do not vary greatly in these counties. In Clearfield County, apiaries usually range from 6 to 30 hives, but occasionally contain 80 or more. At Rockton there has not been a complete failure in twenty years; but small yards give best results. The soil is naturally acid, and lime must be applied to obtain a good growth of white clover. The native or indigenous honey plants are the familiar species which have been previously mentioned. In Jefferson County good results are obtained with small apiaries; but the seasons differ so widely that honey production is pursued chiefly as a side line. This is true over nearly all of the state, says Rea.

In the southwest portion of the state, in the vicinity of Pittsburgh, a region of oil-fields, mines, and great steel-mills, there is pasturage for only a few bees. "I lived at Pittsburgh," writes a beekeeper, "and it is the poorest place I ever saw for bees." Not much buckwheat or clover is sown, and bees are compelled to depend almost entirely on wild flowers. While white clover is abundant, it yields only about one year in three. During the bloom of apple-orchards it is often rainy, but this is true of all orchard country.

NEW JERSEY

Total area, 8224 square miles. According to its physical features New Jersey may be divided into two natural regions: A northern mountainous and rugged section bounded on the south by a line running from Staten Island to Trenton; and a southern section known as the Coastal Plain which is comparatively level. The Raritan and Millstone rivers form a good dividing line between the two regions.

The area of the Coastal Plain is 4400 square miles, of which 300,000 acres are tidal marsh bordering the Atlantic Ocean and Delaware Bay. This great plain slopes gently from its center both eastward and westward and contains no rocky elevations, its rounded hills being composed wholly of sand and earthy material. The salt marshes are covered with marsh and spike grasses, sedge, and rush, which produce nectarless wind-pollinated flowers. Over one million acres of this region are covered with pitch pine and oaks.

In the southern half of the state, says E. G. Carr in the American Bee Journal, alsike clover is the main dependence for surplus. But after it has bloomed bees secure enough honey for their maintenance during the winter from late summer and fall-blooming plants. In the eastern portion of Ocean and Burlington counties there is a large area of cedar swamps in which sweet pepperbush and blueberries are abundant and yield considerable surplus. On the cranberry bogs a careful beekeeper would be able to secure cranberry honey unmixed. Along the eastern bank of the Delaware River in Burlington, Camden, Gloucester, and Salem counties there is a belt of land covered with Spanish needles, from which is secured an abundance of golden-yellow honey with a pleasant flavor. Thickets of willows, red maple, sweet gum, and woodbine are common on the banks of the streams.

Southern New Jersey is a center for growing asparagus, cantaloupes, and watermelons. Large acreages of fruit trees are located in Cumberland, Gloucester, and Burlington counties. One project comprises one thousand acres of fruit trees in one lot. They are not important sources of nectar, according to Carr, but each year there is a greater tendency for fruit-growers to rent bees for the pollination of their orchards. The apiaries of the Coastal Plain seldom contain a large number of colonies or secure a large surplus; but more than 200 pounds of extracted honey has been obtained from a single colony at Mt. Holly, Burlington County. Usually the apiaries range from 10 to 30 colonies.

The northern half of the state is more broken and diversified than the southern, becoming mountainous in the northwest. The land is generally rich and productive, consisting of clay soils well suited to farming, but many ridges are so covered with loose stones or bare rock that cultivation is impossible. As the result of the high cost of labor, farming is languishing, and sumac is becoming common along the fences and on the hillsides. "There is such a variety of major nectar-producing plants in the northern zone," says E. G. Carr in the American Bee Journal, "that there is scarcely a time from June first until frost that the bees do not get more than a living." In the spring pollen and nectar are gathered from the maples, dandelion, and fruit bloom. A surplus is gathered from white clover, alsike clover, white and yellow sweet clover, buckwheat, goldenrod, and aster. Should one or more of these plants fail to yield, there is still an opportunity to get a profitable crop. Notwithstanding poor wintering and uncontrolled swarming, good crops are secured.

There are two pronounced honey flows, the earlier coming from the clovers, the later from buckwheat, goldenrod, and aster. As New York and Philadelphia offer good markets for milk, a large number of New Jersey farmers are engaged in dairying and have largely planted alsike clover. In Morris and Hunterdon counties sweet clover is a valuable honey plant. Present indications are that its cultivation will steadily increase. In the hilly areas a good flow of nectar in favorable seasons may be expected from buckwheat, of which there are 8000 acres in the state. The largest number of beekeepers is found in the northern part of the state, but the apiaries are usually small. Sussex, Hunterdon, Morris, and Warren counties, writes E. G. Carr, offer the best opportunities for beekeeping, although the agricultural counties possess many advantages. In Sussex County, where an abundance of buckwheat is grown in the Delaware Valley, there is a great and unoccupied field. In Hunterdon County, although it contains the largest number of colonies of bees and of beekeepers, there is also much good territory. As the winters are mild in the larger part of New Jersey and much of the territory is understocked, beekeeping as a side line might be profitably pursued more extensively than at present.

In the northeastern corner of the state, in Hudson County, suburban to New York, and in parts of Bergen, Passaic, Essex, and Middlesex counties, there is no early source of honey. Much of the land is lying idle, and is every year becoming more densely covered with goldenrods and asters. The goldenrods, which are very abundant in Middlesex County, begin yielding nectar about the first of September. If the weather is favorable a beekeeper may secure from it a surplus of one hundred pounds. The asters continue to bloom until they are killed by frost. Heartsease is also to a limited extent a dependable source of nectar. At Lyndhurst the apiaries range from 10 to 15 colonies.

"There is on file in the Department of Agriculture a list of three thousand names of persons in the state who keep bees. From inspection records and statistics it is estimated that there are approximately thirty thousand colonies of bees in the state. The local demand for honey is such that the greater part of the production of these colonies is consumed within a few miles of the point of production. More than this, tons of honey are brought into the state for home use."

DELAWARE

Delaware, which, with the exception of Rhode Island, is the smallest state in the Union, has a total area of 2370 square miles. The larger portion of its surface is a low sandy plain bordered on the southeast by wide marshes along Delaware Bay and the ocean, but becoming hilly in the extreme north. On the southern boundary, extending into Maryland, there is a great cypress swamp covering 50,000 acres.

Delaware is an agricultural state; and beekeeping, which is pursued chiefly as a side line, offers few inducements to the specialist. The production of honey is largely dependent on white clover, alsike clover, fruit trees, berry bushes, vegetables, and buckwheat. Approximately the same results are obtained both in the northern and central sections. White clover and alsike clovers are best adapted to the clay soils of the north. There is a belt of crimson clover in the central part, and buckwheat is grown in the south. Apples, pears, peaches, and strawberries are generally cultivated throughout the entire state, and in its central section there is a dense acreage of muskmelons and watermelons.

In the north, near Wilmington, the apiaries are small. This is also true around Marshallton. Besides the clovers, dandelion, locust, poison ivy, sweet clover, smartweed, goldenrod, Spanish needles, milkweed, and aster are the best-known honey plants.

In the central part of the state, near Dover, where the soil is sandy, a surplus is obtained from crimson clover, white clover, buckwheat, and goldenrod; but buckwheat is reliable only about once in two or three years. A specialist at Dover has distributed his bees in apiaries which average 65 colonies. At Kenton the yards range from 15 to 25 colonies. In the south, in Sussex County, the apiaries contain from 1 to 20 colonies; but at Milford there is one of 28. Buckwheat, of which there are some 4000 acres, is the chief dependence of the beekeeper; but white clover is also valuable. The tidal marshes are not suitable for beekeeping.

MARYLAND

Total area, 12,210 square miles, of which 9860 square miles are land. The eastern three-fifths of the state form a low sandy plain which is divided by Chesapeake Bay into two portions, the East and West Shores. The soils are mostly sandy loams suitable for the cultivation of vegetables, melons, strawberries, tomatoes, and peach and apple trees. This section, also known as Tidewater Maryland, is often reported as not well adapted to beekeeping; but the number of colonies maintained at present is probably less than the honey flora would support.

The East Shore, which is less than 100 feet above sea level, is deeply indented by estuaries and bordered by river swamps and broad tidal marshes which cover about 276,000 acres. The average per colony secured by beekeepers located in different parts of the region is about 40 pounds of honey, and the average size of the yards is 20 colonies, spring count. The honey flora of the East Shore will not support large apiaries.

The West Shore is likewise a sandy plain, but it has an elevation of 100 to 300 feet above sea level. The conditions of beekeeping are very similar to those on the eastern side of the bay. The average per colony is about 50 pounds of honey, and the average size of the apiaries, spring count, 10 colonies. The largest amount of honey is secured in Charles and Prince Georges counties. The honey flora of the Coastal Plain consists of white clover, sweet clover, goldenrod, aster, locust, tulip tree, and fruit bloom. There is a large acreage of tobacco on the West Shore, but it is harvested before it blooms. The western edge of the Coastal Plain is sharply defined by the "fall line," or the rise to the Piedmont Plateau.

Between the Coastal Plain and the Appalachian Mountains lies a hilly, rolling country, known as the Piedmont Plateau, which has an area of 2500 square miles. It gradually rises from a height of 450 feet in the eastern portion to over 900 feet in Frederick County. The soils range from sandy loams to heavy clays, and are especially well adapted to growing grasses, corn, wheat, and tomatoes. In this region are included the larger parts of the counties of Howard, Baltimore, Harford, Montgomery, Carroll, and Frederick. A large percentage of the honey reported in the state is produced in Baltimore County. The beekeepers of Baltimore City and Baltimore County form the largest single group in the state. Outside of this county, Carroll, Frederick, and Montgomery counties show the largest number of beekeepers, with the largest number of colonies present in Montgomery County.

A beekeeper living near Baltimore City writes that he does not know of any location where a man could make a living entirely from bees. The crop of honey is at times ruined by a flood of honey-dew. For instance, an apiary of 9 colonies, just outside the mile limit, increased to 17 colonies and stored 675 pounds of surplus, consisting largely of honey-dew gathered from Norway maple. The bee pasturage is composed of the usual honey plants found in low-lying sections of the state.

Excellent locations for medium-sized apiaries occur in the western part of the Plateau in Montgomery, Frederick, and Carroll counties, and in the northern parts of Baltimore and Harford counties. There is a larger number of tulip trees in this territory than in the other counties in Maryland, and the clover yield is larger and more constant. There is also an early honey flow from fruit trees, and a little flow, beginning about September 1 and lasting until frost, from goldenrod, asters, and smartweed, all of which are abundant. Between the two honey flows there is an interval when strong colonies are compelled to consume a part of their stores. The more extensive planting of sweet clover would bridge over or shorten this gap.

The more noteworthy honey plants are white clover, alsike clover, tulip tree, black locust, red maple, sugar maple, sumac, persimmon, sweet clover, blue thistle, goldenrod, and aster. Basswood has been frequently planted in the cities and villages, and is common in the forests of the Piedmont uplands and in the mountains. There are three million fruit trees, and thousands of acres of strawberries and raspberries under cultivation. The black locust is common in the western Piedmont, and in some years, as in 1920, yields a heavy flow of nectar, but very often it is unreliable. The Piedmont was never covered by the great ice sheet, and the soils are derived from the decay of the underlying rocks of granite, gneiss, schist, slate, shale, and sandstone. They contain sufficient lime for the growth of corn and the grasses, but usually not enough for the clovers and other legumes. Neutral or acid soil is shown by the predominance of chestnut and oak trees, and by the abundance of huckleberries, Rhododendrons, and Kalmias. As the temperature in Maryland is usually unfavorable for the secretion of nectar by white clover, it is a reliable source of honey only one year in three, although it is very abundant in this region.

In Maryland, above the "fall line" on the Piedmont Plateau, tulip tree is sufficiently abundant to yield a crop of honey regularly. On the Coastal Plain it never furnishes a surplus. Formerly in the central part of the state it was one of the main surplus-honey plants, and very important in Montgomery County. But it has been so largely cut for pulp wood that there has been a great decrease in the quantity of honey obtained. When the flowers are late in opening, and the weather is warm and dry, the honey flow is very much heavier than when the bloom is early. Under such conditions there are few if any better honey plants than tulip tree. A large quantity of honey is stored, even when the trees are scarce, but when they are abundant there is little danger of overstocking a location with bees. "Tulip tree," says Phillips, "is perhaps exceeded by no other plant in reliability of yield, and few other trees furnish as much nectar as a tree of this species."

West of South Mountains are the counties of Washington, Allegany, and Garrett, constituting the Appalachian mountainous region. This section is crossed by numerous mountain ranges, 2000 to 3000 feet in elevation, between which are very fertile valleys. The winters are cold and the summers cool, the average annual temperature being 50 degrees F. This region was formerly covered by an extensive forest, more than one-half of which still remains. In Washington County is the broad Hagerstown Valley, resting upon limestone, a part of the great Appalachian Limestone Valley which extends from eastern Pennsylvania to central Alabama. Allegany County is the seat of important coal mines, and agriculture receives less attention. The average surplus is small. A local beekeeper describes the honey flora as follows: "White clover yields very little nectar. Not much alsike clover is grown here. In spring the fields are yellow with dandelion blossoms. Red maple and redbud bloom in great numbers. Then comes the tulip tree. Sumac is very abundant, and yields well. Raspberries, blackberries, and huckleberries abound on this the 29th day of November. Chestnut, oak, walnut, and hickory furnish pollen, but no nectar. The cultivation of buckwheat is largely confined to Garrett County, which, in 1919, reported 5000 acres. There is also a large acreage of fruit-trees of bearing age. In the mountains beekeeping is pursued chiefly by farmers as a side line, and improved apparatus and methods are required in order to obtain the best results."

Maryland, says E. N. Cory, is not considered a beekeeping state, and the average yield per colony is extremely low; but there are a number of beekeepers who have made excellent records with a limited number of colonies. The honey produced is almost entirely retailed within this state, which enables many to make a success which would not be possible if they had to ship to a wholesale market. There are many attractive opportunities for small beekeepers with only a few colonies, as the growing season is long and the honey plants varied. There are several beekeepers in the state who are making an excellent living from honey and the sale of bees and queens. It may be that the possibilities in honey production are underrated, particularly in the Piedmont Plateau section, and in the mountain valleys of Washington and Allegany counties.

VIRGINIA

Total area, 42,627 square miles. Virginia, like Maryland, may be divided into three provinces: The Coastal Plain, the Piedmont Plateau, and a western mountainous district. In the Coastal Plain, or Tidewater Virginia, an area embracing about 11,000 square miles, there are comparatively few beekeepers. Along the coast are extensive marshes, while inland the soil is sandy, and covered with a forest of long-leaved pine. Few counties contain more than two or three hundred colonies, nine reporting less than one hundred each; but the counties of eastern Virginia are of small size. In Westmoreland and Essex counties the yards range from 1 to 20 colonies, and on an average obtain about 40 pounds of surplus. The chief honey plants are alfalfa, clover, black locust, gallberry, cowpeas, goldenrod, and asters. At Center Cross, Essex County, the soil is a red clay, and the principal crops are corn and wheat. This locality will not support more than 25 colonies in one apiary. The seasons are not reliable, and it is a country without a history so far as beekeeping is concerned. At Achilles, Gloucester County, on the coast, the extent of the bee pasturage is greatly reduced by the large area of marshland. In an apiary of 30 colonies an average of about 50 sections is obtained. Partridge-pea, which begins to bloom about July 10, was formerly the most important honey plant; but it has become much less abundant of late years, owing to truck crops replacing wheat and oats.

In the southeast corner, around Norfolk, large tracts of land are devoted to market gardens, and the growing of watermelons, peanuts, and sweet potatoes. From one to three crops are produced annually. Usually little honey is gathered away from the swamps. The principal honey plants are gallberry, tulip tree, gum trees, blackberry, sumac, huckleberry, motherwort, sweet pepperbush, goldenrod, and asters. There are near Norfolk about twenty small yards, which average not far from 25 colonies.

The great Dismal Swamp is not a desirable location for the production of honey. The growth of plant life is so luxuriant and dense that it is difficult to penetrate, the crop in some seasons is a failure, and foul brood is prevalent. While large areas of the swamp are dry, except for a few farmers in the marginal portion it is uninhabited. Most of the surface consists of recently formed peat. The predominant types of vegetation are black tupelo, white tupelo, red maple, sweet bay, and a variety of shrubs. The shores of Lake Drummond, near the center of the swamp, are bordered by moss-hung cypress-trees; but the cypress, white cedar, and pine growing in this region are of no value to the beekeeper.

The Piedmont Plateau occupies about 18,000 square miles in the center of the state, rising at the foot of the Blue Ridge to an elevation of 700 to 1200 feet above the level of the sea. It was formerly covered with oak, which is now partly replaced by a second growth of pine. Thousands of acres of tobacco are grown in the central and southern portions of this section. The only honey plants which yield a large surplus are tulip tree and sourwood, the latter being less valuable here than among the mountains. Tulip tree blossoms in May, when the weather is often cold and rainy. Sourwood is likely to be a total failure one year in four. Other honey plants are white clover, alsike clover, sumac, black locust, catalpa. goldenrod, asters, maples, cowpeas, persimmon, blue thistle, buckbush, and buckwheat.

The bee yards in the northern part of the Piedmont Plateau are mostly small, averaging from 30 to 50 pounds of honey. In the eastern counties of this section the beekeepers as a rule give little attention to either apparatus or methods, using box hives or hollow logs.

The best part of the Piedmont Plateau is the region directly east of the Blue Ridge Mountains. There are several commercial apiaries in Pittsylvania County, and the average surplus in a good year is 40 or 50 pounds. There are from 12 to 15 smaller yards within a radius of five miles, but the locality is not fully stocked. A large quantity of honey-dew is gathered in some years, especially from the young shoots of pine trees. In the foothills sourwood yields a fine flow in July. Buckwheat, which blooms in the same month, is also reliable. One of the best sections of Virginia for beekeeping is the Piedmont Plateau in the vicinity of the sourwood ranges, where there are also white clover, sumac, and sometimes buckbush and blue thistle. In Halifax County, on the south border, conditions are less favorable, and the fall flow is apt to be completely ruined, the weather being either too wet or too dry. However, in 1925, Henry W. Weatherford, who lives in this county, was reported to have taken off an average of about 60 pounds of honey previous to June 8.

The mountainous region covers about 12,900 square miles. Its eastern boundary is the Blue Ridge, an imposing chain of mountains which crosses the state from northeast to southwest. West of the Blue Ridge lies the Great Limestone Valley of Virginia, about 20 miles wide, embracing nearly 7500 square miles, with a fertile clay loam, the most productive section of the state. The northern part of this valley, drained by the Shenandoah River, is known as the Valley of the Shenandoah. Apples, peaches, and pears flourish in the greatest abundance throughout the valley, especially in the northern counties, and a super of apple honey is obtained in favorable seasons. Blue thistle, or viper's bugloss (*Echium vulgare*), was formerly very common, and an important source of nectar; but, although still valuable, it is steadily disappearing with the extension of cultivation.

While there are a good many small or moderate-sized apiaries in Frederick and Clarke counties, there appear to be very few beekeepers who specialize in this industry. European foul brood, introduced from New York, has helped to destroy the colonies. The surplus, which comes from white clover, blue thistle, and sumac, averages in a good year about 30 pounds. Forty years ago this country had a great quantity of white clover, blue thistle, sumac, and aster; but with more intensive

Southwestern Virginia is a rich farming section with well-timbered mountains. cultivation the reapers and mowers confine these plants to waste places and the hedgerows. The minor honey plants are silver and red maple, willow, peach, pear, and apple trees, dandelion, black locust, honey locust, persimmon, catalpa, white clover, sweet clover, and aster. In Warren County there is a large number of bees, and the average crop is about 50 pounds per colony. Farther southward sourwood becomes important, but the other honey plants are essentially the same. The apiaries seldom exceed 20 colonies, and the average surplus is 40 to 50 pounds. No very favorable reports from this region have been received.

The extreme southwest is crossed by numerous mountain ridges, between which are smaller limestone valleys. At Olinger, in Lee County, in the extreme southwest, the honey is gathered from tulip tree, sourwood, basswood, black locust, white clover, and goldenrod. At times there is a large amount of honey-dew. The apiaries seldom exceed 50 colonies, while the average size is about seven colonies. Sweet clover and alfalfa are being introduced. It is one of the most promising sections in the state for beekeeping.

Of beekeeping in Virginia, W. J. Schoene, State Entomologist, writes: "Beekeeping is of minor importance in Virginia. With very few exceptions the bees are kept by farmers who give them no care or attention beyond taking away the honey and hiving the swarms. It is to be doubted if there are more than ten commercial beekeepers in the state who make an important part of their living from this industry. Very likely there are many places where bees could be kept with success, provided the right persons were interested."

WEST VIRGINIA

Total area, 24,170 square miles. The eastern one-third of the state is a mountainous area belonging to the Appalachian region; the western two-thirds, known as the Alleghany Plateau, is a rugged country in which the numerous streams have worn deep and narrow valleys. Much of the land is better adapted to grazing than to farming; but along the Ohio River the soils are fertile clay and sandy loams. Fairly good conditions for beekeeping prevail over the entire state, each section having its advantages and disadvantages.

The mountainous region consists of a series of parallel narrow limestone valleys containing the most fertile agricultural lands in the state. Here the old-time log hive is still common, largely because of the difficulties of transportation to the remote farms. At Martinsburg it is estimated that there are about 10 apiaries averaging 30 colonies each, and within a radius of five miles 50 small yards. An entire failure does not often occur, and in a good year 40 pounds of section honey are obtained. This is a fruit country with a large acreage of apple trees. The surplus comes largely from the clovers. Blueweed, or blue thistle, is common, and is reliable. Jefferson County lies in the Shenandoah Valley, where the land is nearly all in farms, and under cultivation. Apple bloom builds up the colonies in the spring; but in the fall the sources of nectar are few. On an average, 15 pounds of section honey, chiefly from white clover, are obtained annually. In Mineral County the yards are small, ranging from 2 to 12 hives. Very little modern equipment is used, and the crop is a failure about once in four years. The honey plants are dandelion, blue thistle, white clover, milkweed, and goldenrod, which yield a fair surplus.

Among the barren mountains of Tucker County beekeeping is not profitable except as a side line. The winters are severe, the temperature falling at times to

10 degrees below zero. The winds are high and the summers cool. A beekeeper at Thomas writes that he has kept from 20 to 60 colonies, and that in some years he gets no honey, in others as much as 60 pounds per colony. In Barbour County the honey plants are wild blackberry, tulip tree, black locust, sumac, basswood, sourwood, white clover, fruit bloom, and aster. Many farmers have from 1 to 6 hives. This is a poor country for commercial beekeeping. There are hundreds of locust trees in the pastures; and, while they bloom in some years profusely, they bloom irregularly. In the southeast corner of the state most localities in Greenbrier, Monroe, Summers, and Fayette counties would sustain probably from 20 to 50 colonies. Sixty pounds per colony is regarded as a good crop, 40 fair, and 20 poor. In Mercer, McDowell, and Wyoming counties the honey flora and the number of colonies are about the same as in the above counties.

In Marshall County, in the Panhandle, the largest apiaries range from 20 to 50 hives. The surplus comes from white clover and basswood; but fruit bloom, black locust, tulip tree, aster, and sometimes buckwheat, are all of considerable value. Near Wheeling poor results are reported. At Pennsboro, Ritchie County, is located Grant Luzader, West Virginia's largest commercial beekeeper; he has never failed to produce a crop of honey.

In the eight counties along the Ohio River there are many thousand colonies of bees, which in a single year produce several hundred thousand pounds of honey. In Wood County, around Parkersburg, there are about 35 yards, the largest of which contains 50 colonies; but the apiaries are usually of small size. Small apiaries are numerous around Charleston, ranging from 1 to 40 colonies. Heartsease and white clover are abundant, and usually reliable in the valley of the Kanawha River, but goldenrod does not yield well. Of Gilmer County, W. D. Zinn, one of the most progressive farmers of West Virginia, writes: "Never have I seen a country where white clover grows more luxuriantly." According to the replies of many beekeepers, small or medium-sized beeyards are numerous in West Virginia, but there is comparatively little commercial beekeeping.

The disappearance of many farm apiaries during the past 15 years has resulted in an annual loss to the state of thousands of dollars from the imperfect pollination of fruit bloom. "No farmer," says Reese, "can afford to be without five or more colonies of Italian bees; and no livestock on the farm will pay a greater percentage of profit." If the bees receive proper care there can be no question that they are a great benefit. Unfortunately they are often wholly neglected.

The honey flora of West Virginia is varied, and there is a nearly continuous flow from the first spring flowers until frost. The whole state lies within the deciduous-leaved or hardwood forest belt, and 16,000 square miles are covered with forest. There is a wealth of shrubs, as Rhododendrons, Kalmias, and Azaleas. The more important nectar-producing trees are basswood (3 species), tulip tree, sourwood, sumac, black locust, redbud, maples, black gum, holly, willows, and a great variety of wild and domesticated fruit trees.

"It seems that no plant stands out as the most important nectar-producer anywhere in the state," says C. A. Reese. "In some localities, especially along the Ohio River, the different clovers are probably most important. White clover, where it occurs in quantity, produces an excellent honey in color and flavor. Alsike clover is now being generally grown, and is an excellent nectar-yielding plant. Sweet clover is becoming more abundant every year, especially in the extreme eastern counties. Sourwood is rather widely distributed, and yields nectar in large amounts. Blue thistle is important, especially in the valleys of the eastern counties."

Asters are common throughout the state, and are increasing in abundance. Raspberry is abundant on burnt-over areas, and the huckleberry is common in the mountains. Swamp milkweed (*Asclepias incarnata*) thrives in the valleys everywhere. Some 30,000 acres of buckwheat are grown in the northwestern portions of the state, which yield well in favorable seasons. In the valley of the Ohio there is an immense acreage of apples.

Charles A. Reese, formerly State Apiarist, writes as follows: "The best beekeeping territory in the state is located in the headlands of the Elk, Gauley, and Greenbrier rivers, in Webster, Pocahontas, Randolph, and Nicholas counties. The area is noted and quite famous for the excellent-flavored honey that it produces. It commands a very attractive price in the smaller towns located near by. Unfortunately roughness of the country and the lack of proper transportation facilities prevent any extensive commercial operations."

KENTUCKY

Total area, 40,598 square miles. Kentucky is an agricultural rather than a manufacturing state, possessing a mild and wholesome climate, and over 22,000 square miles of fertile soils, derived largely from the decomposition of limestone, on which white clover, alsike clover, and sweet clover thrive in great luxuriance. In the north-central and western portions 420,800 acres are covered with tobacco — a larger acreage than is found in any other state. Apples are grown in great abundance in all sections, and pears and other fruits to a less extent. The eastern and central tracts are covered with a hardwood forest similar to that of Tennessee. Tulip tree was at one time very common; and, while merchantable trees have been largely cut for timber, they still furnish a part of the surplus in the less thickly settled sections of the state. Young trees are also rapidly springing up and beginning to bloom.

The eastern section of the state, comprising 10,000 square miles, is a rugged, or mountainous region belonging to the Appalachian coal-fields, and containing rich deposits of coal and oil. In the narrow fertile valleys there are many small or medium-sized beeyards maintained by the farmers and fruit-growers. The principal honey plants are tulip tree, basswood, white clover, goldenrod, and aster, while sourwood is not uncommon. "The reason," says Garman, "why little emphasis is placed on the mountainous eastern part of Kentucky as a honey-producing section is because it is relatively thinly peopled, and because of the backward condition of its beekeeping. Though many settlers in the valley have a few hives, the honey produced is so dark and unattractive that it could hardly be sold anywhere else. Much of it does not find its way to market; but there are also some good beekeepers to be found in this section. The steadiness of the honey flow in the mountains is apparently due to the large area of untilled land, on which a variety of native plants flourish throughout the summer. In the central and western portions of the state the land is more closely cultivated."

The best section of Kentucky for beekeeping is the famous blue-grass region, or Lexington Plain, which occupies about 10,000 square miles in the north-central part of the state. Its eastern limit is Portsmouth, on the Ohio River, and its western the mouth of Salt River beyond Louisville, while its center extends southward half way across the state. It is a limestone area, well drained, and free from swamps, but with its surface broken by many rounded hills and knobs. Fifty years ago tobacco was largely grown in Pendleton, Bracken, and adjacent counties; and gradually the shallow fertile soil was washed away by heavy rains, and the eroded, often gullied farms became unproductive. Farm after farm was abandoned, and in many instances sold for taxes, and more than one-third of the population of Pendleton County moved away. Then sweet clover was introduced, and, finding a congenial home on the limestone hills, spread in every direction. With the renovation of the soil the farmers returned, and dairy farming and the sale of sweet clover seed brought prosperity and comfort to northern Kentucky. Both the yellow and white species are grown extensively. Beginning with the first of June there is a continuous honey flow until fall, and 75, 100, and even 200 pounds of honey per colony have been obtained. In Pendleton and Bracken counties, the average size of the apiaries is about 60 colonies; and in each county there are more than 3000 colonies. In addition to sweet clover, white clover, black locust, tulip tree, fruit bloom, milkweed, catnip, motherwort, and aster are common. In Franklin County, 30 miles southwest, conditions are not equally good, and the apiaries contain only about ten colonies.

Around Lexington there are some 1600 square miles on which there is a luxuriant growth of the well-known Kentucky blue grass (*Poa pratensis*), so called from the blue color of the seed-vessels. White clover is most abundant in this region; but it is too far south for the secretion to be reliable. Black locust is common in the villages and on farms everywhere in Kentucky, but is less abundant in the forests. Other honey plants are sweet clover, dandelion, alsike clover, goldenrod, and aster. The blue-grass region is not as good a location for beekeeping as Pendleton and Bracken counties, as they have an advantage in growing large quantities of the two sweet clovers. According to Garman, the spring and fall honey flows are apt to be separated by a severe drought in August and early September that often causes a period of famine for the bees. If their surplus is taken away completely they suffer during this interval from lack of stores. When white clover fails the drought is especially trying to bees, since they can get little nectar until the blooming of the goldenrods and asters in the fall. At Stanford, in Gerrard County, at the southern extremity of the Lexington Plain, there are only a few bees, and the apiaries are small. White clover is a comparative failure every alternate year; but field asters and goldenrod never fail to furnish winter stores. The Lexington Plain, or blue-grass region, is surrounded by a ridge, or a series of knobs, 200 feet high, the boundary of the Highland Rim Plateau, in which this great plain was formed by erosion.

South and west of the blue-grass region extends the Highland Rim Plateau covering fully one-half of the state. The southern portion of this plateau, drained by the Cumberland River, and the southwestern portion, drained by Green River, have a limestone soil suitable for the cultivation of cereals and grasses. In Edmonson County, the seat of Mammoth Cave, the limestone is full of caverns and subterranean rivers. Away from the Cumberland hills, clover, locust, and aster are the most valuable honey plants, supplemented by dandelion, maples, willows, fruit bloom, tulip tree, raspberry, horehound, catnip, milkweed, smartweed, persimmon, redbud, honey locust, ironweed, and motherwort. But ten miles eastward, among the "Cumberland knobs," buckwheat, sourwood, and asters yield the surplus, which is colored dark by the buckwheat. In Lincoln County, clover is reliable about four years in six, locust one year in five, and aster every year, if the weather is not cold and rainy. Buckwheat yields heavily some years in the southern part of this county; and in Pulaski County, near the south border, there is another large buckwheat-growing area.

In western Kentucky there is a second coal-field, 4000 square miles in extent, parts of which have a very fertile soil. In general the eastern, southern, and southwestern portions of the state are most backward in honey production, and furnish a dark honey, sometimes injured both in color and quality by a mixture of honeydew. The central, northern, and northeastern portions are best adapted to commercial beekeeping, where sweet clover, white clover, and black locust are most abundant, and yield a light-colored honey. Beekeeping is badly neglected in much of the state, and there are many colonies in log gums and box hives.

Todd County, on the south border, is only a fair locality, as white clover is not abundant and does not secrete nectar freely. Crimson clover grown for seed yields most freely. Not much honey is gathered during the spring and summer, although maples, sumac, buckbush, and in some years black locust, carry the bees through the first half of the season. The fall honey flow is most important, and 80 per cent. of the crop is gathered between August 15 and October 1 from boneset, smartweed, goldenrod, and aster; but at least 75 per cent. comes from boneset (*Eupatorium serotinum*), which yields a white or light amber honey of excellent quality. In Trigg County the apiaries have only 6 to 12 colonies.

West of the Tennessee River there is a lowland area covered with loessial soil which belongs to the Gulf Coastal Plain. Forests of cypress cover the river swamps, and on the higher land there is a dense acreage of tobacco. Much of this section is not well adapted to bee culture; but along the Mississippi River and at Paducah, where the Tennessee flows into the Ohio, there are numerous colonies of bees. Owing to the greater moisture of the soil, Spanish needles (*Bidens aristosa*) and frost flower (*Aster ericoides*) sometimes yield here, when in other portions of the state it is so dry that there is no fall honey flow.

TENNESSEE

Total area, 42,022 square miles. According to its surface features the state may be divided into three natural regions: Eastern Tennessee, or the Cumberland Plateau, including the eastern valley of the Tennessee River and the rugged mountainous section bordering on North Carolina; Middle Tennessee, consisting of the Nashville Basin, a limestone valley surrounded by the Highland Rim Plateau; and Western Tennessee, or the Mississippi Slope. The surface ranges in altitude from less than 350 feet on the low alluvial bottom-lands along the Mississippi River to 6000 feet in the mountains of East Tennessee. The Cumberland River crosses the northern part, while the Mississippi forms the western boundary. Tennessee lies in the great Atlantic forest region, and more than 2000 species of flowering plants are listed by Gattinger in his Flora of Tennessee.

The mountainous district is crossed along its eastern border by a disjointed range of mountains broken into irregular short chains and valleys, and farther west by the lower ridges of the Cumberland Plateau. Through the center of this region the Tennessee River and its tributaries have eroded a deep valley known as the Valley of East Tennessee. This valley is one of the best locations in the state for beekeeping, as the fertile soils are derived partly from limestone, and white clover, alsike clover, and black locust, the chief sources of nectar, are abundant. The apiaries are most numerous in Sullivan, Washington, Hawkins, Hancock, Claiborne, Union, Grainger, and Knox counties. Apple, peach, and pear trees and strawberries are extensively cultivated in the southern part of the valley. Other honey plants growing either in the valley or on the neighboring mountains are alsike clover, sweet clover, sumac, tulip tree, persimmon, basswood, sourwood, locust, redbud, sugar maple, red maple, cowpeas, goldenrod, and aster. In the mountainous counties of Cooke, Sevier, Blount, Monroe, Knox, Hawkins, Campbell, and

Anderson there are also good opportunities for beekeepers. At an elevation of 2000 feet or more, tulip tree, basswood, and aster are the chief sources of nectar.

Another fairly good region for the production of honey is the central valley around Nashville, where most of the commercial apiaries in the state are located. As in the eastern valley of the Tennessee River, the honey plants are white clover, alsike clover, black locust, basswood, and aster. A few commercial apiaries contain 100 to 125 colonies; but a large number range from 25 to 75. As a rule, in the central section of the state it is not advisable to place more than 25 to 50 colonies in one yard, and the yards should be five miles apart. As the land is valuable, and in a high state of cultivation, few nectar-bearing weeds are permitted to grow. The surplus obtained in different years varies greatly. At Springhill, in the northern part of Maury County, in 1913, according to John M. Davis, it is doubtful if 500 colonies in one yard could have gathered all the nectar available; but since then on an average not one-fourth of a crop has been obtained, the season of 1919 being a complete failure. A bumper crop may be expected about once in ten years. Queen-rearing is much more profitable than honey production; and, owing to the unreliability of white clover, beekeeping should be combined with general farming or some other industry.

The Central Valley around Nashville is surrounded by the Highland Rim Plateau, which attains an altitude of 1200 to 1400 feet. This section produces very little clover; but there is an abundance of sourwood, tulip tree, aster, basswood, and black locust. The soils of this section are stony or clay loams, and the percentage of land in farms is much less than in the Nashville Basin. In Franklin, Lincoln, Bedford, and Marshall counties, in the south-central portion of the state, there are large areas of crimson clover, grown for seed, which yields heavily, while white clover and black locust are also abundant.

The Mississippi Slope comprises the gently rolling country between the Tennessee and Mississippi rivers. There are cypress swamps along both these large rivers, and the sluggish mud-colored streams are bordered by lagoons and swampy land. In the cypress swamps and lowlands are found the swamp locust (*Gleditsia monosperma*), the gums (*Nyssa sylvatica* and *N. aquatica*), the mountain sweet pepperbush (*Clethra acuminata*), frequent in the mountains but rare in middle and west Tennessee; the American holly (*Ilex opaca*), the swamp holly (*I. decidua*), two species of buckthorn, and the rattan-vine (*Berchemia scandens*). Cotton is grown over a large part of the Slope, especially in the vicinity of the Mississippi, and yields the larger part of the surplus. Along the north border there is a dense area of tobacco; and boneset, aster, and white clover near Reelfoot Lake are abundant. Along the Tennessee and Mississippi rivers smartweed, bitterweed, goldenrod, and aster are good honey plants.

NORTH CAROLINA

Total area, 52,426 square miles. North Carolina, which extends from the coast 500 miles westward to the Appalachian Highlands, consists of three natural provinces: The Atlantic Coastal Plain, or Tidewater North Carolina, a low and nearly level sandy plain, also known as the "flatwoods"; the Piedmont Plateau, a central hilly section; and a series of mountain ranges in the west, which reach an altitude of over 6000 feet. The western and eastern sections of the state are best adapted to bee culture, and contain large areas of land that are not and probably never will be reclaimed. The middle portion of the state has been more largely cleared for agricultural purposes, and has a poorer honey flora.

The Coastal Plain, or Tidewater North Carolina, is a vast area of low land comprising 20,000 square miles. It is about 125 miles in width, and the eastern half is not more than 20 feet above sea level. Swamps and lakes abound, and the coast line is deeply and irregularly indented by broad tidewater rivers, bays, and sounds. Its western boundary is the "fall line," which extends from the point where the Great Pedee River crosses the south boundary line, to Weldon on the Roanoke. The soil is generally sandy, but between the rivers it is a black peaty loam. The climate is mild; and the southeast corner, where the palmetto, magnolia, mock orange, and sweet bay flourish, is sub-tropical. A large area of the uplands, known as the pine barrens, is covered by a pine forest.

The chief honey plants are the gallberry, tupelos, or gum trees (*Nyssa biflora* and *N. aquatica*), huckleberry, blackberry, holly, tulip tree, and maple. Of less importance are rattan, persimmon, cotton, cowpeas, sumac, asters, goldenrod, wild cherry, willow, and blueberry. Gallberry (*Ilex glabra*) covers thousands of acres on the uplands near the swamps and along the rivers, blooming from May 10 to June 1, and yielding a honey which is excellent, but rather thin, and inclined to ferment unless well ripened. The tupelos extend up into Virginia and are abundant all along the North Carolina coast line. The white tupelo is more abundant in the southern part.

The Coastal Plain is the leading honey-producing region of the state, and the possibilities of honey production on a commercial scale are great. In 1919, the poorest year known for the production of honey in the Coastal Plain, the southeastern counties of Wayne, Onslow, Duplin, Sampson, Pender, and Bladen reported only a small amount of honey; but in 1919 these same counties produced much greater crops. A beekeeper near Wilmington writes that he obtains on an average 45 pounds of extracted and 35 of section honey per colony. A few miles distant there is much vacant territory.

In the Coastal Plain the best locations are in the swamp lands along the rivers and creeks that flow into the Atlantic Ocean, and near the many sounds and bays which extend far inland. The semi-arid lands of the western states require irrigation; but in North Carolina great areas need drainage. Both propositions require state and government aid. Adapted to this wet sour land are a number of plants which yield an immense amount of honey. In the southeastern swamps the gum trees are very abundant, while on the higher lands grow the gallberry, blackberry, and huckleberry. "But for these plants," says E. R. Root, "the whole southeastern section of the United States would be without table honey, for on the sour soil of the Coastal Plain neither sweet clover nor white clover will thrive. The gallberry is a most valuable honey plant, covering much land that will probably never be reclaimed for farm purposes. It offers great opportunities to the prospective beekeeper, and in the near future much more gallberry honey will probably be sold than at present. The blackberry and huckleberry are also very common, and are an ever reliable source of honey for stimulating brood-rearing. The animal life of the swamps is likely, however, to deter the timid beekeeper from entering them, for they abound in mosquitoes, redbugs, venonous snakes, and other enemies of the human race. But the more adventurous apiarist can successfully protect himself from most of these dangers and annoyances; and happily, with few exceptions, the swamp land is said to be free from malaria."

The middle section of the state, or the Piedmont Plateau, extends from the "fall line" to the Blue Ridge, which rises abruptly above the plateau to a height of two or three thousand feet. Many large rivers take their rise in the valleys, which are several hundred feet lower than the intervening upland. Agriculture is the chief industry. The surplus honey is obtained chiefly from tulip tree, sourwood, crimson clover, white and alsike clover. Other plants of value are willow, dandelion, goldenrod, aster, persimmon, holly, fruit bloom, maple, black gum, boneset, sumac, and cowpeas. The average size of the apiaries some years ago in 16 counties was about 40 colonies. Eastward there are no specially favored localities. The cotton belt, as cotton bloom seems to be nearly nectarless in this state, is considered a poor location. In 1915 a beekeeper near the center of the state increased the number of his colonies from 80 to 125, and secured 5000 pounds of bulk comb honey. In the western portion of the Piedmont, which attains an elevation of 1200 feet, sourwood becomes abundant; but otherwise conditions are about the same as in its eastern half. A crop is obtained usually every other year. The larger part of the surplus comes from tulip tree and sourwood. There are many good locations unoccupied in this region.

In the western mountainous region the Blue Ridge and the Great Smoky Mountains cross the state from northeast to southwest. These two ridges are connected by many cross-chains, separated by deep fertile valleys. The forest is composed of a great number of trees and shrubs. The leading honey plant is sourwood. It seldom exceeds 30 feet in height, and blooms profusely from July 1 to 21, yielding a better honey than any other plant in the state. In the extreme northwest the larger part of the surplus comes from sourwood; but in the southwest it is gathered partly from other sources. Other honey plants in this section are basswood, tulip tree, locust, clovers, asters, maple, apple bloom, wild cherry, sumac, buckwheat, and goldenrod.

The species yielding a surplus are chiefly trees and shrubs, and are few in number. The four most important honey plants are sourwood, tulip tree, gallberry, and crimson clover. Sourwood is restricted chiefly to the mountains and the rugged portion of the Piedmont upland. The clovers are most abundant in the central region, while in the Coastal Plain gallberry and the gum trees are most important. Tulip tree is found over the entire state, except in the extreme eastern lowlands. Black gum is common from the center of the state to the coast. The white tupelo is most abundant in the southeastern swamps. Maples, willows, persimmon, sumac, and goldenrod grow throughout the state. Asters are also widely distributed, and near Raleigh add largely to the winter stores. Basswood occurs in the north coves of the mountain ravines. Redbud, locust, wild plum, smartweed, boneset, white clover, and Spanish needles are found chiefly in the western part of the state, while rattan-vine, holly, blueberry, huckleberry, and gallberry occur eastward. Some thousands of acres of buckwheat are grown in the northwest corner. In the north-central section, extending southeastward into the Coastal Plain, there is an immense acreage of tobacco. In the south-central portion, and in many of the counties of the Coastal Plain, cotton is very extensively grown; but it is reported to yield little nectar in North Carolina. Apples are most common in the mountains, and pears and peaches in the central region. The forest in the western part of the state is composed of many kinds of trees intermingled, and everywhere presents a variety comparable with almost any portion of the tropics. There are 112 species of trees, and twice that number of shrubs. The blueberry, grape, cherry, and cranberry are more or less common. The mountain-sides and the valleys are adorned by the profuse bloom of the mountain laurel, the Rhododendron, and the flame-colored azalea. In no other state are there native so many medicinal herbs.

The beekeepers of North Carolina are unevenly distributed, and there are many opportunities for establishing new apiaries. It has been estimated that tons of nectar are going to waste annually. In the past apiculture has been much neglected in North Carolina, but to-day new methods and improved apparatus are being introduced. A state association has been formed, and the work of education and encouragement has been placed in charge of a specialist in beekeeping.

SOUTH CAROLINA

Total area, 30,989 square miles. The state consists of two main divisions: The Coastal Plain and the Piedmont Plateau, the "fall line" running from Cheraw, in the northeast corner, through Camden and Columbia to Augusta on the Savannah River. Along the coast there is a broad border of salt marsh, beyond which there are 10,000,000 acres of level country with a sandy soil partly covered with pine trees, known as the "flatwoods," or pine barrens. Gum trees and bay trees are common in the river swamps, while near the coast the palmetto occurs, and South Carolina is often called the "Palmetto State." Figs and pomegranates flourish, and oranges can be grown in the subtropical climate of the southeastern portion.

In the state as a whole beekeeping is not well developed; there are many small, poorly managed apiaries with the bees mostly in box hives. Here and there are small but well managed modern apiaries, which yield good returns. During the last three or four years there has been a rapidly increasing interest in beekeeping throughout the state. In many counties, bees are being transferred to modern hives, and there is a strong movement toward better methods and the organization of the beekeepers. There are active beekeepers' organizations in five counties with a total membership of about 175. Anderson County, with a membership of about 50, publishes a four-page monthly bulletin. Other counties will probably organize at an early date.

Bee diseases are reported to be absent, and certainly are not prevalent in the state.

The Piedmont Plateau is a rolling or rugged country, becoming mountainous in the northwest, where an elevation of 2000 feet is attained. The surface soil ranges from a gray sandy loam to a clay loam, a stiff red clay subsoil predominating. In the numerous valleys along the streams there is a great variety of hardwood trees. Apples, peaches, and pears are abundant and cotton is cultivated over the entire section. Blackberry, boneset, horehound, goldenrod, and asters are common herbs and considerable clover and hairy vetch are grown. The winters are mild and comparatively short. The bees are seldom confined to their hives long at a time, and brood-rearing begins early in the spring. The main honey plants are tulip tree, persimmon, sourwood, holly, black willow, sumac, clover, white tupelo, goldenrod, aster, and cotton. Beekeeping has been more highly developed in the Piedmont section than on the Coastal Plain, and for this reason much more honey is produced here than in the latter section.

Beekeeping in the Coastal Plain is at present in a very undeveloped condition; but it is the opinion of E. S. Prevost that this section offers the best opportunities in the state for commercial beekeeping. The chief sources of surplus honey are gallberry, tupelo, holly, rattan, and titi. In the Coastal Plain the apiaries are for the most part small, and the problem of successful management is in a transitional state. The best locations are on the margins of the river swamps, as along the Pedee, Black, Cooper, Edisto, Cambahee, and Savannah River swamps.

No accurate figures as to the surplus obtained in the state are available, but yields of 75 to 100 pounds per colony are not unusual. This could be greatly increased by better methods of beekeeping, for fully 80 per cent. of the nectar of this region is lost through want of strong colonies to gather it.

GEORGIA

Total area, 58,665 square miles. Georgia is divided by the "fall line," which passes from Augusta on the Savannah River through Milledgeville and Macon to Columbus on the Chattahoochee River, into Southern Georgia, or the Coastal Plain, and Northern Georgia, comprising the Piedmont Plateau and a series of mountain ranges and limestone valleys belonging to the Appalachian Mountain System.

Southern or Low Georgia, which includes more than one-half of the state, has an area of 35,000 square miles, the highest altitude of which is less than 400 feet. The sea coast, which is 128 miles in length, is bordered by a chain of low-lying fertile islands which are separated from the mainland by narrow lagoons. Along the coast there is a strip of marshy alluvial land where a nearly subtropical climate prevails, which is favorable to the growth of the orange, lemon, sweet potato, banana, peanut, sugar cane, magnolia, and palmetto. The great Okefinokee Swamp in Ware and Charlton counties is estimated to have a circumference of 128 miles, but it is without inhabitants. However, the territory surrounding the swamp offers excellent locations for beekeeping. The coast belt is succeeded by the Coastal Plain, a level or gently rolling section with a sandy soil, which was originally densely wooded with long-leaf yellow pine.

The most interesting and promising portion of Georgia to the commercial beekeeper, according to E. R. Root, is the southeastern region, where the gallberry occupies a large area, and the black gum and white tupelo are abundant in the swamps. Here are located the most advanced beekeepers, who operate thousands of colonies of bees. At Waycross, Ware County, is located the most extensive beekeeper in the United States. The principal honey plants are spring titi, gallberry, black gum, white tupelo, holly, clethra, chinquapin, velvet bean, snow-vine, and cotton. The site of an apiary in a swampy region near the headwaters of the Suwannee River is described by C. P. Dadant as a flat country from which all the large trees have been removed, and which to-day is covered with stumps, straight slim pines, and a thick undergrowth of saw palmetto, several kinds of huckleberries, and a great abundance of gallberry. The tupelos also are found in the cypress swamps. The apiaries are in the brush, and at some distance from the settlements, and range from 50 to 75 colonies, and occasionally number 100, the average being about 60. Wayne County, a little farther north, is a good location; and an apiarist at Doctortown reports 1800 colonies in a few yards. The crop is usually reliable, and in a good year a surplus of 100 pounds per colony is obtained. In January the maples yield well; spring titi blooms in March and April; black gum yields heavily in April, and the willows are also important. In May there is a good flow from white tupelo and gallberry, the latter surpassing all the other honey plants, and in June snow-vine (*Cissus arborea*) has been known to yield 100 pounds of extracted honey. All the farmers have usually a few bees. Bitterweed yields nectar in a few places in late summer, injuring the quality of the honey with which it is mixed. At Valdosta, in Lowndes County, on the south-central border, apiaries vary from 50 to 80 colonies. In a good season colonies store about 30 pounds from tupelo, 30 to 40 from gallberry, and from cotton 40 to 50 pounds. Sweet bay (*Magnolia glauca*), which is abundant in the swamps, is also reported to yield from 15 to 20 pounds. The flow from cotton lasts for about 90 days. A beekeeper in Berrien County writes that within a radius of 70 miles there are several large commercial apiaries, which average about 60 pounds per colony, but about one year in three there is a short crop. There are also about 12 small yards in a radius of five miles. In Coffee County the yards are smaller. There are in this locality many box hives. Cotton in Ben Hill County was formerly an excellent source of honey; but since the advent of the boll weevil the cotton-fields do not supply much nectar. A beekeeper at Fitzgerald reports that he has obtained very little surplus during the past three years.

The southwest counties are less suitable for beekeeping than the southeast portion. Miller County, in the southwest corner, is a fair location. One beekeeper reports that there is no great profit in beekeeping there, but S. V. Brown writes that some conspicuous successes have been made in that area. The boll weevil has reduced the crop from cotton, and the fall flow is sometimes injured by honey from bitterweed. In Decatur and Early counties the apiaries are usually small, and chunk honey is chiefly produced, which is sold at home. Lee County, farther north, is regarded as an ideal location for the small beekeeper. Brood-rearing begins early, and the swarms store from 50 to 75 pounds of honey. Around Cordele, from 40 to 75 colonies are kept in one yard, and there are about 20 apiaries in this vicinity. From Cordele northward to the "fall line," or the city of Macon, the counties contain few colonies of bees.

Effingham and Screven counties, which are bordered on the east by the Savannah River, offer fair opportunities. At Oliver, Screven County, there are several small apiaries in modern hives. During six successive years a medium crop of honey has been secured. The surplus comes chiefly from gallberry, which begins blooming about the 10th of May. Farther westward, in Emanuel and Laurens counties, gallberry continues to be the most important honey plant; and while 50 pounds per colony is a fair average, strong colonies sometimes store 200 or even 300 pounds of surplus. Bibb County lies on the border line between the Coastal Plain and the Piedmont Plateau, and in this area more fruit, grain, clover, alfalfa, and buckwheat are grown than southward. Near the city of Macon there are two large commercial apiaries which report an average surplus of about 35 pounds per colony.

Northern Georgia comprises the Piedmont Plateau and the northwestern mountainous region, the dividing line extending from Habersham County, in the northeast, through Bartow County to Haralson County on the west border. The Piedmont Plateau, which has an elevation of about 1000 feet above sea level, is a land of hills and valleys, with a healthful climate, mild winters, and cool nights in summer. Its area is about 15,000 square miles. The surplus comes chiefly from cotton, tulip tree, locust, sumac, goldenrod, aster, persimmon, blackberry, willow, and sweet clover.

A beekeeper at Temple, Carroll County, writes: "I have been interested in beekeeping for about 35 years, operating for the past ten years 25 hives. The season is colder here, and the land is more extensively cultivated than in south Georgia. On an average I obtain 25 pounds to the hive." In Douglas County, conditions are similar. Fifty pounds is considered a good average surplus, and this is secured only two years out of three. At Lula, Hall County, in the northeast, there are no specialists. Seventy-five per cent. of the bees are in old-fashioned hives, which supply very little honey; but there are large apple and peach orchards in addition to the wild flora, offering a good opening to an expert beekeeper. At Bogart, Oconee County, not more than 20 or 30 colonies can be kept in one yard; yet one apiarist has more than 30 yards with over 800 colonies. His lowest average has been 56 pounds per colony, and his highest 86 pounds. According to F. C. Pellett, about 25 pounds per colony are stored from blackberry in April, following which there is a light flow from tulip tree, sumac, and gum. There is a little sourwood in this locality, which gives a light flow. Cotton and cowpeas follow sourwood, and about 12 pounds per colony are obtained from aster in the fall. In general, throughout the Piedmont Plateau the apiaries are small and the honey flora meager.

Northern Georgia is crossed by the great Appalachian Chain of mountains known as the Blue Ridge, west of which is the Great Limestone Valley with a fertile clay soil. Extensive areas of this region are covered with forest. The most valuable honey plants are sourwood, basswood, tulip tree, persimmon, blackberry, goldenrod, and aster. Honey-dew is occasionally so abundant that it furnishes a surplus. There is almost no commercial beekeeping; but nearly every farmer has from one to twelve colonies, and obtains about 10 pounds of surplus per colony. Box hives or "gums" are the rule, and section honey is an "absolute novelty and mystery" in many localities. The rough mountain area is thinly settled, and produces no cultivated crops of value to the beekeeper; but the fertile limestone valleys in the northwest are suitable for the cultivation of fruits and many other farm products. The winters are never severe, and the ground is seldom covered with snow.

Throughout the Piedmont Plateau and in the upper part of the Coastal Plain there is a dense acreage of cotton, but reports vary greatly as to its value as a honey plant. It secretes nectar best in the southern portion, yielding in some places as much as all other sources combined. Gallberry, the most important honey plant in the state, grows in nearly all parts of the Coastal Plain except on limestone soil. "There has never been a total failure of gallberry bloom within the memory of the oldest inhabitant," writes S. V. Brown in June, 1925. It is most abundant in the "flatwoods." Tulip tree occurs over the larger part of the state, and is usually a reliable source of honey. Sumac (*Rhus copallina*) is also common, and in a few localities in north Georgia is the main source of surplus. Sourwood is abundant principally in the northern part of the state. Saw palmetto (*Serenoa serrulata*) is confined almost entirely to the wire grass or Altamaha grit region, where it is very abundant, especially near the coast. Titi (*Cliftonia monophylla*) is very abundant along the streams of the "wire grass" region, blooming in March and April. It is called spring titi to distinguish it from summer titi (*Cyrilla racemiflora*), which has a wider range, extending up into middle Georgia, and blooming later in June. It is not so common, nor does it yield so abundantly as spring titi. Snow-vine is another honey plant of southern Georgia. It blooms from June to September. Sweet pepperbush grows over nearly the whole Coastal Plain and along the border between Georgia and Florida. White and black tupelo are often very abundant along streams in the southern part of the state. Common holly (*Ilex opaca*) is widely distributed over the state, but is seldom very abundant. It blooms in April. White holly (*Ilex myrtifolia*) grows only in cypress swamps and in other wet places in the Coastal Plain. It blooms in May at the same time as gallberry, and the honey is never obtained unmixed with that from other plants. Swamp gallberry (*Ilex coriacea*) grows in damp shady places, and is far from being as abundant as gallberry. It blooms about two weeks earlier. Blackberry, as has been stated, is in some localities of great value. Several species (*Rubus cuneifolius, R. allegheniensis,* and *R. trivialis*) are common over a large part of the state. They bloom in April and May. Willow (*Salix nigra*) grows along streams over a large area. It blooms in March, and in a few places yields a small surplus. Asters (*Aster adnatus* and *A. squarrosus* are the most common species) are found over the entire state, and in many places are the main reliance for a fall surplus and for winter stores. Goldenrod is also of value. There is a dense area of peaches in the northwest corner and in the west-central part around Macon. Partridge-pea furnishes a surplus in only a few localities. Sweet clover is reported to be a good honey plant in Clarke County, and is spreading rapidly in Middle Georgia; but, as it requires a limestone soil, it is rare south of the "fall line." In central and southwest Georgia there are thousands of acres of watermelons. Several other honey plants of some importance are found along the coast and in the Costal Plain, among which are Mexican clover, rattan-vine, and yellow jessamine.

FLORIDA

Total area, 58,666 square miles. The entire state forms a part of the Coastal Plain; and, except for a central ridge and a hilly region in the northwest, is nearly level. Geologically it is the youngest of the states. The foundation of the whole peninsula is a series of coral reefs, the upper half of which did not rise above the waters of the Gulf until the Tertiary and the lower half until the Quarternary Period. Gradually its wave-washed surface became covered with a deposit of gravel, white sand, and loam. The underlying rock is thus limestone, in which there have been eroded numerous caverns and channels by subterranean streams which often rise to the surface as "boiling springs."

The entire length of the state, from St. Mary's River to the most southern key, is 450 miles. The vegetation of the southern portion of the peninsula naturally differs widely from that of the northern, and offers a convenient basis for dividing the state into three sections: North, Middle, and South Florida. North Florida is bounded on the south by a line running from Cedar Keys to Fernandina, and has a climate similar to that of the Gulf States. Middle Florida lies between this section and a line passing from the mouth of the Caloosahatchee River to Indian River Inlet. The climate is semi-tropical, and adapted to growing oranges and other citrus fruits. South Florida has a nearly tropical climate, and its vegetation has come largely from the West Indies. Of 247 species of plants growing here, 187 are common to tropical America. Common tropical plants which grow in this region are the banana, cocoanut, royal palm, custard apple, mango, guava, and all kinds of citrus fruits.

Northern Florida, or the region north of a line extending from Cedar Keys to Jacksonville, does not have a climate suitable for growing citrus fruits, as the temperature at times drops to 16 degrees below freezing. The surplus honey is secured chiefly from gallberry, tupelo, titi, partridge-pea, summer farewell, and chinquapin, as in southern Georgia, and similar methods of bee culture are followed in both localities. The northwestern counties offer along the streams and coast moderately good opportunities for beekeeping, but a large area of hilly upland is timbered with pine, oak, and hickory. The soils are sandy loams or sands underlaid by red clay subsoils.

The most famous portion of the state for the production of honey is the swamp-land along the Apalachicola and Ocklocknee Rivers, where the white tupelo and spring titi are very abundant, and about 15,000 colonies of bees are maintained. From the Georgia line to the mouth of the Apalachicola River the country is a great swamp and the river has low banks, the main channel breaking up into small streams which wind through the marshland. This bottom-land may be 10 miles across, and is covered with a luxuriant growth of white and black tupelo. As it is overflowed in the rainy season, it is necessary in many cases to place the hives on platforms six to ten feet high. In one instance as many as 360 colonies in one apiary were thus raised above the ground. Apiary sites are leased along the banks of the river and its tributaries, and, as they are accessible only by water, are visited by motor-boats. Foggy weather and light showers prevent the early withering of the flowers and favor a large yield of honey. As there is a shortage of pollen in this locality after the flow from the tupelos is over, the bees are moved up the river for the remainder of the summer.

The season opens with the blooming of black titi (*Cliftonia monophylla*), an evergreen shrub common in the swamps, in March and April, and is followed by the black tupelo and white tupelo in April and May. Two hundred and fifty barrels of extracted honey were once secured in 26 days by A. B. Marchant. E. R. Root, who visited this region some years ago, states that vast quantities of nectar go to waste, and that it is doubtful if the country is ever overstocked with bees. A beekeeper at Apalachicola writes: "There are 20 apiaries in this locality numbering from 100 to 330 colonies. In a good year the average surplus is 150 pounds. A full crop is obtained three years in five; but there is never less than one-quarter of a crop." Much nectar is also gathered from the white bloom of the snow-vine, which heavily festoons the trees in the swamps and along the edges of the river. Many do not care, however, to live in a desolate, unwholesome swamp where malaria and mosquitoes are prevalent. Other objections are the shortage of pollen and the absence of late-blooming honey plants to maintain the strength of the colonies in the fall. In recent years the Apalachicola region for some reason has not shown as good results in honey production as it did formerly.

Throughout the north-central portion of the state on the high pine land there are in July and August thousands of acres of partridge-pea in bloom. Nectar is secreted not by the yellow flowers, but by extra-floral nectaries at the base of the leaf-stems. The average surplus obtained from this source is about 30 pounds per colony. There is also a large area covered by a little-known plant which blooms late in the season, called summer farewell (*Kuhnistera pinnata*). It grows about two feet tall, and the spikes of white flowers yield a large amount of fine-flavored white honey. In Columbia County there is much bush chinquapin. In Hamilton County, on the north border, the apiaries do not contain more than 7 to 10 hives, mostly in boxes and log gums. Gallberry yields the larger part of the surplus; but soft maple, holly, black gum, sweet gum, and prickly ash are of value.

At Gainesville, in Alachua County, a fairly good surplus comes from many plants; but the most important are gallberry, which grows commonly in the flatwoods, partridge-pea, and summer farewell in the pine lands. There is also a large acreage of peanuts and of velvet beans. An average surplus of 50 pounds per colony has been obtained for a number of years, and it may exceed 90 pounds. In Florida there are thousands of acres of peanuts grown, both for oil-mills and for fattening hogs.

In the extreme northeastern portion of the state, in which are the cities of Jacksonville and St. Augustine, the counties of Nassau, Duval, Baker, Clay, and St. John's are of little importance for beekeeping. The apiaries are small, and the average surplus is only about 20 pounds. Most of the honey is apparently gathered in the fall from goldenrod; but a portion of it comes from gallberry, cabbage palmetto, and chinquapin.

Middle Florida, which is the great citrus-growing region of the peninsula, as the temperature seldom falls below 43 degrees F., extends southward as far as Lake Okeechobee. Throughout the central portion there are innumerable lakes, and hammocks varying in size from one to a thousand acres. The hammocks support a dense growth of oak, magnolia, sweet bay, elm, laurel, and cabbage palmetto; but there are thousands of acres of land, covered with long-leaved pine and pitch pine which are only moderately fertile, and produce few nectar-bearing plants. Flowers are not as abundant as the name Florida would suggest. "By far the larger part of Florida," says J. J. Wilder, "will not support bees," or, more definitely stated, beekeeping on a large scale is not profitable over a large area. Bees fly on pleasant days throughout the year and brood-rearing continues more or less during the winter months, so that a much larger amount of stores is required than in the

North. The best locations are along the rivers, or in the vicinity of swamps, or near the coast.

The three areas in Middle Florida in which beekeeping is most profitable are: The orange-groves of the highland district; the region along Indian River; and Manatee County. The largest groves of oranges are in the highland district in the counties of Marion, Volusia, Lake, Orange, Hillsborough, and Polk. The apiaries range from 25 to 50 colonies. The average surplus at Deland, in Volusia County, in a good year is about 40 pounds of honey per colony. Orange bloom comes in March and April, and lasts for about two weeks; and to secure a satisfactory harvest it is usually necessary to feed the bees in January. Gallberry and scrub palmetto succeed orange bloom; but the cattle-men burn over once a year a large area of woodland, killing or injuring both plants. On the margins of the hammocks, along the streams and in the flatwoods, or low pine lands, which partially overflow during the rainy season, gallberry yields a pound per hive each day, but it is not reliable every year. Other honey plants growing in Middle Florida are cabbage palmetto, Mexican clover, cowpeas, wild pennyroyal, purple-flowered mint, peanuts, Andromeda or wicker bloom, wild sunflower, and a large number of minor species. But on the high pine lands as a rule little honey is stored after July, and it becomes necessary to feed more or less during the fall and winter. A crop is obtained two years in three, and then the average surplus does not usually exceed 100 pounds; but occasionally 300 pounds or even more have been reported. On the west coast the counties of Lafayette, Levy, Citrus, Hernando, and Pasco offer few inducements to the beekeeper.

In the counties of Brevard and St. Lucie, along the Indian River, there are thousands of acres of citrus fruits which are the source of a large amount of orange honey. Beekeeping in this locality is young and very profitable. The orange flow is followed in about six weeks by a flow from scrub palmetto, which, three years in five, gives a good surplus. In the fall partridge-pea in the pine woods seldom fails to yield ample winter stores.

The third center in Middle Florida for the production of honey is in Manatee County on the west coast, sixty miles south of Tampa. In the vicinity of Bradentown, Palmetto, and Manatee a large surplus is obtained from orange, gallberry, and palmetto. Near Bradentown there are several apiaries which together number about 1500 colonies, and a surplus of from 65 to 75 pounds is secured nearly every year. There are also many small yards in this vicinity, which report a surplus of 50 or more pounds of honey four years out of five. Other important honey plants in this vicinity are wild pennyroyal and (along the coast) black mangrove and sea grape (*Coccolobis uvifera*). The saw-grass flats of this region should not be passed over without mention. They cover hundreds of acres, and for a large part of the year are under water; but in late summer they support a luxuriant growth of saw-grass, smartweed, thoroughwort, and wild sunflower. In August a wealth of bloom unfolds sufficient to maintain many colonies for months.

Tropical or Southern Florida, with its vast Everglades or saw-grass swamps, its almost impassable Big Cypress Swamp in which silence and perpetual twilight dwell, its dense mangrove swamps fringing the coast and sea-islands, and its numberless coral keys crowned with spreading palms and extending far into the Gulf of Mexico, is a land of remarkable interest and mystery, but much of it is forbidden territory to the beekeeper. The vegetation is largely tropical and has been derived from the West Indies. The honey plants of chief importance are the black mangrove and several species of palms. The land remains largely in its wild and primitive condition and there are few roads or other means of communication. The towns and villages are located along the east and west shores where a narrow strip of land has been brought under cultivation.

The southeastern counties of Palm Beach, Brevard, and Dade comprise about four million and seven hundred thousand acres of land of which only two hundred thousand are in farms. The larger portion of these counties is covered by the Everglades, which serve as an outlet to Lake Okeechobee, and have an area of 6,400 square miles. The land is only about 18 feet above sea level and forms a great saw-grass marsh which is traversed by many water channels. It has the appearance of a vast level field of grain. The soil consists of brown fibrous peat of a spongy character composed of partially decomposed saw grass. At a depth of 3 to 12 feet the soil is underlaid by a great sheet of limestone. During the rainy season it is from six to eight feet under water. There are thousands of islands with an area ranging from one to one hundred acres, but their combined area is only about 150 square miles. The swamp itself is devoid of honey plants, but there grow on the islands wild orange, lemon-trees, willows, *Magnolia glauca,* woodbine, persimmon, buttonbush, prickly ash, and cabbage palmetto. Except along the coast this section offers little inducement to the beekeeper. On Biscayne Key and on other keys there are extensive apiaries that produce some large averages, mostly from black mangrove, cabbage palmetto, and the cocoanut palm.

On the Gulf side of the peninsula the great county of Lee occupies two and a half million acres, of which less than 5000 are in farms. In the center of this county is the Big Cypress Swamp, about which little is known. There are many apiaries along the coast, as at Fort Myers, Naples, and Marco Pass, which are dependent chiefly on the black mangrove. The larger portion of northern Monroe County, at the southwestern extremity of Florida, is covered by the Everglades and the southern portion by an extensive mangrove swamp to the exclusion of nearly all other vegetation. Black mangrove also is common on most of the keys, in some instances forming a girdle around them. The flow from this tree comes in midsummer, and it is one of the most dependable and heaviest yielders of nectar in the state. Several extensive beekeepers are located on the keys, particularly at Biscayne Key and Largo Key. Black mangrove suffered a severe freeze in 1895, but is now rapidly coming back.

Reports of remarkable crops of honey come from Tasmania, a new town near the west shore of Lake Okeechobee. Two hundred pounds of honey per colony are in some seasons gathered from gallberry, scrub palmetto, pennyroyal, orange, wild sunflower, red bay, and huckleberry. At Okeechobee cabbage and saw palmetto are very abundant, but they do not yield every year.

The following lists of the honey plants of the Peninsula and the Keys were prepared by J. C. Goodwin, State Bee Inspector:

CHINQUAPIN (*Castanea pumila*).—A very common shrub in Suwanee County, where it yields a heavy flow of nectar. The honey is dark colored, resembling molasses, and has a strong flavor. Good for feeding back.

CORAL-VINE (*Antigonon leptopus*).—A perennial vine, with bright red flowers, very common in central and southern Florida. It is in bloom nearly eight months. Bees gather considerable dark honey and also some pollen from the flowers.

GALLBERRY (*Ilex glabra*).—Also called inkberry. An evergreen shrub with small leathery leaves, blooming in May. One of the best honey plants, and the source of a large surplus of light-colored honey of good body. It does not granulate. An acid-soil plant, not found on limestone soils. Central Florida and northward.

GOLDENROD (*Solidago*).—Many species occur in the state. The goldenrods are very common in the central section, but are very unreliable yielders. They secrete well in some parts of the state but not in others—secretion being apparently influenced by the type of soil. A heavy, golden-colored honey, which should be well ripened.

JAMAICA DOGWOOD (*Ichthymenthia piscipula*).—Sometimes called sumac, but it belongs to the pulse family, and is not at all related to the genus *Rhus*. The bark is very poisonous, and if macerated and thrown into streams and ponds it will stupefy fish. It blooms in spring, and the white flowers are very abundant nectar-producers. Bees swarm on the bloom. The peculiar odor of the nectar can be detected in the hives. The honey is superior to that of mangrove and does not granulate. This tree is very common on the Keys.

MANGROVE, BLACK (*Avicennia nitida*).—An evergreen shrub or tree growing in sandy soil overflowed by the tide. Abundant along the lower coast and on the Keys. Up to 1895, when it was killed to the ground by frost, it yielded wonderful crops of honey. Since the "big freeze" it has been a poor honey plant on the Peninsula, but it is now becoming valuable again. It blooms from June to August on the Peninsula; on the Keys, from April to July. The honey is light and thin and often has a brackish taste. On the Keys and Thousand Islands it has always been productive, as it was not destroyed by the freeze of 1895.

MEXICAN CLOVER (*Richardia scabra*).—The name is a misnomer as it is not a clover. There are two species, known as the high bush and the low bush. The former is the better honey plant. Grows along roadways and in old fields. The honey is light amber and has a mild flavor.

ORANGE (*Citrus Aurantium*).—The waxy white blossoms open from the middle of February to the middle of March, the date varying with the weather. They furnish a good surplus in the middle and southern portions of the state. The honey is light in color, of heavy body, and has a fine flavor. As many other plants are in bloom during the orange flow, pure orange honey is hard to get.

SCRUB PALMETTO (*Sabal megacarpa*).—A low shrub with long crooked stems, growing in the flatwoods and on the edges of heavy damp hammocks. The small, stalkless white flowers are borne in a large many-branched spike. The honey is of good grade.

CABBAGE PALMETTO (*Sabal Palmetto*).—A large tree, widely distributed in the southern two-thirds of the state. The stalkless white flowers are produced in great profusion in branched spikes. It blooms in July or August, and nectar secretion is greatly influenced by the weather. Rain will cause the flowers to blight. The honey is of good grade, but foams and ferments if taken off unripe. Even after it is sealed it will often foam as though fermenting.

SAW PALMETTO (*Serenoa serrulata*).—Closely resembles scrub palmetto, but has a wider distribution. Common on dry ground in the Gulf Coast region. It blooms from April to May. The honey is of good quality even though taken off before it is sealed. It is said by some to be the best honey in the state.

PENNYROYAL (*Satureja rigida*).—A square-stemmed plant belonging to the mint family, growing on pine lands in the southern part of the state. It blooms from December to March. Weather conditions make the amount of surplus uncertain, but it is good for building up in early spring. The honey is light in color and good in quality.

POISONWOOD (*Rhus Metopium*).—A small tree, blooming in February and March. An excellent source of nectar. The sap is very poisonous to the touch. It has been confused with manchineel, but the two trees belong to entirely different families.

SUMMER FAREWELL (*Kuhnistera pinnata*).—This plant, which grows on the high pine land, is rapidly becoming known in Florida as an excellent source of nectar. The small white flowers appear in late summer and fall, when there is little else in bloom on which bees can work. The honey is light-colored and granulates quickly.

SUNFLOWER.—The name is rather loosely applied to various species of Compositae growing abundantly in the lower half of the state, especially in low land in Seminole and Volusia counties. There are thousands of acres. They are in bloom from September to the early part of November, but are not reliable honey plants. The honey is not classed as first grade.

TITI (*Cliftonia monophylla*).—A small shrub with shining leaves and white flowers. Common in swamps and along streams in the Apalachicola River section of west Florida. Sometimes 30 feet tall. The honey is red and strong, but good for building up colonies for the tupelo flow.

TUPELO (*Nyssa aquatica*).—Large tree, very abundant along the Apalachicola and Flint rivers. The small white flowers in good weather are wonderful nectar-yielders. Blooms in April. The light-colored honey has an excellent flavor and does not granulate.

WONDER-PLANT (*Penstemon laevigatus*).—Commonly called smooth beardtongue. A perennial herbaceous plant growing in sandy soil and blooming in late summer. The reddish white tubular flowers have yielded a good surplus during the three years it has been under observation.

THE HONEY PLANTS OF THE FLORIDA KEYS

Agave decipiens.—Lower Keys. March to May. No surplus.

Avicennia nitida.—Black mangrove. All Keys. May to June. Surplus not dependable.

Crysobalanus Icaco.—Cocoa plum. April to November.

Citrus Limetta.—Lime. Lower Keys. February to May. A small surplus.

Coccolobis uvifera.—Sea-grape. All Keys. March and April. Surplus.

Conocarpus erecta. Buttonwood. All Keys. July and August. Surplus.

Crotalaria pumila.—Upper Keys. January, February, and sometimes later. Pollen and nectar.

Elaphrium Simaruba.—Gum elemi. Gumbo-limbo. Lower Keys. April and May. Surplus.

Ichthyomenthia piscipula. Jamaica dogwood. Lower Keys. March, April, and May. Surplus.

Laguncularia racemosa.—White mangrove. All Keys. May and June. Surplus.

Lysiloma bahamensis.—Wild tamarind. March and April. Lower Keys.

Melothria crassifolia.—Creeping cucumber. Upper Keys. All summer.

Metopium toxiferum.—Poisonwood. Coral-sumac. Doctor-gum. All Keys. February to April. Surplus.

Mikania cordifolia. Climbing hempweed. Upper Keys. May to December. Surplus.

Pithelobium guadalupensis.—Black bead. Upper Keys. October and November. Surplus.

Portulaca oleracea.—All Keys. Summer and autumn. Surplus.

Portulaca phaeosperma.—Purslane. All Keys. Summer and autumn. Surplus.

Sabal Palmetto.—Cabbage palmetto. Upper Keys. June and July. Good surplus.

Serrenoa serrulata.—Saw palmetto. Lower Keys. March to May. Good surplus.

Sophora tomentosa.—All Keys. October and November. Surplus.

Tamarindus indica.—Tamarind. Lower Keys. February to April.

Thrinax microcarpa.—Brittle thatch. Upper and Lower Keys. March and April. Good surplus if the weather is favorable.

Vitis.—Many species of grape. In bloom throughout the year.

Yucca aloifolia.—Spanish bayonet. All Keys. April, May, and June. No surplus.

ALABAMA

Total area, 51,998 square miles. Alabama is often popularly divided into four sections, differing in surface features, climate, soil, and vegetation: The Cereal Belt, which occupies the valley of the Tennessee River in the northern part of the state; the Mineral Belt in the mountainous section; the Black Belt, or agricultural region, a strip 60 or more miles wide south and west of the mountains; and the pine barrens in the southern part of the state. Divided more strictly according to its physical features, Alabama consists of the Coastal Plain and a rugged region in the northeast, where the great Appalachian Range terminates in a series of mountainous ridges and low hills. The line separating the two regions extends from Columbus on the east border to Wetumpka, about 20 miles north of Montgomery, thence northwesterly to Tuscaloosa, whence it runs northward to Tuscumbia, on the Tennessee River, which it follows to the west boundary. Physiographers recognize four sections in northeastern Alabama, but their areas and boundaries do not here call for special description.

The Tennessee River crosses the northern part of the state in a broad irregular curve. Westerly the land bordering the river is level or rolling; but easterly it becomes rugged and mountainous. The soil is very productive, and there is a large acreage of corn, wheat, rye, and oats, whence this section is known as the Cereal Belt. A beekeeper at Grant, Marshall County, describes beekeeping in the Tennessee Valley as follows: "I am a planter of hay, corn, and cotton on the Tennessee River. My records show that my average crop of honey since 1912 has been a little over 50 pounds per colony. Previous to that it was 10 to 15 pounds larger. The decrease I attribute to the logging of tulip tree, which was formerly one of my most important sources of honey. Other honey plants are sourwood, which blooms in June, huckleberry, persimmon, rattan-vine, and basswood; but the last is very uncertain. There were formerly many small beeyards in this vicinity, but the unfavorable winter of 1919 destroyed them." In the mountain valleys of Jackson County in the northeast corner of the state there are more than 5300 colonies of bees.

The northeast-central portion of the state is traversed by a series of mountain ridges and intervening limestone valleys produced by erosion. The soils are largely sandy loams, and the forest is composed chiefly of oaks, hickory, and scrub pines. From the vast coal measures, covering 5500 square miles, and the immense deposits of iron, this region is often called the Mineral Belt. "While there are in northern Alabama many colonies of bees," writes a farmer who has lived in this section for a lifetime, "scattered about in small yards, there is little commercial beekeeping. Both white and black farmers usually have a few colonies in box hives, and both handle them in an equally successful manner—that is, as little as possible, and both have the same 'luck.' But I believe beekeeping would pay well here if properly managed. At times a hundred pounds of honey are obtained from a 'gum,' and colonies can be bought for two or three dollars each." Another beekeeper, at Munford, Talladega County, writes: "I have kept bees for 20 years, and for 15 years have used frame hives. Part of the time I have made fair crops, but commercially have obtained only limited results." The honey flora in the mountains consists of tulip tree, sourwood, persimmon, black locust, chinquapin, sumac, black gum, holly, blackberry, cowpeas, horsemint, bitterweed, goldenrod, and aster. Cotton does not yield a surplus, but furnishes a part of the winter stores. In the north-central portion, except along the streams, where the nectar-yielding trees mentioned above occur, there are few honey plants. The farms are small, and scattered here and there are a few fruit trees.

At Brierfield, Bibb County, there are some small beeyards. About ten per cent. are in modern hives, the remainder in box hives. In this locality no surplus is stored after the middle of June. During a period of five years one yard has paid only expenses.

The east-central portion of the state lies in the Piedmont Plateau. In Chambers County the apiaries range from 1 to 20 colonies, but there are few over 10.

The Coastal Plain includes the southern half of the state and a belt of land along the northwest border about 50 miles wide. It may be divided into two tracts of land: The Black Prairie and the pine barrens. The coast region is nearly level, and its barren sandy soil is covered with a vast forest of southern pine; but in the river valleys are live oak, magnolia, sweet bay, black gum, yellow jessamine, bitterweed, wisteria, yaupon, azalea, gallberry, and velvet bean. Rice, oranges, figs, and other sub-tropical plants flourish in the southeast corner in the rich mucklike soil near the waters of the Gulf. The titi swamps yield well early in the season, and a little later gallberry. Bay trees fringe the streams, and blackberries abound in the swamps. But in many localities, after titi and gallberry have bloomed, the bees must be moved or they will starve, as there are no later sources of honey, and hundreds of colonies have thus perished in the past.

In Houston County the yards are mostly of small size, although near Cottonwood there is one containing 100 colonies. In the spring they sometimes average 60 pounds of honey from titi, gallberry, and cotton. Notwithstanding the large number of colonies of bees found in this area, it is considered much better adapted to queen-rearing than to the production of honey. The hives are mostly box hives, or "log gums," and the bees receive very little attention except at "robbing time." The best locations are in the vicinity of swamps. Four-fifths of the present acreage of cotton in Alabama are found in 10 counties in the southeastern portion of the state.

In the southwestern section of Alabama, in Clarke County, which is bounded on the west by the Tombigbee River and on the east by the Alabama River, there are nearly 6000 colonies of bees — a larger number than is found in any other county in the state. The main crop of honey comes from white tupelo, holly, blackberry, and velvet bean. Sometimes a surplus is obtained from boneset. At Mount Pleasant, on the east side of the Alabama River, the honey plants bloom approximately at the following dates: Redbud, Feb. 1.; plum, Feb. 17; huckleberry and yellow jessamine, March 13; crab apple and haw, March 20; blackberry, April 1; holly, April 11; persimmon, April 29; black tupelo, April 24; white tupelo, April 28; gallberry, May 24; velvet bean, Sept. 26. An average surplus of 70 pounds per hive has been obtained for seven years. There is, perhaps, no better location in the pine barrens than this.

North of the pine barrens, and south of the Mineral Belt, is the Black Prairie, the famous sweet-clover country of Alabama. This tract is 50 to 60 miles in breadth, comprising 13,000 square miles, extending into the state of Mississippi and northward along the border nearly as far as Corinth. The underlying rock is largely Cretaceous limestone from which there has been derived a black fertile soil well adapted to growing cotton and sweet clover. The cotton crop of Alabama is produced chiefly in this belt, little being grown on the pine barrens or in the mineral regions to the north. Six counties—Montgomery, Dallas, Lowndes, Marengo, Wilcox, and Bullock — together report more than one-fifth of the total acreage of cotton. Thousands of acres of sweet clover formerly flourished here; but since 1922 it has been partially destroyed by cattle-raising. The sweet-clover region begins near Union Springs, about 50 miles from the Georgia line. At Columbus, on the Chattahoochee River, no honey is stored from this plant. The apiaries in this section are much larger than in any other part of the Coastal Plain, and range from 50 to 200 colonies, or rarely to 500 in a single yard.

In the vicinity of the city of Montgomery, where more package bees and queens are reared than in any other equal area in the United States, and where sweet clover is abundant, the farms have been greatly improved, and there are many evidences of prosperity. The average size of the apiaries is about 75 colonies. The surplus honey comes mainly from sweet clover; and during three years there is usually one big crop, one medium, and one poor. The main income is from bees and queens shipped north. The honey flow lasts from 40 to 50 days, and the honey granulates as soon as the weather becomes cool. A beekeeper who had 600 colonies reports that the crop was only about two tons one year, but in the next it was 19¾ tons. In a radius of 25 miles from the city of Montgomery there are several commercial beekeepers raising bees and queens, and about 100 smaller apiaries, which contain nearly 4000 colonies. The main crop comes from sweet clover, and an average of 80 pounds per colony is stored once in three years. A beekeeper who has been doing a large business in raising package bees writes: "I have been in the business for 27 years, and have now 500 colonies. I started with 75 colonies, and have acquired $20,000, practically all from bees." At Fitzpatrick, in 1917, there were reported 900 colonies in 11 yards, which were devoted chiefly to raising queens and bees. At Hayneville, Lowndes County, there are apiaries in every direction, located about 3½ miles apart. The commercial apiaries number about 25, and contain a total of 2500 colonies. Their main business is raising package bees and queens. A crop of 75 to 100 pounds per colony is obtained in a good year, 90 per cent. of which comes from sweet clover. The crop is never a complete failure. A beekeeper writes: "I started beekeeping 18 years ago with three colonies. I now have not far from 700 colonies, run for package bees in the spring, and for honey in June and July. In addition I have 600 nuclei for queen-rearing. This year (1920) I have sold 800 two-pound packages of bees, and will probably produce 8000 queens." The honey plants of this region, in addition to sweet clover, are tulip tree, black tupelo, fruit trees, black locust, chinquapin, basswood, maple, rattan-vine, blackberry, buttonbush, snow-vine, redbud, partridge-pea, cotton, velvet beans, cowpeas, bitterweed, China-berry, goldenrod, and aster. Bitterweed is a very important source for winter stores, and those who raise package bees consider it a great help.

Westward in the Black Belt conditions are fairly good, but are not improving. At Demopolis, on the Tombigbee River, the honey gathered from cotton is poor in quality, and in some years there is none at all. Bees in this locality will not work on cotton if there is any other plant in bloom — even bitterweed. For a hundred miles north of Demopolis, along the border, there are many good locations. At Sumterville, Sumter County, there are several apiaries which are reported to contain 100 to 300 colonies. Sweet clover is the only honey plant of consequence. At Cochrane, Pickens County, there are nine apiaries which average over 80 colonies each. But for sweet clover and Spanish needles the bees would die of starvation. The high price of hay has caused the farmers in this county to cut the Johnson grass (*Sorghum Halepense*) three times a year, which kills the sweet clover, as it has no chance to produce seed. In Lamar County, north of Pickens County, beekeeping does not receive much attention. The sweet-clover belt extends into Mississippi, where we shall again meet with it.

The Black Prairie, with its thousands of acres of sweet clover, is undoubtedly the best section of Alabama for beekeeping. All of the sandy land is inferior in fertility, and probably does not afford an average surplus of over 40 pounds per colony. Yet the large number of colonies in the southeast counties, and in many of the northern counties, would indicate that the possibilities in these sections may be much better than would appear from the actual number of pounds of honey produced. Unfortunately, throughout the Black Belt cattle-grazing injures the sweet clover, and prevents it from blooming. The spraying of cotton with poisonous solutions has also been injurious to the beekeeper, and many colonies have perished from this cause.

The northern beekeeper who thinks of moving to the South should not fail, first, to visit the section in which he proposes to settle, and become familiar with its climate, soil, and people. There is a large negro population who live in the country and till the soil, and few northern men care to live among the blacks. The white population is confined chiefly to the towns. There is also in low or swampy areas, and in many river valleys, danger from malaria and fevers. The methods and apparatus of modern bee culture have been much neglected.

In a general way Alabama, particularly the territory around Montgomery, easily ranks first in the production of bees and queens for shipment to the North early in the spring. This is made possible by the heavy growth of sweet clover in the black muck soil and by the ever-present bitterweed that is at its best in this locality, though scattered all over the state. The plant is so bitter that nothing will eat it. While the honey is too bitter for table use, it makes good winter stores. If the honey were of good quality the beekeeper would take it away; but, left in the hives, it makes the colonies strong in the spring, which is so necessary for early shipment of bees.

MISSISSIPPI

Total area, 46,856 square miles. The entire surface of the state is included in the Gulf Coastal Plain, and the soils are formed largely from the rocks of the younger geological formations. There are a great number of soil types, as many as twenty occurring in a single county. The state may be divided into a rolling or hilly upland, comprising five-sixths of its area, and a low flat alluvial bottom-land lying between the Yazoo and Mississippi rivers, known as the Yazoo Delta. Most of the state lies below an elevation of 500 feet; but in Union and Tippah counties there are ridges which rise 1000 feet above sea level. Cotton is the most important crop, and is cultivated throughout the state except in the southeast corner. The densest area is in the Yazoo Delta. The best locations for commercial beekeeping are in the sweet clover belt and the Yazoo Delta; but fair opportunities may be found in isolated areas in all parts of the state near swamps and in the river valleys. No section of Mississippi is overstocked with bees.

The extreme northeastern portion of the state is a somewhat rugged area, lying between 400 and 600 feet above sea level. The soil is chiefly a fine sandy loam, varying in depth from five to fifteen inches. The principal crops are cotton and corn. East of the Northeast Highland is the Northeast Prairie, a narrow belt of land varying in width from ten to twelve miles in the northern part to about double that width in the southern part. The typical soils of this section are derived from the weathering of a soft marly chalk or limestone known as the Rotten Limestone or Selma Chalk. The black soils and large level fields are very attractive features, but these lands are not continuous over the entire area, as there are isolated patches of sandy soils of different origin and value. About one-third of this area is included in the famous Black Prairie Belt of Alabama, which enters Mississippi at Noxubee County and extends northward to the Tennessee line, including parts of Noxubee, Lowndes, Oktibbeha, Clay, Monroe, Chickasaw, Prentiss, and Alcorn counties. Throughout this section white sweet clover is abundant, but it is not common outside of it. Cotton is also grown extensively in all these counties. Other honey plants are black willow, fruit trees, black locust, redbud, blackberry, black gum, partridge-pea, Spanish needles, and goldenrod. In August and September a surplus is often obtained from bitterweed, which is usually left for spring feeding. This plant is very important where package bees are produced. The larger apiaries usually range from 50 to 100 colonies. One beekeeper has 500 colonies, and another does a large business in package bees. At West Point, in Clay County, sweet clover is disappearing in consequence of the land being pastured or seeded to alfalfa. Smaller crops are reported to be obtained to-day than in past years. A beekeeper writes that he has abandoned the attempt to produce honey, and devotes his attention to selling bees and queens, thus making a fair living.

But from Booneville, Prentiss County, the report is more favorable. No complete failure from sweet clover has occurred during the past ten years. An average surplus in a good year is perhaps 50 pounds; but with proper attention it could be increased probably to 100 pounds. There are in this locality some 100 apiaries, the largest containing 50 colonies. Of the Black Prairie, R. B. Willson, formerly the State Specialist in Beekeeping, writes: "The honey flow from Melilotus (the natives call all sweet clover by its generic name) and bitterweed are late, but many very light flows during the early part of the season make this section well adapted to the queen and package business." At Mayhew, Lowndes County, is located a beekeeper who, in 1925, had 2700 colonies of bees and 3000 nuclei which were used in producing package bees and queens.

Between the Yazoo and Mississippi rivers lies the Yazoo Delta, a low alluvial region 160 miles long, occupying 6000 square miles, the larger part of which a long line of levees protects from the overflow of the Mississippi at high water. "To me it is the most beautiful part of the state," says Willson, "despite the fact that it is generally flat. Plantations of from one thousand to ten thousand acres are common, and, because of the extreme fertility of the soil, the people are wealthy and cultured, and, as a consequence, living conditions are excellent. The population is approximately 90 per cent. black; but this is merely an indication that the white man is the lord of the manor still, as he was in the days before the Civil War in most parts of the South."

The honey flora consists of white tupelo, black gum, holly (*Ilex opaca*), blackberry, black locust, and white clover. Brunnichia (*Brunnichia cirrhosa*), a climbing vine which produces in June great sheets of bloom, is believed to yield a surplus. A large amount of honey is also stored from holly. Nowhere is there to be found an acreage of cotton more dense than in Yazoo County. "Definite information as to the value of cotton as a honey plant," says Willson, "is difficult to obtain; but in recent years the severe losses from poisoning by calcium-arsenate dust, used to control the boll weevil, would indicate that bees work cotton heavily."

Tupelo, though generally present, does not appear to be one of the surplus-making plants, because few beekeepers in the Delta secure a surplus before June, and the species of *Nyssa* bloom earlier. Willson states that there were on his mailing-list the names of approximately 100 beekeepers in the Delta. Of these, ten were commercial honey-producers with apiaries averaging 120 colonies. The remaining 90 might average eight colonies to the beekeeper. The Delta is bordered on the east by a high bluff or escarpment, made of a solid silty soil known as loess.

The Loess Soil, or Brown Loam area, extends through the entire state along the eastern border of the Mississippi River flood plain, or the Yazoo Delta. It has an average width of about thirty miles. The land is high and well drained. The principal soil is a brown loam derived from the weathering of the loess, which varies in color from a light yellowish brown to a dark brown. It is moderately rich in mineral plant foods, ranking third in the state as a soil type, being surpassed only by the soils of the Delta and the Northeast Prairie. The crops most extensively grown are corn, cotton, and trucking crops. It is not a section well adapted to bee culture. In the southern part of this area, at Port Gibson, in Claiborne County, on the Mississippi River, there is an apiary of 200 colonies. A crop has been obtained, on an average, three years out of four, during twenty years. But in 1917 and 1918 there was no rain for seventy days, while in 1919 and 1920 there was rain every two or three days with the result that during all four years there were very short crops.

In the north-central portion of the state, between the Black or Northeast Prairie and the Loess Soil area, lie two belts of land known as the Flatwoods and the Shortleaf Pine area. The principal soil of the Flatwoods is a clay or silt loam derived from the weathering of a heavy clay. It is deficient in lime. The soils of the Shortleaf Pine belt are largely sandy loams, and as a rule are not rich in the mineral plant foods. The counties in this section report few colonies of bees, as the only good locations are found along the river and creek bottoms. The honey plants are bitterweed, holly, black gum, blackberry, persimmon, black locust, white clover, and tulip tree. At Grayport, Grenada County, the surplus comes wholly from holly (*Ilex opaca*), which is reliable unless it rains constantly while it is in bloom. Strong colonies have stored 17 pounds in a single day; but there is difficulty in getting the colonies strong in time for the flow, which comes late in April or at the beginning of May, and lasts for about three weeks. The flavor of holly honey is excellent; and it is still liquid after two or three years. In this locality, but for holly it would not pay to keep bees.

The south-central part of the state is crossed by a narrow raised tract of land, covered principally with clay soils derived from the weathering of marl and limestone. There are also small areas of sandy soils. The black lime soil is well adapted to the growing of cotton, corn, alfalfa, lespedeza and other clovers. Melilotus grows wild here and attains a vigorous growth. Alfalfa yields an average of four cuttings. A beekeeper at Hillsboro reports that, while his surplus varies from 5 to 75 pounds, it averages about 35 pounds. The prolonged light flows in the fall encourage breeding, and this locality is better adapted to the production of bees than honey. Many more bees could be kept, but no large commercial apiaries would succeed. The honey plants are tulip tree, wild crab apple, buttonbush, persimmon, and boneset. There is a large acreage of cotton, but its value as a honey plant is doubtful. Bitterweed often yields well in the fall.

South of latitude 32 degrees, or a line extending from near the northern boundary of Wayne County to the city of Vicksburg, on the Mississippi River, the land is generally level, sandy, not very productive, and covered almost entirely with long-leaved pine except in the river valleys. Wherever the pine has been cut, gallberry (*Ilex glabra*) grows in great profusion. Along the streams and in the marshes black and white tupelo and titi are found in great abundance. Scrub palmetto grows throughout this region also, and in places is so dense as to make the land practically worthless for agricultural purposes. Holly occurs locally, and the tulip tree is confined largely to the sides of the creeks.

Along the coast there is a strip of sandy land about ten miles wide where the mean annual temperature is about 68 degrees, and the Satsuma orange grows without protection. This region has been called the poorest section of the state for beekeeping; but the observations of R. B. Willson convinced him that anywhere along the larger streams in this region commercial honey production could be carried on with decided success. The honey is generally of fine quality and flavor, and much of it is from light amber to amber in color.

Climatic conditions in this section are ideal for living, except for the severe Gulf storms which come perhaps once in seven or eight years, and for hordes of mosquitoes which infest the coast whenever a southwest wind blows from the marshes along Lake Ponchartrain in the very early spring. There is no foul brood along the Gulf Coast, and roads are good and the land cheap. At Biloxi, on the shore of the gulf, the largest apiary numbers 25 colonies. Forty to sixty pounds of honey are at times stored by strong colonies; but the average is about 25 to 30 pounds. The surplus honey is gathered from gallberry, gum trees, and blackberry.

The following plants are known to yield a surplus: Bitterweed (*Helenium tenuifolium*), white tupelo (*Nyssa aquatica*), black gum (*Nyssa biflora*), sweet clover (*Melilotus alba*), gallberry (*Ilex glabra*), holly (*Ilex opaca*), rattan-vine (*Berchemia scandens*), tulip tree (*Liriodendron Tulipifera*), partridge-pea (*Chamaechrista fasciculata*), basswood (*Tilia pubescens*), Spanish needles (*Bidens aristosa*).

More or less surplus is also obtained from the following species: Sumac (*Rhus glabra*), black willow (*Salix nigra*), white clover (*Trifolium repens*), persimmon (*Diospyros virginiana*), bush tupelo (*Nyssa acuminata*), scrub palmetto (*Sabal megacarpa*), sourwood (*Oxydendrum arboreum*), swamp gallberry (*Ilex lucida*).

Less important honey plants often visited by bees: Horsemint (*Monarda punctata*), haw (*Crataegus spp.*), mallow (*Hibiscus moschatus*), redbud (*Cersis canadensis*), buckthorn (*Rhamnus lanceolata*), spindle tree (*Euonymus americanus*), Ogeche plum (*Nyssa ogeche*), black alder (*Ilex verticillata*), white titi (*Cyrilla racemiflora*), brunnichia (*Brunnichia cirrhosa*).

The species of *Nyssa* are widely distributed throughout the state. Holly (*Ilex opaca*) occurs locally; but gallberry (*I. glabra*) and swamp gallberry (*I. lucida*) are found generally south of latitude 32 degrees. Black locust is found locally, scattered throughout the northern and southwestern part of the state. Willow grows chiefly along the Mississippi River and its tributaries. The Rubus group occurs in all parts of the state. Sweet clover is confined chiefly to the northeast prairie, although found to a small extent in the Delta and on the loess formation south of the Delta. Basswood is found largely in the counties along the Mississippi River, through Bolivar County, in the southern part of which there is a thick growth.

LOUISIANA

Total area, 48,500 square miles, of which 3000 square miles are water surface. According to their elevation and origin, Louisiana may be divided into lowlands and uplands; but no part of the state has an altitude of more than 500 feet, and the average elevation is less than 75 feet. Each of the larger streams, as the Mississippi River, the Red River, and their tributaries, flows through a belt of bottom-land liable to overflow at times of high water. These alluvial bottom-lands occupy nearly one-third of the state and include the territory best adapted to bee culture. Along the coast there is a low swampy region traversed by slight ridges which extends inland from 20 to 60 miles. The coast swamps and the bottom-lands along the rivers comprise about 20,000 square miles, or nearly one-half of the total area of the state. The remainder consists of uplands of prairie and forest occupying the northern and northwestern portions of the state and a small area south of the

State of Mississippi. Louisiana has a semi-tropical climate, with a mean annual temperature of 70 degrees F., and an average rainfall of 55 inches.

The sections of Louisiana best adapted to beekeeping, according to E. C. Davis, Specialist in Bee Culture, are the bottom-lands along the Red and Mississippi rivers and the Atchafalaya River basin, which is as fine a country for honey production as can be found in the United States. These areas are subject to overflow at times, although this may not occur once during an interval of ten years. After the levees are finished by the government this trouble will be eliminated. There are spots all over the state which are fine for commercial beekeping, but the above locations are the best. Some of them contain apiaries which have even 500 colonies and produce on an average 100 pounds of honey per colony annually. In parts of the Red River section, where white clover is very abundant, sometimes as much as 200 pounds per colony is secured.

The salt-marsh region, including the Mississippi Delta, is a belt of land 20 to 60 miles wide, extending along the coast, and irregularly indented or divided by numerous large lakes, tidal bays, and lagoons. Of this area, 3,500,000 acres are subject to tidal overflow. On the inner margin of this belt, rice is extensively grown; and in the Delta there is a large acreage of sugar cane and orange trees. In the counties of Lafourche, Terrebonne, Assumption, St. Mary's, and Iberia there is an extensive cypress swamp, much of which would be inundated by the Mississippi at high water were it not protected by dykes. Snow is almost unknown, and frosts occur only two or three days in a year.

In the extreme south, in Terrebonne County, Dulac, 18 miles south of Houma, is situated at the end of a cypress tract, and on the east, south, and west is surrounded by lakes and marshes as far as he eye can reach. The best-known honey plants are white tupelo, willow, red haw, white clover, buttonbush, pepper-vine, thoroughwort, and heartsease. The spring honey flow extends through March, April, and May, and is always dependable. If there is not too much rain, there is a fall honey flow from flowers on pastured land. The earlier honey is lighter colored, and has the better flavor. "At my home yard," writes a beekeeper, "I secured 10 gallons per colony, and have a surplus to take off early in the spring." Bees are not abundant along the coast.

The bottom or alluvial lands which border both the larger and small streams offer excellent opportunities for beekeeping. The flood plain of the Mississippi above Baton Rouge is 50 miles wide, and in the Delta it has nearly double this width. The valley of the Red River has an average width of 25 miles, and the valleys of the Atchafalaya and Ouachita have an average width of 10 miles. Considerable white clover is grown for seed and not a little honey comes from the plant; but this southern white clover honey is darker than the northern. Each one of the smaller streams has its own narrow belt of bottom-land. The rivers flow through their flood plains on ridges built up by their own deposits, and their banks are much higher than the country back of them, which is low and swampy, and is overflowed at high water unless protected by dykes. There are many crescent-shaped lakes and bayous along both the Red and Mississippi rivers, portions of former channels now abandoned. The alluvial soils are extremely fertile, and support a large acreage of cotton, corn, and other farm products.

Of the Red River counties, Avoyelles is considered by many the most suitable for beekeeping. The average surplus does not usually exceed 60 pounds per colony; but some phenomenal yields have been reported. This region is also excellent for raising bees, although too much rain often causes a heavy loss, and interferes with the honey flow. The honey flora consists of white tupelo, willow, haw, persimmon, rattan-vine, water locust, hundreds of acres of magnolia, snow-vine, horsemint, heartsease, white clover, bitterweed, boneset, goldenrod, and aster. The counties of Carroll, Madison, Tensas, and Concordia in the Mississippi Valley contain only a few colonies of bees, but southward there are many good locations in Iberville, West Baton Rouge, and Pointe Coupee parishes. In Iberville Parish, in five apiaries of 200 colonies each, an average of 10 gallons per colony is secured. Little attention is at present given to beekeeping among the orange groves; but this may prove to be a most profitable region.

Fair opportunities for beekeeping are reported along the Black River, a tributary of the Red River, and northward on the Boeuf River, Bayou Macon, and Tensas River. "Along the Atchafalaya River from its source at Red River to the Gulf," writes E. C. Davis, "is a veritable paradise for bees. There are numerous places where as many as 500 colonies could be kept in one yard. One apiarist in this territory having 250 colonies of bees, spring count, produced in one season more than $5000 worth of honey, and at the same time increased his colonies to 329." The region south of Lake Pontchartrain is also reported good. The alluvial lands of Louisiana belong to the Quarternary Period, and are of recent origin.

The uplands, including the pine barrens in the north-central portion of the state, the southwest prairies, and an area south of the State of Mississippi, are, as a whole, poorly adapted to bee culture. Between the valleys of the Red and Ouachita Rivers the land is largely timbered with short-leaved pine and oak; but farther southward there is a large forest of long-leaved pine growing on a thin sterile soil. While there are many small beeyards throughout this section, little effort is made to produce large yields of honey.

South of Mississippi, in the southeast part of the state, there is another tract of long-leaved pine. Much of the pine has been cut for timber, and the open range has been used for pasturing livestock so that many flowers grow in protected places. Most of the counties contain very few bees, and the surplus often does not exceed 20 pounds, but there have been yields of as high as 100 pounds. In the spring the bees gather nectar from maple, tulip tree, holly, gallberry, rattan-vine, persimmon, willow, tupelos, and palmetto; but after June 1 the active season closes, as there are few fall flowers. Box hives and "gums" are very common. Bees commence work about January 15 and continue until frost. Large quantities of honey-dew are gathered.

The prairies of the southwest are treeless except along the streams, and are well covered with grass, which is grazed by herds of cattle and horses. A large area has in recent years been brought under cultivation, and the land greatly improved. Five parishes in the southwestern corner report four-fifths of the total acreage of rice, of which Calcasieu alone reports more than one-third. Like all the other grasses, the bloom of rice is nectarless. There are no commercial apiaries, and a good crop at Welsh is obtained about one year in five.

The principal honey plants of Louisiana are willow (*Salix nigra*), maple (*Acer rubrum*), white tupelo (*Nyssa aquatica*), black gum (*Nyssa biflora*), white clover, carpet grass (*Lippia nodiflora*), cabbage palmetto, water locust (*Gleditsia aquatica*), orange, snow-vine, (*Cissus arborea*), persimmon, partridge-pea, rattan-vine (*Berchemia scandens*), buttonbush, pepper-vine, velvet beans, heartsease, boneset, climbing boneset, goldenrod, and cotton. The honey flow in the central part of the state may be divided into three periods, but they merge one into the other, so the bees always have some flowers to work on. The spring crop is secured from willows, white tupelo, and white clover. Tupelo is more abundant than the other two, and is the main source of honey, but willows cover a large area in the swamps and often yield a large surplus. The honey from tupelo comes in a veritable flood, and the bees work from dawn until long after sunset, but it is of short duration. The summer honey flow continues for several months, and is gathered from a great variety of flowers; but a considerable part of it is honey-dew. The summer honey is red in color and very poor in quality, with a rank flavor. Cotton is the staple field crop. It yields no honey in the hill section of the state, but does furnish considerable on the lands along the rivers. Goldenrod, horsemint, boneset, smartweed, and asters are the sources of the fall flow, which begins by September 10 and lasts for about a month. The honey is amber-colored, but has a good flavor.

ARKANSAS

Total area, 53,335 square miles. A line drawn from Sevier County, on the southwest border, to Clark County, and thence northeasterly to Randolph County on the northeast border, divides the state into a northwestern highland section and a southeast lowland section. The northwestern division belongs to the Ozark Plateau, and is rugged and mountainous. The southeastern division is part of the Coastal Plain, and consists of the bottom-lands of the Mississippi River and its tributaries and the rolling inter-stream uplands. In eastern and southern Arkansas the honey crop is dependable annually, and it is also reliable in a few mountain counties where the soil is of limestone origin.

The northern part of the Ozark uplift is occupied by the rugged Boston Mountains, and the west-central portion by the Ouachita Mountains. Much of the soil is sandy and sterile; but there are large areas which are underlaid by limestone, and are suitable for growing the clovers. Northwestern Arkansas, like southwestern Missouri, is well adapted to fruit-growing; and in Benton County, in the northwest corner, there are more apple trees than can be found in an equal area in any other part of the United States except in Oregon. Peaches are abundant along the entire western border, and there is a large acreage of strawberries in Benton County. Several other counties in the northwestern part of the state are also devoted to fruit-growing to an almost equal extent. But apple trees are so frequently sprayed when in bloom that at present they are a doubtful benefit to bee culture. The principal honey plants are white clover, sumac, sweet clover, horsemint, black locust, basswood, raspberry, fruit-bloom, buckbush, redbud, honey locust, persimmon, heartsease, cotton, bitterweed, and goldenrod.

The number of colonies of bees in the highlands is about the same as in the lowlands. The poorest portion of the state for beekeeping is found in the mountains, where the land is too rough for cultivation, and the sour soils are heavily wooded with oak and pine. Where there are limestone valleys the crop of honey is more dependable. Benton County, in the northwest corner, is a great fruit-growing country; but the apiaries seldom exceed 30 colonies, and the surplus is often only 10 or 12 pounds.

The largest number of colonies of bees in the Ozark region is found in the west-central part of the state in the Ouachita Mountains. At Harris, Washington County, a surplus has been obtained every year for 32 years, except once, when, owing to continued wet weather, the season was the poorest for the production of honey ever known in western Arkansas. The middle-west border is a poor location for the production of honey, but is fine for queen-rearing. Persimmon blooms in May; basswood among the hills in June; sweet clover in June and July. White clover does not always produce nectar, and cotton also is not always reliable.

The northeast counties of the upland region, as Randolph, Lawrence, Independence, Sharp, Izard, and Stone, contain relatively few colonies; but in Independence County there are a number of apiaries which contain from 50 to 100 colonies. The surplus comes chiefly from sweet clover, the flow continuing from six to eight weeks. Blackberry, raspberry, and cotton are less important sources of nectar, and in some years large quantities of honey-dew are gathered.

While the number of colonies of bees in the lowlands does not at present much exceed those in the mountainous northwestern section, the future development of beekeeping will take place chiefly in the eastern and southern portions of Arkansas. It is the opinion of J. V. Ormond, based on extensive travel in this state, that these sections are dependable, and offer opportunities for engaging profitably in beekeeping on a commercial scale. According to H. H. Bennett, the alluvial bottom-lands of the Mississippi, Arkansas, and Red Rivers represent one of the richest and most important areas of land in the world. Probably no land used by man exceeds them in fertility and productiveness. The very fertile bottoms of the Arkansas River have produced 1½ bales of cotton per acre without fertilizers in the best years. Cotton secretes nectar very freely on these fertile bottom-lands.

In the northeast portion of Arkansas in the vicinity of the swamps of St. Francis River beekeeping remains in a primitive and undeveloped condition. There are many small apiaries in which very little care and attention are given to the methods of bee culture, and gums and boxes of various sizes are used for hives. The average surplus does not much exceed 20 pounds per colony. But in Poinsett and Mississippi counties conditions are better. Of the honey plants in this locality the buttonbush covers large areas of the marsh, willows are abundant, and heartsease and Spanish needles are common. A little farther westward in Independence County larger yards and crops are reported. On the higher land are found persimmon, white clover, sweet clover, sumac, and goldenrod. The dry upland on which white clover occasionally yields well is a poor location.

The winters in the southeast portion of the state are very mild, and this is an excellent location for raising bees for sale. Along the Ouachita River, in the southern portion of the state, holly is the principal honey plant, blooming in May. Not much honey is stored later in the season, although at times there is a flow from bitterweed in August and September. Other honey plants are maple, rattan-vine, persimmon, basswood, black gum, cotton, goldenrod, Spanish needles, heartsease, and thoroughwort.

"In the southwestern portion of the state," writes a beekeeper of Columbia County, "I believe that scientific, intelligent bee culture would succeed, although it it not considered a favorable locality." Holly is the most valuable honey plant. It yields a clear white honey, and the flow is not greatly affected by the weather. In July and August the bees store to a small extent from cotton, and sometimes also from goldenrod. Sumac, sweet clover, and rattan-vine also deserve mention.

Near the center of the state on the Arkansas River are Pulaski and Jefferson counties, in which there are 163,000 acres of cotton under cultivation. On the fertile bottom-lands cotton secretes nectar abundantly and is the source of thousands of pounds of honey. If there is sufficient rain in August a fair crop is obtained chiefly from cotton. Other honey plants are buckeye, redbud, black gum, blackberry, rattan-vine, basswood, white clover, sumac, cow-itch (*Cissus incisa*), and bitterweed. All are more or less affected by extremes in the weather.

Between the Arkansas and White Rivers are the counties of Lonoke, Prairie, and Arkansas, which consist largely of low marshy land. On this wet soil more than one hundred thousand acres of rice are grown. In the neighborhood of the swamps and on the dry land near them there are many honey plants, as swamp holly (*Ilex decidua*), white tupelo (*Nyssa aquatica*), dwarf palmetto (*Sabal glabra*), sour gum (*Nyssa sylvatica*), buttonbush, hackberry, snow-vine, redbud, red maple, cow-itch, persimmon, honey locust, white clover, bitterweed, blackberry, and rattan-vine. In Lonoke County farkleberry (*Vaccinium arboreum*) is very abundant. It blooms in May and the white bell-shaped flowers are visited by hosts of bees. While there are a large number of small yards in this marshy region, commercial beekeeping receives almost no attention.

The best locations for beekeeping in Arkansas are found on the bottom-lands in the river valleys in the eastern and southern portions of the state. In the Mississippi, Red, St. Francis, Arkansas, and Ouachita valleys white clover grows in profusion, and secretes nectar well in favorable seasons, since the soil is very fertile. Holly in the southern portion yields a large surplus of honey, taking the place of gallberry in North Carolina and Georgia. Beekeeping is in an almost wholly undeveloped state, employing crude apparatus and methods, and, strictly speaking, there are not more than two or three commercial apiaries in the state.

TEXAS

Total area, 265,780 square miles. Texas, which is 740 miles in length and 825 miles in breadth, is nearly the same size as France, and its area is about one-twelfth of that of the entire United States. Its soils, rainfall, temperature, surface features, and flora vary widely in different portions of the state. From the low coast marshes along the Gulf of Mexico, covered with reeds and cypress-trees, the land gradually rises in altitude until on the high treeless plains in the northwest it reaches an elevation of 4000 feet and in the Rocky Mountains west of the Pecos River a height of 9000 feet. From over 60 inches in the northeastern Coastal Plain the annual rainfall decreases to 25 inches in the center of the state and to less than 10 inches at El Paso. The varied and dense vegetation of eastern Texas, consisting largely of a mixed forest and many prairie flowers, gives place in the almost rainless desert of the Trans-Pecos region to cactus, yucca, and agave. According to a conservative estimate about one-half of the land, or 85,000,000 acres, can be cultivated; but at present less than one-third of this area is under tillage. Irrigation is practiced to a limited extent in many counties; but, except in connection with rice-growing along the coast, the most extensive irrigated areas are in he valleys of the Rio Grande, Pecos, and Nueces rivers.

Notwithstanding its immense area and great diversity of physical features, Texas may be divided into six fairly well-defined physiographic regions: Northeastern Texas, or the east Texas timbered region; Southeastern Texas, or the Rio Grande Plain; the Black Prairie, or the cotton belt; Central Texas; the Staked Plains, or high plains; and the Trans-Pecos, or mountainous region west of the Pecos River.

Northeastern Texas, or the east Texas timbered region, is an agricultural section, but about one-third of its territory is covered with a mixed forest of long-leaved pine, short-leaved pine, loblolly pine, and post oak on the higher land, and a variety of hardwood trees on the river-bottoms. This forest is the southern termination of the great mixed forest of the Gulf states, its farther extension being checked by the decreasing rainfall. In the timberland along the rivers the following trees and shrubs secrete nectar: black tupelo, white tupelo, honey locust, sweet bay, willow, persimmon, redbud, red maple, papaw, catalpa, basswood, sumac, buckthorn, plum, cherry, gallberry, dwarf palmetto, holly, rattan-vine, and yellow jessamine. In the low flat lands of the coast in the underthicket are found holly and gallberry, and on the dryer land blueberry. The banks of the Neches River are lined with black tupelo and white tupelo, the belt of timber being one to two miles wide. In the Coast Prairies there occur almost impenetrable thickets of holly, thornbush, chinquapin, and magnolia. There are many peach and pear trees in the northeastern counties, and in the vicinity of Galveston a large acreage of orange trees.

Of herbaceous plants sweet clover, partridge-pea, heartsease, cotton, tie-vine, cow-itch, horsemint, boneset, and aster are reported as good honey plants in various localities. Bitterweed blooms from June until frost, but bees work on it only in dry seasons. Broomweed and crownbeard often yield in the fall an amber-colored honey, but in dry seasons they are of little value. No one plant is of paramount importance, and there are large areas without a sufficient honey flora to make beekeeping profitable.

Northeastern Texas is bounded on the south by the San Antonio River and on the west by the Black Prairie, the dividing line running from Paris, Lamar County, to Seguin in Guadalupe County. The counties are mostly small in size, ranging from 800 to 1000 square miles, many of them reporting less than a thousand colonies of bees. The soils are sandy loams and the honey plants are widely scattered, none of them yielding a large surplus. Beekeeping is pursued largely as a side-line and usually only crops of moderate size are obtained. Old-fashioned box hives are largely used and chunk honey produced.

While the large yields of southeastern Texas are seldom secured in the northeastern section, there is seldom an entire failure. The chief difficulty in this part of the state is that none of the honey plants produce a large surplus, and they are so widely scattered that, in order to secure a fair crop, it is often necessary to move the bees several times.

Southern Texas, or the Rio Grande Plain, comprises the area between the San Antonio River and the Gulf of Mexico on the east and the Rio Grande River on the west, and south of the north line of Kinney, Uvalde, and Medina counties. In climate and flora it belongs to the Gulf region of Mexico. The woodland, according to Bray, embraces 70 to 80 species of small trees and shrubs, not one of which appears in the Atlantic forests of east Texas. It is a semi-arid region with many days of intense sunshine, a loose soil destitute of vegetable mold, and a water level so deep as to be beyond the reach of all plants except perennials with long roots. Except in the river valleys there are no large trees; but vast areas are covered with a chaparral of low thorny bushes and small trees from 2 to 15 feet tall. It is noteworthy that 30 per cent. of the species, and a far greater percentage of individuals, belong to two sub-families of the pulse family, the Mimosas or Acacias (Mimosae), and the Cassias (Caesalpineae). Two small trees, huisache (*Acacia farnesiana*) and retama (*Parkinsonia aculeata*), are distributed throughout this section, and approximately determine its limits. Until comparatively recent times the Rio Grande Plain was a grassland. There are many who remember when hundreds of acres now in brushland were in grassland.

Southeast of San Antonio is a vast forest of mesquite trees, 10 to 15 feet tall. On the rich low flats this tree reaches its maximum growth of 20 feet or more. The open grass floor is close set with white brush (*Aloysia ligustrina*) and prickly pear. On gravelly slopes there grows a straggling shrub (*Parkinsonia texana*), and on ridges and bluffs that most valuable honey plant huajilla. Huisache is most common on clay soils near the coast. Northward there is a "black chaparral" of which 60 to 75 per cent consists of black brush (*Acacia amentacea*), which is chiefly valuable for pollen. The more important honey plants of the Rio Grande Plain are as follows: granjeno, agarita, two species of Texas ebony, eight species of acacia, huisache, mesquite, Leucena, retama, five species of mimosa, palo verde, redbud, coral bean, Eysenhardtia, guaiacum, colima, three species of sumac, two species of Texas buckeye, two species of Brazilwood, prickly pear, Texas persimmon, coma, Adelia, anaqua, and white brush. (For descriptions of these plants see Part III.)

The thorny chaparral of southern Texas secretes nectar much better in a dry season than in a wet one, as much rain blasts the bloom and causes it to fall from the stems; but there must have been sufficient rain during the preceding fall and winter to enable the plants to store up the required food material. The time of blooming is also greatly influenced by the occurrence of rain. If there is sufficient wet weather, mesquite, soapbush, and Brazilwood may bloom twice in the same season. In Uvalde County white brush blooms after every heavy rain; but the flowers last only five or six days. In Bee County mesquite blooms about April 1, and again about June 20, but it may fail to bloom entirely, or it may bloom continuously from April to July. Catsclaw may also bloom twice in a single season.

When there is sufficient moisture and an abundance of bloom this is the most important section of Texas for beekeeping; but in 1917 and 1918 there was a long drought, and in many localities fifty per cent. of the bees died of starvation. Even where they were fed with sugar there was not enough pollen obtainable to permit of brood-rearing, and in 1918 thousands of colonies died for want of pollen. The resulting loss and disappointment have caused many specialists in honey production to move away. But in favorable seasons it is almost impossible to overstock this region, and immense crops of honey are obtained. "It is in this region," says Pellett, "that commercial beekeeping has reached its highest development in Texas. In several counties there are more commercial beekeepers than are found in whole states in other sections of the country." Over a large part of this section mesquite, huajilla, and catsclaw are the chief honey plants, the average surplus ranging from 25 to 100 pounds, according to the locality.

Cameron County is situated in the extreme southern part of Texas. The surface is a nearly level alluvial prairie, and the soils are deep and fertile. Along the Rio Grande River irrigation is practiced extensively, and more than 50,000 acres are under cultivation. At Brownsville are found the most southern apiaries in the United States, the largest of which contains 60 colonies. The honey plants are mesquite (February to July); Texas ebony (June); anaqua (February to September); tenazza (April to September); Brazilwood (September to November); horsemint (April to October); Gaillardia (March to July); cow-itch (June to September); and yerba dulce (September to October). The blended honey has a fine flavor, is amber-colored, and does not granulate quickly. Nearly all the plants bloom annually, and several of them after each rain. An average of only about 25 pounds per colony is obtained, since a large amount is consumed in brood-rearing, and bees swarm as late as December. This is an excellent location for raising bees, but a rather poor one for honey production.

In Brooks County the gently rolling surface is largely covered with mesquite, and on hundreds of acres there are no bees to gather the nectar. At Falfurrias there is an apiary of 200 colonies. In 1911 and 1919 the crop was a complete failure; but since 1908 there have been three very large crops. In Nueces, Jim Wells, and San Patricio counties there are many large ranches on which thousands of cattle graze. Mesquite is the most common tree, but other common honey plants are huajilla, white brush, catsclaw, horehound, horsemint, and broomweed. The average rainfall is about 26 inches. Beekeeping is pursued chiefly in connection with diversified farming, and the yards range from 10 to 50 colonies, with a few larger ones.

In Bee County there is a large area of prairie land; and, as the rainfall is over 30 inches, fruit-growing and truck-farming are developing rapidly. During 19 years at Beeville a beekeeper has averaged 65 pounds of surplus per year, but in 1916 and 1917 the crop was a failure, and 50 per cent. of the bees died from starvation. The sources of honey are wild currant, or agarita, catsclaw, horsemint, mesquite, and Brazilwood. At Oakville, Liveoak County, there are several commercial apiaries. In Goliad County, through which flows the San Antonio River, more than half of the area is prairie with a black sandy loam soil. The Collier apiaries contain over 1000 colonies of bees. W. C. Collier writes that in his opinion a beekeeper could keep bees profitably along most of the rivers of southern Texas. At Goliad the sources of surplus are mesquite, Brazilwood, and catsclaw.

In Atacosa County about 3000 acres are under irrigation, and fruit-growing is profitable. From Jourdanton a beekeeper writes: "We have over 750 colonies of bees which are run for extracted and chunk honey. In a good year the average surplus per colony is over 125 pounds. Our bees begin to breed in January, and in an ordinary season are gathering nectar from huajilla, catsclaw, and white brush by April." In Zavalla County during recent years irrigation projects taking water from the Nueces River and artesian wells have made rapid progress. From Indio a beekeeper reports that he has 500 colonies in four yards along the Nueces River, where in the past he has secured very large crops of honey. Many portions of this section are wholly without bees.

Uvalde County produces a large quantity of honey, 200 pounds per colony being not unsual, and as much as 400 pounds has been obtained from a strong colony. There were at one time not far from 15,000 colonies in the county; but in 1917 and 1918 it was so dry that half the bees died or were moved away. D. C. Milam, who has lived in Uvalde County for 28 years, states that during this time there were two years in which he removed honey from the hives every month except January. A beekeeper living at Sabinal has eleven apiaries which contain from 60 to 100 colonies each. In a favorable season the average surplus is 60 pounds per colony. Near Uvalde the apiaries in some cases number 200 or more colonies. The surplus comes from huajilla, catsclaw, and mesquite, while Brazilwood, white brush, and soapbush are common, but they are not reliable every year. Soapbush blooms about the first of April, if the weather is dry, yielding a heavy white honey. White brush blooms for five or six days after every heavy rain. Large crops are sometimes obtained in the fall from broomweed. In the adjoining county of Medina the general surface is rolling, becoming mountainous in the northern part. The average rainfall is 26 inches. In the vicinity of La Coste there are about 50 apiaries which contain 2380 colonies of bees. There are also over 100 small yards in which there are from 2 to 30 colonies. In 1918 the gross returns from 360 colonies was $3075, and in 1919 the gross returns from 400 colonies was $2620. For several years past Wilson, Atascosa, Live Oak, and Trio counties have been among the greatest honey-producing counties of the state. A sharp escarpment midway between San Antonio and New Braunfels marks the dividing line between the Rio Grande Plain and the Black Prairie.

West of Northeastern Texas there is a narrow belt of land known as the Black Prairie, which has a dark, waxy, very fertile soil not easily eroded. It is underlaid by Upper Cretaceous limestone. It is a grassland country, and is bare of trees except for groves of mesquite and oak; but in the river valleys are oaks, pecan, cottonwood, and various shrubs. It is an agricultural region with a larger population than any other portion of the state, and beekeeping is usually pursued in connection with general farming. The most dense area of cotton in the United States is found on the Black Prairie. Millions of acres of corn are grown, and fruits and vegetables are generally cultivated. The principal honey plants are cotton, mesquite, and horsemint; but sumac, broomweed, goldenrod, sunflower, and marigold are valuable. An apiary should be located where the bees can gather from more than one plant. There are many excellent locations in the Black Prairie, where in favorable seasons large crops may be secured.

Within a radius of 20 miles of New Braunfels the Scholl apiaries contain 1500 colonies in 31 yards. Here and northward to Waco and Waxahachie cotton is the main dependence for honey. "In an average season," according to Scholl, "a good yield may be expected from cotton in the black-land districts and the river valleys. Under favorable conditions it is not excelled by any other nectar-yielder in the cotton-growing belt. On poor soil and on sandy land it does not secrete nectar plentifully, and in some sections or under certain weather conditions, not at all." Nectar is secreted most freely when the air is warm and damp. On the bottom-lands of the Brazos River there are cotton plantations which are several thousand acres in extent. It is the only source of nectar, and averages about 75 pounds of bulk comb honey annually. One season the surplus exceeded 100 pounds per colony. The honey is very light in color, with a very white comb, and has an excellent flavor when well ripened. With the beginning of September many fall plants begin to bloom, as broomweed, which is very abundant, boneset, and rosinweed. Throughout the Trinity and Brazos valleys heartsease, or smartweed, has become thoroughly established, and the bees work on it steadily.

In Caldwell County, at Lockhart, there is an apiary of 500 colonies, which during six years has never experienced a total failure. The best crop was over 100 pounds per colony, the poorest 23 pounds per colony. Near Austin, in Travis County, there are five apiaries which average about 100 colonies and obtain a good crop three years out of five. There are also more than 60 small yards. A beekeeper at Marlin, Falls County, who operates 200 colonies, reports that he has had only one failure in 20 years, which was in 1917, the dryest year ever known in this county. There are three apiaries containing 25, 300, and 450 colonies respectively at Heidenheimer, Bell County, which have obtained a crop every year except during the dry seasons of 1917 and 1918. At Belton the honey plants are horehound, mesquite, milkweed, prickly ash, China-tree, horsemint, cotton, broomweed, and goldenrod. During droughts bees often obtain sufficient honey-dew from live oak for winter stores. A crop has been obtained every year, except one, during 31 years.

At Waco, McLennan County, near the center of the cotton belt, about 2000 acres of the Brazos bottom-lands have been reclaimed from overflow. The soils vary from black alluvial loams in the river valleys to black waxy and sandy loams on the uplands. Attention is given chiefly to agricultural pursuits, and 239,000 acres of cotton are grown. Farther northward around Waxahachie there are reported to be 2000 colonies of bees. There are 274,000 acres of cotton. A beekeeper writes: "I began with one colony in 1903 and have increased to 700. I average 60 to 70 pounds per colony annually. There is a good market for all the honey I can produce within a radius of 25 miles. I ship many pounds of bees in packages, am making money, and like the business." There is ample room for many more bees in this part of the cotton belt. Around Dallas, Dallas County, there are 700 colonies in commercial apiaries. During six years the smallest average was 30 pounds and the largest 140 pounds. The county is estimated to contain 6000 colonies of bees.

On the north border of the state south of the Red River are Grayson, Fannin, Lamar, and Red River counties. Grayson County is reported to be a poor bee country and to contain no yards with more than 50 colonies. Feeding is necessary in order to obtain a surplus of honey. In Fannin County the apiaries are also small. Cotton and sweet clover are usually reliable, but horsemint yields only about every other year. At Paris, Lamar County, there are 600 colonies in two apiaries, and a failure of the crop has not been known for many years. Rattan, cotton, and horsemint are the sources of honey; but, as in Fannin County, bees must be fed in spring, which is rainy and backward. The farmers are sowing fields of 10 to 30 acres of sweet clover, and at Roxton within a radius of one mile there are 100 acres. In Red River County the valley has a heavy red soil suitable for general farming and for growing cotton and corn. A beekeeper from Manchester writes:

'I have 300 colonies in a line of apiaries. The surplus is gathered from huckleberry and cotton. Huckleberry begins to yield nectar about the middle of May, and the bloom lasts for about three weeks. Then there is a dearth of nectar until cotton begins to yield early in July. There are very few beekeepers here. The unoccupied territory extends down the Red River for 75 miles to the vicinity of Texarkana." The Red River Valley is considered a promising region.

West of the Black Prairie lies the vast territory of central Texas, extending westward to the sharp escarpment of the Staked Plains, and in the extreme southwest to the Pecos River, and from the Red River on the north to the Rio Grande on the south. It is a semi-arid region with a rainfall of 20 to 25 inches. The soil is loose and sandy, and chaparral and mesquite woodland are common. The climate favors grassland, and sheep and cattle raising are important industries. The northern portion is relatively level, and supports a large acreage of cotton and corn. The middle portion is known as the Texas Hills Region. Deep canyons have been worn in the surface by the streams, along which there is an abundant nectar-yielding flora. There are many colonies of bees here.

The four northern rows of counties of Central Texas contain very few colonies of bees, seven counties reporting none or less than ten. This region has a rainfall of about 26 inches and is drained by the Wichita and the northern tributaries of the Brazos River. The land is largely divided into ranches. Dwarf mesquite occurs in nearly all the counties. The central portion of the Texas Hills Region, which is drained by the Colorado River, is an excellent section for beekeeping. Many counties contain over 2000 colonies of bees, and three (San Saba, Coleman, Llano) report over 3000 colonies. A beekeeper at Llano, Llano County, writes that a good crop is obtained every year unless it is too dry. At Brady, in McCulloch County, there is an apiary of 280 colonies. In 1917 and 1918 the crop, owing to drought, was a total failure, and 85 per cent. of the colonies were lost. But usually a fair crop is obtained which ranges from 40 to 60 pounds. The honey plants are mesquite, beeweed, cotton, prickly pear, and broomweed.

The southern portion of Central Texas is occupied by the Edwards Plateau. The soils are of limestone origin, underlaid by the Lower Cretaceous, but are badly eroded, and the surface is broken by deep gorges and steep bluffs. The western half is almost devoid of streams, and, except in the extreme east, the plateau is not adapted to beekeeping. The land is covered with an open mixed forest and dense thickets of shin oak and mountain cedar. Among the trees valuable to the beekeeper are mesquite, wild cherry, Texas persimmon, Brazilwood, Acacia, hackberry, sumac, and coral bean. Other species of value are kinnikinnick, chittam, wild China, wild plum, agarita, catsclaw, redbud, and horehound. In the Nueces Canyon, Edwards County, which is drained by the east fork of the Nueces River, there are numerous commercial apiaries which contain approximately 1400 colonies of bees. On an average about 67 pounds per colony are obtained. Considerable honey-dew is gathered from pecan and oak, which also furnish much pollen. The western part of the plateau, which extends across the Pecos River, is very rough, and is a stock-raising country.

The Llano Estacado, or Staked Plains, the southern termination of the Great Plains, is a level region with few streams, but in the southern portion there are numerous lakes, part of which are saline. There are great extremes of heat in summer and of cold in winter. The soils are mostly sandy loams, and, as there is an abundance of underground water, many thousand acres are irrigated from wells. This region, known as the Panhandle, is a grassland, where cattle-raising is the principal industry. It is subject to high winds, which "would blow bees off the earth." The honey flora is not sufficient to support bees, and the few efforts which have been made to keep them have failed.

The eastern portion of the Trans-Pecos, or Rocky Mountain region, consists of broad level plains, covered with sotol, or beargrass, yucca, cactus, and agave. The rainfall is from 10 to 12 inches, and there are no streams, except a few creeks leading to the Pecos River. The western portion is broken by numerous short mountain ranges, the summits of which are covered with pine and mountain cedar, but the valleys and plains are treeless. At Fort Stockton, near the center of Pecos County, are marvelous springs with a flow of 55,000,000 gallons daily, most of which is used for irrigation. Approximately 8000 acres are watered and are planted with alfalfa and fruits. There are several commercial apiaries at Stockton which range from 50 to 300 colonies, besides about a dozen apiaries with from 5 to 20 colonies. A surplus of 30 to 100 pounds is stored annually from mesquite, catsclaw, and alfalfa. In the northwest corner of Pecos County 20,000 acres are under ditch and 8000 acres in cultivation.

At Barstow, Ward County, there are several apiaries which range from 50 to 600 colonies. A full crop is obtained about once in four years, but a partial crop is obtained during the off years. A fair average is 60 pounds per colony. Alfalfa is the most important honey plant, but a good flow is often obtained from mesquite and catsclaw. More than 15,000 acres are under irrigation in this locality. There are several small apiaries. A beginning has been made in raising yellow and white sweet clover under irrigation. A surplus in spring is obtained from mesquite and catsclaw, but the most bountiful flow is from alfalfa. The bloom does not always yield nectar. Large fields within a mile of an apiary have been found destitute of bees while other fields were yielding a surplus. If the atmosphere is not too dry there is a good flow from cotton in July and August.

In the vicinity of El Paso, El Paso County, there are 15,000 acres under irrigation and cultivation. The Elephant Butte Dam, a government enterprise, provides for the irrigation of 50,000 acres. A beekeeper at South Clint writes: "There are three beekeepers in this locality who operate 1250, 250, and 100 colonies respectively. From alfalfa and mesquite we obtain from 85 to 100 pounds of honey annually. We have 11 yards within a radius of 12 miles. They contain from 100 to 150 colonies each, and are about three miles apart, fully stocking the valley with bees. Alfalfa is a sure crop, and we have not had a failure in 10 years."

OHIO

Total area, 41,040 square miles. The surface of Ohio is diversified by numerous hills, and broad valleys eroded by the streams, but it is nowhere mountainous. The northwest portion is more level than elsewhere, as here the original valleys have been filled by glacial drift. A range of hills extending from Trumbull County on the east to Darke County on the west divides the state into two unequal slopes. The smaller northern slope contains many square miles of swamps and marshlands, especially in the northwest, in which aster and other fall flowers are very abundant. Away from the swamps not much honey is gathered in the fall. Nearly 6,000,000 apple trees of bearing age are grown in all parts of the state. The orchards are most numerous along the eastern border of the state, but the greatest acreage of apples is in the southern part. A great variety of fruits, especially peaches, thrive on a belt of land south of Lake Erie. Buckwheat is grown by the thousand acres in the northeast. Of the great hardwood forest which once covered the entire state there still remain small areas of basswood, locust, tulip tree, cherry, maple, catalpa, and buckeye, and a few other nectar-producing trees which are locally of value. From the abundance of the buckeye, Ohio is known as the "Buckeye State."

West of a line extending from Sandusky through Columbus to the east line of Adams County the underlying rock is limestone, except in the northwest corner, in the counties of Williams and Fulton and the northern part of Defiance and Henry counties, where the rock is black shale. East of this line, including a little over one-half of the state, the rock is largely black shale and sandstone, from which the surface soils have been derived by weathering. In some of this territory the farmers, urged by the county farm agents, are putting lime into the soil. Where this is being done the clovers are being grown. In the southeast section a very productive soil of limestone origin is found in Belmont, Monroe, Noble, and Morgan counties. The central and western portions of the state have been heavily glaciated, and the surface soils are derived directly from the drift; but in the southwest corner, where the land is very level, the clay and muck soils are the deposits of an old glacial lake. The poorest and thinnest soils are found on a belt of land extending north and south through the center of the state, derived from the decomposition of the shales. Beekeeping is only moderately successful in this section, many counties containing only a few colonies each. The excellent soils of the western portion of the state are largely of limestone origin, and are well adapted to support a luxuriant growth of clovers. While tobacco may be grown successfully on a great variety of soils, it favors a soil containing lime, and it is for this reason that it is so largely cultivated in the western counties. The southwestern corner of the state has a surface soil of brown silty loam of moderate fertility overlying heavy beds of glacial till. Clover does well on the southern drift, but secretes less nectar than in the northwest counties because of the higher summer temperature.

The best section of the state for beekeeping is the southern slope of the valley of the Maumee River in the northwest and the territory directly south of it. Nearly every farmer in this section is growing alsike clover for seed, and there are many farmers with 10 to 20 acres, and a few with 30 to 60 acres under cultivation. Red clover is grown, and sweet clover has become well established on ditch-banks, roadsides, and in waste places. Basswood was formerly much more abundant than it is to-day; but, even where the trees are common, it is far from dependable. A beekeeper in Wood County writes that, although he has thousands of basswood trees near his apiaries, during the past four years his bees have stored only a little basswood honey. Of secondary importance are fruit bloom, dandelion, catnip, heartsease, Spanish needles, goldenrod, and aster. On the southern slope of the Maumee Valley commercial beekeeping is at its best. There are more bees, and more honey is produced, than anywhere else in the state. Conditions are especially favorable around Tiffin, Defiance, and Delphos. The Miami Valley, farther south, according to State Bee Inspector Chas. A. Reese, is rapidly developing into one of Ohio's best beekeeping areas.

Several counties in the northeastern part of the state furnish some clover and considerable buckwheat. Sweet clover has made a good start. There is likewise a good fall honey flow from a variety of sources, chiefly swamp plants. There is an area in the northeast corner in which apple orchards are very abundant, while pears, peaches, cherries, strawberries, and bush fruits are common south of Lake Erie. The mixed honey from fall honey plants is amber-colored, of good body, and fine flavor. The apiaries contain from 6 to 10 colonies. In Ashtabula County and the northern part of Trumbull County there are several commercial beekeepers, and a large crop of honey is often secured. It is one of the few areas in the state that are blessed by two distinct honey flows, says Reese. But in most of the counties in this corner of the state the number of colonies is limited. In Lorain County the only sources of nectar are white clover and alsike clover, but there are several commercial beekeepers, especially around Oberlin. It is remarkable that in Medina County, where the soil is semi-acid and the conditions for beekeeping are not the best, there should have been established the largest industry for the manufacture of beekeepers' supplies in the world. Had it not been for the great abundance of basswood near Medina in 1870, The A. I. Root Company would probably never have come into existence. There was then always a certainty of a surplus from clover or basswood. Even without basswood The A. I. Root Company produced from about 600 colonies nearly three carloads of honey in 1924.

The counties in the southeastern section of the state, as Belmont, Monroe, Noble, Morgan, and Athens, have several thousand colonies of bees. Over the limestone, fertile clay soils prevail, while over the sandstone there are poorer sandy soils in this section. A part of this area is so hilly and rocky that it is unsuitable for tillage. In the extreme south-central part of the state there are many small apiaries, but the average surplus per colony is small. The honey plants are the clovers, sumac, basswood, tulip tree, locust, redbud, and aster.

In the southwest corner around Cincinnati the limestone rocks are overlaid by a thick layer of glacial drift, which is thinly covered with a moderately fertile silty soil. White clover and alsike clover are abundant; and, while there are few commercial apiaries, there are many small yards, from which the average yields are small. According to a local apiarist there are only fair opportunities in this region.

In the production of honey Ohio ranks about with Pennsylvania, Indiana, and Illinois, but lower than the states of New York, Michigan, Wisconsin, and Minnesota. The soils of Ohio very generally lack lime; and in consequence the growth of the clovers is retarded. Sweet clover, owing to the recommendations of the county agents and extension specialists, is rapidly increasing in many portions of the state. Reese says there are approximately one hundred and fifty thousand acres in the northwestern part of the state alone, and that it is improving beekeeping conditions even in the northeastern section where lime is being put into the soil.

MICHIGAN

Total area, 57,980 square miles. Michigan is situated amid the Great Lakes, and is divided by the Mackinac Straits into two natural provinces, the Upper and Lower peninsulas. The larger portion of the Lower Peninsula lies between Lake Huron and Lake Michigan; the Upper Peninsula is bounded on the north by Lake Superior and on the south by the state of Wisconsin and lakes Michigan and Huron. The effect of the great bodies of water on the climate is very marked. During the winter the temperature of the lakes falls to the freezing-point, or below it, and in the spring the development of vegetation along the shores is retarded and thus escapes the injurious effects of cold eastern waves. In the fall the water, which has become moderately warm during the summer, keeps the lake shores free from frosts until late in the season. The temperature is much more equable than in the states west of Lake Michigan. During the summer the shores of the lakes are comparatively cool and are the resort of thousands of tourists. The mean annual temperature of the Southern Peninsula is forty-six degrees, and of the Northern Peninsula forty-two degrees. The average temperature from June to September inclusive in the Northern Peninsula is about sixty degrees, so that conditions are favorable for the growth of crops. In the northern part of the Southern Peninsula frosts do not, as a rule, occur later than May 15, nor earlier than September 30, although there may be a frost on the high elevations any month during the summer. In the Northern Peninsula frosts may also occur in the interior during any summer month, but they are rare along the lake shores.

The rainfall in the Northern Peninsula averages about 34 inches, a large part of it coming during the growing season. There is an exceedingly heavy snowfall in the western part of the Peninsula. The average annual rainfall in the Southern

Peninsula is about 30 inches. Long periods of dry weather injurious to vegetation seldom occur. Sunshine prevails during about one-half the number of hours possible, and the percentage is much higher in the summer months than in those of winter. During the greater part of the year the winds are westerly.

The climate of Michigan, of which a brief outline has been given, is very favorable to agricultural pursuits. Due to its northern latitude, it enjoys a longer period of daylight than such states as Indiana, Kentucky, and others farther south, which, in the opinion of B. F. Kindig, is of some consequence when we consider

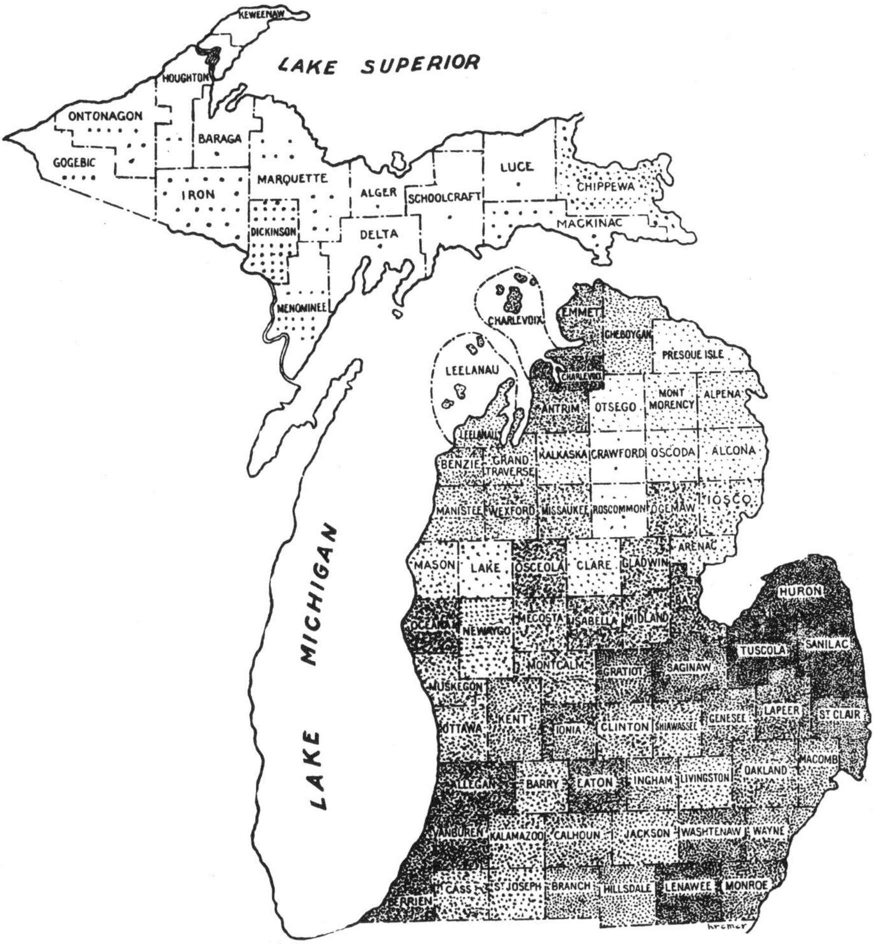

Fig. 129.—Map showing bee population of Michigan

that this extra daylight is longest during the months of June and July. The state is noted for its enormous annual production and the great variety of its crops. The "Fruit Belt" along Lake Michigan is especially well adapted, as to climate, for growing fruit, on account of cool springs which retard the opening of the flower buds, the moderate summers with much sunshine, late falls free from frosts, and mild winters with much snow. The equable climate also gives a great variety to the flora. More than 1600 species of flowering plants have been listed, of which 165 are trees and shrubs.

"The absence of extremely hot weather in summer and of extremely cold weather in winter is a matter of tremendous importance to beekeeping," says Kindig. "The moderate heat of summer is sufficient to produce a very heavy flow of nectar, but does not dry the soil excessively. This results in uniformly heavy honey flows of long duration — one of the outstanding characteristics in Michigan beekeeping. The cooler summers also cause the production of a very mild-flavored honey, as compared with honey from the same sources produced where the temperature is higher. This feature of our climate insures very high-grade honey. Outdoor wintering of bees is being successfully practiced in all parts of the state."

The soils of Michigan consist of glacial drift, and are very variable within short distances. They are very closely related to the production of honey in all parts of the state. There were two or more invasions of Michigan by the great continental ice-sheet, separated by long intervals. The oldest invasion came from the northwest across the basin of Lake Superior. A later invasion was from the northeast, and brought the material which to-day forms the upper layer of the soil. The glacial drift which was deposited by the melting of the ice varies greatly in thickness; on the border of Lake Michigan it is more than 600 feet thick, in other places it is very thin or absent, the underlying rock coming to the surface. It consists of a mixed mass of boulders, gravel, sand, and clay, forming stony loams in the moraines, gravelly loams in the river terraces, and sandy loams in the sand plains. There are also clayey soil plains, which were deposited by the waters of ancient glacial lakes. The soils of each of the two peninsulas will be described separately.

THE NORTHERN PENINSULA

The western half of the Northern Peninsula is rugged, or along the western shore mountainous, attaining an altitude of 1443 feet above Lake Superior in the Porcupine Mountains. The rocks are folded granites, gneisses, and schists, which have weathered in localities to rounded knobs and sharp ridges. The eastern half of this province is more level, and the rocks consist of strata of sandstone and limestone which remain in an almost undisturbed position. Along the southern coast bordering Lake Michigan and Lake Huron there is a tract of limestone, thinly covered with drift. The average elevation of the Northern Peninsula is probably about 250 feet above the lakes.

It is estimated that about 60 per cent. of the land can be used for farming, leaving about 40 per cent. of light soils, rocky areas, and swamps difficult to reclaim. The swamps and lakes cover about one-quarter of the Upper Peninsula. They are found largely in Schoolcraft and Luce counties, although small swamp areas occur in most counties. They do not offer desirable locations for beekeeping, according to Kindig, but in some years aster, goldenrod, boneset, willow-herb, and other honey plants are abundant, and afford a valuable late flow of nectar. Extensive rocky areas occur at the surface or at a very slight depth in Marquette, Houghton, and Keweenaw counties. These rugged areas are unsuitable for farming, but large tracts of raspberry and fireweed are of interest to the beekeeper. Heavy soils occur in Houghton and Keweenaw counties, on which much white clover, alsike clover, and some sweet clover are grown. An added attraction is that the Keweenaw Peninsula has a larger population than any other portion of the Upper Peninsula on account of the mining industries located there.

A large area is covered with glacial moraines which have a rolling or hilly surface. The soil is variable, ranging from a heavy clay to a mixture of boulders, gravel, and sand. Usually there is sufficient clayey material present to render it very productive. Among the moraines there occur sandy to clayey loam plains, a part of which make excellent farm lands. Outside of the moraines there are large areas of outwash sand plains, which were laid down by the water escaping from beneath the glacier. On a portion of these plains the honey flora is so poor that they are of little value to the beekeeper. They are often covered by a pine forest. The moraines and sand plains cover about 42 per cent. of the Northern Peninsula. Lake clay soils are found at the eastern end of the Upper Peninsula in Chippewa County, in the central portion in Delta County, and at the western end in Baraga, Houghton, Ontonagon, and Gogebic counties. These areas are generally level and the soils very fertile. They are well adapted to the growth of alsike clover and white clover, and the possibilities of extensive beekeeping in these counties are very great. They occupy about 3 per cent. of the Upper Peninsula. The limestone soil covering the tract of limestone along Lake Huron and Lake Michigan is one of the best soils in the state. It is filled with fragments of limestone and fine calcareous material produced by the grinding of the ice upon the underlying rocks. Alsike clover and white clover are very abundant in a part of this area, the soil of which is also well adapted to growing sweet clover.

While there are not a great number of honey plants in the Upper Peninsula, there are a small number which are very common and yield a great amount of nectar. The long, clear days and cool nights are very favorable to a profuse secretion of nectar. Maples, which are especially abundant in the hardwood sections of the western half of the Peninsula, are helpful in building up the colonies in early spring. Following maple comes dandelion bloom with its ample supply of pollen. Wild red raspberry abounds on the uncultivated hardwood land. It occurs in nearly all parts of the Peninsula, and almost every good location for beekeeping is within range of large areas of this shrub. It is a dependable source of nectar and is second in importance only to alsike clover. White clover and alsike clover grow with the greatest luxuriance on the lake clay soils and on the limestone soils. In Ontonagon and Chippewa counties the flow from alsike clover is described as remarkable both for its length and abundance. It has been widely scattered through the hauling of hay, and is by far the most valuable source of nectar in Upper Michigan. After a forest fire willow-herb, or fireweed, springs up and is an important honey plant for several years, until it is driven out by raspberry and other hardier plants. It blooms from July until frost and yields well at a low temperature. Basswood was formerly much more common than it is to-day; but there are a few locations in which occasionally it is the source of a heavy flow of nectar. Goldenrod is found chiefly on the heavy soils and often gives a good late flow. The main honey flow, beginning with the opening of raspberry bloom, lasts for about six weeks.

In most of the counties of the Upper Peninsula the number of colonies of bees is small; but in Chippewa County, where there is a luxuriant growth of the clovers on the lake clay soils, there are about 1200 colonies, and a surplus of honey for exportation is obtained; but in some years several hundred colonies are moved in from localities where the season is not satisfactory. Extracted honey is chiefly produced, as the cool nights are unfavorable for the production of comb honey. Most colonies of bees are packed and wintered outdoors. Snow comes early and the hives are soon deeply buried beneath it and well protected from wind and snow. In 1924 colonies were wintered in a cellar with better results. Occasionally there comes an open winter; and, when the temperature falls to 35 degrees below zero, many colonies perish. The honey is white or very light colored and of the best quality. The average surplus per colony is high. There are a great number of excellent unoccupied locations. The many small cities and towns afford good social and educational advantages, and are yet near large areas of good beekeeping territory. The main roads between the cities are paved, or hard and well graveled, making transportation from one end of the Pensinsula to the other very easy. Of the ten and a half million acres only a small percentage is devoted to agriculture, and there will long remain an almost unlimited territory for beekeeping.

THE SOUTHERN PENINSULA

The rocks of the Southern Peninsula consist of a series of limestone, shale, and sandstone beds, all of which lie in nearly horizontal strata with a gentle dip toward the center of the Peninsula. The glacial drift, which covers so deeply most of the rock surface, was brought from the highlands of Canada by a glacier which moved southwesterly across the state. The average thickness is about 300 feet, but in the interior of the northern part of the Peninsula it may exceed 1000 feet. The average altitude of the Peninsula above the lakes is about 255 feet. The highest elevation is in Osceola County, where there is an area of perhaps two square miles which is 1000 feet above Lake Michigan, or approximately 1600 feet above sea level.

The northern and central portions of the Southern Peninsula are occupied by narrow concentric belts, or oblong broader areas, of drift known as moraines, between which there are narrower belts of boulder clay formed under the ice sheet. The moraines have a rolling or hummocky surface and are composed of soils ranging from stony material to a heavy clay. Northward they consist largely of sand. As farming lands they are fair to very good. In the southwestern and western portions of the Peninsula there are extensive sand plains which are of very little value to beekeeeping. In the north-central portion comprising parts of Antrim, Otsego, Oscoda, Crawford, Kalkaska, Roscommon, and Iosco counties, there is a great sandy plain, which was formerly covered with pine.

About one-third of the Peninsula is embraced in sand and gravel areas. Lakes and swamps occupy about one-ninth of it. "The swamp lands become more and more valuable as sources of nectar," says Kindig, "as we approach the southern boundary of the state. The yield from the swamp flowers during the months of August and September often constitutes as good a crop as may be secured from the clovers and other summer flowers. In the four southern tiers of counties, Spanish needles, goldenrod, boneset, verbena, and asters are very valuable sources of fall honey."

The lake clay area is a highly productive belt of land extending around Saginaw Bay, the "Thumb," and southward along Lake Huron to the Ohio line. It is an old lake bed, in which occasional tracts of sand occur. The clovers grow in great abundance and luxuriance on the lake clay soils, which constitute the best beekeeping territory in the state. "While lake clay soils," says Kindig, "are distinctly clover soils, there are yet many other plants growing there which are of great value to the beekeeper. Goldenrod, Canada thistle, basswood, sweet clover, raspberry, dandelion, and many trees grow on the clay soils or on their borders. There are few beekeepers in the heavy clay areas that do not secure surplus honey from one or more plants in addition to the clovers." Good limestone soils occur in Alpena, Presque Isle, and in parts of Cheboygan, Emmet, and Charlevoix counties. The dunes form a narrow strip of desolate land about a mile in width along Lake Michigan.

The Lower Peninsula is divided into two sections by the channel of an old glacial river, the most interesting physical feature of this part of the state, which extends from Saginaw Bay on the east through the valley of Maple River to Grand River on the west. Saginaw Bay was once a large glacial lake extending westward to Gladwin and eastward to Cass City. Its former limits are approximately shown by the lake clay soils surrounding the Bay and covering the "Thumb." The section south of this channel has been largely cleared of forest, and nine-tenths of the land converted into prosperous farms. The wealth and population of Michigan are largely in the southern part of the state, where there are numerous growing towns and cities. Detroit alone has a population of over one million, and affords a market for the entire wholesale crop of honey of this section. Agriculture is highly developed and dairy farming is extensively practiced. More than 65 per cent. of the crop consists of cereals, hay, and forage. The remaining 35 per cent. includes vegetables, fruits, and forest products. As would be expected, the majority of the colonies of bees are located in the southern half of the Lower Peninsula. There are over 300 beekeepers in the state whose apiaries average more than 100 colonies, and more than twenty beekeepers who produce honey in carlots. It is estimated that about 80 per cent. of the honey crop is produced by 20 per cent. of the beekeepers, most of whom are operating commercial outfits.

The best beekeeping region in the Southern Peninsula includes the "Thumb" and the southeastern counties, where, on the belt of lake clay soils, alsike clover and white clover are excellent honey plants. On the Ohio line it extends westward to include Lenawee County. Dairy products are important in nearly every county. In Huron, Tuscola, and Sanilac counties at the apex of the "Thumb," extending westward to Gratiot County, is the most dense area of bees in the state. Nearly the entire crop of honey is consumed or sold locally at retail; and the larger cities, like Detroit, offer a market for a much greater amount than is offered at wholesale.

Alsike clover in Huron County, according to David Running, is more reliable than white clover, which fails in seasons of severe drought. Sweet clover has become an important honey plant in this region in recent years. The Farms Crop Department estimates that there are 75,000 acres of sweet clover in Michigan, found chiefly in the Thumb district, southeastern and northwestern Michigan. In Huron County, as the results of the efforts of David Running, there are several 40 to 60 acre tracts of solid biennial sweet clover. Probably every county in the state is growing white sweet clover to some extent, and sweet-clover seed is produced in many counties.

In 1919 there were 41,000 acres of buckwheat grown in all parts of the state, but the largest acreage is in the southeastern counties, where Monroe County contains 3000 acres and the neighboring counties about 1000 acres each. It is in Michigan a very unreliable honey plant and only rarely produces a surplus. In 1919, there were 74,000 acres of alfalfa. While it is of very little importance as a source of nectar in Michigan, beekeepers, according to Kindig, have from time to time reported the storing of considerable surplus from this plant. Cucumbers and melons cover some 11,000 acres and yield a light amber honey of good flavor. There are several hundred acres of chicory, an equal area of which is found in no other state. Other honey plants of value which are found in the southern counties are dandelion, fruit bloom, raspberry, basswood, wild cherry, goldenrod, and aster. In Sanilac County large quantities of aster honey are secured occasionally.

Southwestern Michigan is a poor region for clover; and, in fact, most of western Michigan is poor territory for beekeeping. Many of the southwestern counties, however, report from 1500 to 2000 colonies of bees. The southwestern corner is an extension of the Kankakee swamp region, and Spanish needles is abundant in this and every other swamp area. It is a major source of nectar in the southern third of the state. In low lands boneset and purple vervain also add many pounds of honey to the winter stores. In the southwestern counties of Allegan, Cass, Berrien, Kalamazoo, Van Buren, and St. Joseph the bulk of the peppermint crop in the United States is produced. According to Kindig, on many of the large marshes it is not possible to cut the whole acreage for distilling until after the plants have bloomed and a considerable surplus of a white honey with a rich spicy flavor is secured. The temperature of the western shore of the state is greatly modified by Lake Michigan, and the influence of the lake extends many miles inward. There is a wonderful fruit belt along the side of the lake. Apple, peach, plum, and cherry trees are grown almost to the exclusion of other crops. Peaches are grown as far north as Frankfort. While little honey is stored from the bloom, sufficient nectar and pollen are gathered to build up strong colonies for the later clover flow.

The soils of the northern half of the Southern Peninsula consist largely of glacial moraines, in the central portion of which there is a great sand plain, where the honey flora is poor and there are few colonies of bees. But that portion next to the lakes is good for bees, and is a great summer resort for thousands of people who are attracted here by its numberless lakes and trout streams and its delightful climate. The winters are not severe — the average annual temperature of January at Alpina is 19 degrees F.

More than half of the land in the northern part of the Southern Peninsula was formerly covered by red and white pine. As the pine was cut for lumber and the stump land burned over, willow-herb sprang up and flourished for two or three years, when it was succeeded by blackberry bushes. It seldom failed to yield well, and 15 or 20 years ago Hutchinson estimated that there were thousands of acres of this plant in northern Michigan; but as the pine has nearly all been cut, willow-herb in this section has had its day. On the pine barrens raspberries will not grow, or the bushes are small and stunted and the bloom nearly nectarless. But there are also in this section tracts of clay soils on which there were belts of magnificent hardwood forest. As this was lumbered, there sprang up a luxuriant growth of wild raspberry, which completely covered the land and never failed to yield nectar even in cold wet weather. But the raspberries soon succumb to the rapidly growing young trees, and the beekeeper is forced to seek a new location. On suitable soils alsike clover has become extremely abundant, and in many localities there are no bees to gather the nectar. Buckwheat is a very uncertain source of nectar and has yielded a surplus only once (1918) in twelve years. Milkweed is abundant in Emmet, Cheboygan, Charlevoix, Antrim, and Grand Traverse counties, where in localities an average of 50 pounds of surplus per colony is sometimes obtained. It is rapidly extending over a larger area and will in the future be one of the principal honey plants of this region. Other honey plants are willows, maples, dandelion, fruit bloom, sumac, and goldenrod.

According to R. H. Kelty the best territory for beekeeping in the northern part of the Lower Peninsula is a strip of land extending along the lake shore from Alpena to Cheboygan, southward into Charlevoix and Antrim counties. Fertile lake clay soils occur in this region. County agricultural agents are vigorously pushing the cultivation of sweet clover as a soil builder. In the sand-plain region several counties report less than one hundred colonies of bees; but the four counties of Antrim, Cheboygan, Emmet, and Charlevoix at the extreme northwest end of the Peninsula contain more than 10,000 colonies, according to B. F. Kindig. The average size of apiaries among farmers is 5 to 25 colonies; among specialists 200 to 300 colonies. An apiary of Ira D. Bartlett at East Jordan, Charlevoix County, containing 160 colonies, harvested one season 17,000 pounds of wild red raspberry and clover extracted honey, and about 200 pounds of comb honey. One colony produced 300 pounds of honey. In 1923, from 160 colonies he produced 25,000 pounds of almost pure raspberry honey. Some colonies that year were very weak; but those that were strong produced from 400 to 500 pounds, and his scale colony showed 461 pounds of white honey and 100 pounds of fall honey. The best localities are well stocked, but in the remote districts there are still excellent opportunities; but they are open to the objection of isolation, the severity of the winters, the sterility of the soil, and the danger of forest fires.

INDIANA

Total area, 36,350 square miles. The surface is generally rolling, and is well drained by numerous streams. The highest lands, which have an altitude of about

1000 feet, are found in east-central Indiana; and the lowest, which are about 500 feet above sea level, are along the Wabash River in the southwestern section. The northern three-fifths of the state was heavily glaciated by the Wisconsin glacier, and almost all of it, except the Brown County hills, by earlier glaciers. In the northwestern part of the state the dark-colored muck and sandy loams are deposits of extensive glacial lakes. In the central part the soils of the best beekeeping regions have been derived from the glacial till, which is in some places hundreds of feet in depth. The underlying rock of eastern and southern Indiana is limestone. The fertile soils of the southern unglaciated two-fifths of the state have been formed partly from the weathering of the underlying rocks and partly from deposits of loess. About 1400 species of flowering plants have been recorded, of which 110 species are trees.

The Kankakee Valley in the northwest is known as the Spanish-needles region, from the great abundance of this plant in the extensive swamps along the Kankakee River, and is an excellent location for migratory beekeeping and for obtaining a late honey flow. Through the center of this section, which includes Lake, Porter, Laporte, Stark, and the northern portion of Jasper and Newton counties, runs the Kankakee River, bordered by broad swamps which are annually overflowed. In Stark County the river broadens into a long narrow lake traversing the county diagonally. The honey flora of the swamps consists chiefly of Spanish-needles (*Bidens aristosa*), buttonbush (*Cephalanthus occidentalis*), and thoroughwort. On the higher land heartsease, aster, goldenrod, and buckwheat, all late-blooming plants, contribute largely to the fall honey flow. Commercial apiaries range from 50 to 100 colonies, but comparatively few contain more than 100 colonies. Farmers keep from one to six colonies. An average surplus of 50 pounds per colony is frequently obtained, and occasionally it exceeds 100 pounds. White clover is usually unreliable, but may yield once in three or four years. Large apiaries are located near Valparaiso, Hobart, and Hebron.

East of the Kankakee River swamps the three northern tiers of counties form a belt of land 60 miles in width, where the fertile limestone soils have been buried under a thick later glacial deposit of sand and gravel. This porous soil is naturally acid, and white clover is much less reliable than in the counties farther south. There are, however, several good white-clover locations near the north border, as at Middlebury and Lagrange, where there are a few commercial apiaries. A good crop is obtained once in three or four years, the surplus averaging about 60 pounds per colony. In Steuben and DeKalb Counties in the northeast corner of the state commercial beekeeping receives little attention. White clover is the principal honey plant of this region; but buckwheat, basswood, sweet clover, and fruit bloom are minor sources of importance.

The white-clover region of Indiana, which occupies the eastern and central portions of the state, is an irregular triangular area with no sharply defined limits. On the north it is bounded by a line parallel to the Wabash River, extending from Allen County southwesterly to Tippecanoe County. Its extension westward is marked by the counties of Tippecanoe, Montgomery, and Putnam; and on the south it is bounded by a line running from Putnam County southeasterly to Dearborn County on the Ohio River. In the northern part of this region white clover is more reliable than in the southern, as the rainfall is greater, the summer temperature less, and the soil is highly calcareous. "The Miami soils area of eastern Indiana and western Ohio (extending into Michigan)," writes E. F. Phillips, "is as good as any for clovers, and is probably unequaled by any other soil area in the United States. * * * * Unfortunately there are no dependable late honey flows; but an experienced beekeeper can secure a crop almost every year from the clovers." The acreage of alsike clover is steadily increasing, and sweet clover is also becoming common. Basswood is important in localities; and fruit-bloom, dandelion, and heartsease deserve mention. Owing to liberal annual appropriations the state is to-day largely free from foul brood. The white-clover region includes some 38 counties; but the counties of Indiana are small. A beekeeper living at New Ross, Montgomery County, who has 52 colonies, writes that the crop of honey is secured from white clover, alsike clover, heartsease, basswood, and wild aster. Minor honey plants are dandelion, raspberry, fruit bloom, haw, sweet clover, milkweed, black locust, maples, melons, and tulip tree. The clovers yield nectar four years in five; but basswood is unreliable, and heartsease depends on the rainfall in late June and July. The clovers usually bloom from four to six weeks; but in 1915 young clover began blooming in August and continued to bloom until killing frosts. There were no old clover plants that year. Bees were swarming in September.

Returning to Henry County, in the east central portion of this region, we find that three-fifths of the land is covered with cereals and forage, and that fruit trees also occupy a large area. In the central portion of the state 95 per cent. of the land is in farms. Beekeeping receives a fair share of attention. The honey plants are red maple, dandelion, fruit bloom, and black locust (March 1 to June 1); white clover (June 1 to July 20); sweet clover (July 10 to August 30); heartsease and asters (September 1 to October 30). Only small yards can be maintained without overstocking. Around Indianapolis, Marion County, there is a large number of small yards. The average surplus in a good year is 75 to 100 pounds, and at least a partial crop is obtained every year from white clover, alsike clover, and sweet clover. During the past 10 years the amount of honey produced in this county has doubled.

South of Indianapolis the counties in the white-clover region generally report fewer colonies and a smaller amount of honey. A beekeeper at Dillboro, Dearborn County, who keeps from 20 to 50 colonies, in a good year secures from sweet clover, honey locust, white clover, and aster 100 or more pounds of surplus, but in a poor year only about 25 pounds. The summers in southern Indiana are so hot and dry that white clover secretes nectar much less freely than northward. Moreover, white clover is confined largely to land not under tillage, and this area is very small in many southern counties. For example, the area of Fayette County is 138,000 acres, of which, in 1920, 133,000 were in farms; and of 103,000 acres comprising Union County, 100,000 acres were in farms.

Southern Indiana was not glaciated, and the soils are derived partly by the weathering of the underlying rocks, and partly from a thin deposit of fine material known as loess, which largely covers this region. These silty soils are better adapted to growing winter wheat, oats, and grass than to corn, which is by far the most important crop in the central states. Beekeepers usually divide southern Indiana into three parts — a southeast, a south-central, and a southwest section.

In the southeastern section the honey plants are redbud, fruit bloom, black locust, white locust, white clover, sweet clover, goldenrod, and aster. A beekeeper at Madison, Jefferson County, writes: "We have depended largely in the past on sweet clover; but during the last few years it has been disappointing, largely beceause it is pastured too closely." Black locust is so common in this section that it is an important source of surplus. The more important honey plants of south-central Indiana are white clover, alsike clover, sweet clover, goldenrod, and aster; but there are many wild nectar-bearing flowers, as willow, maple, locust, dandelion, raspberry, sumac, and Spanish needles. Along the creek bottoms much nectar goes to waste, and bee culture would be profitable on a small scale.

From Brownstown, Jackson County, a beekeeper writes that his apiary of 45 colonies is the largest in the county. White clover yields a crop only twice in five years, in seasons when the growth has been stimulated the previous summer by much rain. In dry years it yields little nectar. The fall honey flow from goldenrod and aster is uncertain; but usually sufficient honey is stored for winter use. One year, however, 1000 pounds was secured from fall flowers by 23 colonies. In this locality the crop is not large and is unreliable. Good results may be obtained on the banks of the Ohio River, which forms the southern boundary of the state, when, after a flood, it subsides in time.

Thousands of acres of climbing milkweed or bluevine (*Gonolobus laevis*) have, during the past four years, made the counties of southwest Indiana very widely known as a honey-plant region. This vine grows luxuriantly on the bottom-lands of the Ohio River as far east as Spencer County; on the Wabash River as far north as Knox County, and on the White River as far north as Daviess County. The copious secretion of nectar is dependent on hot dry weather, whence this species is called "dry-weather vine." It thrives in the lowland cornfields, matting the ground between the rows, and twining about every cornstalk, until it has become an apothegm, "Shake one cornstalk and you shake an acre." It blooms from mid-July until frost, and yields a fine-flavored, pale pink-colored honey which does not granulate. A surplus of 60 pounds in three weeks and of 80 pounds in two weeks per colony has been obtained. Over 100 pounds per colony may be secured during the entire season. At Vincennes, Knox County, smartweed is the best honey plant, but it fails in dry weather, and the beekeeper depends on climbing milkweed for his winter stores. Sweet clover is also of value in this county. It must not be forgotten that bluevine fails in wet weather, and that the seasons are uncertain.

Along the western border there is a narrow belt of land extending from Knox County to Warren County, where in wet seasons heartsease is the predominant honey plant. A surplus of 75 to 100 pounds of honey per colony is secured about every two years. The honey plants are white clover, alsike clover, sweet clover, heartsease (or smartweed), and asters. Farther northward in Fountain, Warren, and Benton counties the honey flora is poorer, the land is highly cultivated, and a much smaller amount of honey is obtained.

In the case of most states, as in Virginia and North Carolina, the principal physical regions are so clearly defined that no question can arise as to their number and extent; but in Indiana the physical features are so uniform that the honey-plant regions of the state must be based largely on the composition of the soils. Indiana may be divided into two natural sections — a northern glaciated region and a southern unglaciated region. The soils are of very different origin, and a more thorough study of their character and relation to the honey flora is desirable; but the data for such an investigation are not yet available. The honey flora is largely determined by the character of the farm crops. The annual acreage of the cereals is over 9,000,000 acres, none of which (with the exception of 6000 acres of buckwheat) are of value to the beekeeper for nectar. White clover is not likely to become more abundant; but red-clover seed is grown on 16,000 farms, and other clover seed on 1500 farms. The improvement of the honey flora will depend largely on the more etxensive cultivation of alsike clover and sweet clover. There are 7000 acres of mint under cultivation for mint oil, an equal number of acres of small fruits, and five million fruit trees of bearing age

ILLINOIS

Total area, 56,665 square miles. With the exception of its southern extremity, which was originally densely forested, Illinois lies wholly in the great Prairie

Plains of the upper half of the Mississippi Valley. A layer of glacial drift from 10 to 200 feet in thickness extends over all of the state except the northwest corner and the extreme southern counties. The western portion of the glaciated upland along the Mississippi River is deeply covered by a brown silty loam known as loess. The surface of Illinois is a vast grassy plain sloping gently to the southwest, and traversed by the valleys of 275 streams. It is one of the most level states in the Union, with a descent rarely exceeding one foot to the mile, so that swamps are numerous; and it is probable that at no very remote geological period a large part of the land was covered by a shallow lake. The soil is a fine compact back loam, 10 to 15 inches deep, entirely free from stones and gravel, of inexhaustible fertility, and largely underlaid by yellow clay. The absence of trees from the prairies is partly explained by the intense prairie fires which annually burned the dry grasses, and partly by the compactness of the soil which excludes from the roots of trees the oxygen required for oxidation. Referring to the changing color of the soil, J. R. Wooldridge, President of the Illinois State Beekeepers' Association, writes:

"Southward from Effingham the soil loses its black color, becoming a reddish clay, much spotted with white patches, and for about 60 miles is less productive; but from there on to Cairo it produces great crops of peaches and apples. Though the soil on the uplands of this region is red clay, it has proven to be a world-beater when it comes to producing fruits and vegetables. The soil of the valleys is much darker in color, and most of it is very productive also."

The richest and most varied forest in the north temperate zone is found in the lower Wabash Valley in southeastern Illinois, where 107 species of trees occur, and 75 species have been counted in an area of less than a square mile. The flora contains a great variety of flowering plants, but most of the species are of only incidental value to the beekeeper. The larger part of the population is engaged in agriculture.

While the different sections of Illinois vary in altitude, climate, and soil, the whole state is admirably suited to agricultural purposes. In the northern three-fifths of the state, especially in the central portions, there is an immense acreage of corn and oats. This section is often referred to as the "corn belt." A great area in the southern two-fifths of the state is devoted to the growing of orchard fruits, whence it is called the "fruit belt." Apples contribute three-fifths of the crop, while peaches rank next in importance. There is no sharp dividing line between these two belts, as cereals are largely grown in the southern part, and fruit trees in the northern part of the state. But the acreage of corn, oats, and wheat in the central portion of the state is so large that it reduces the extent of the bee pasturage, and is unfavorable to beekeeping.

Very large yields of honey have been obtained in the two northern tiers of counties. The surplus in this section is stored from white clover, sweet clover, and heartsease. Fifty years ago heartsease was hardly known; but now it is important. A little basswood honey is sometimes obtained, but not often. When the temperature and rainfall are both favorable white clover may yield a phenomenal surplus, as at Marengo, McHenry County, in 1913, when 72 colonies averaged 266 sections of comb honey. At Rockford, Winnebago County, there have been four good honey flows from white clover in the past eight years, the largest flow coming after a very wet fall. There are many small apiaries which perhaps average 15 colonies. Around Chicago there is a good dairy region, and as the result of the rapid increase of sweet clover in recent years the production of honey has increased enormously, sweet clover being the main source.

This sweet-clover area extends southward into Will and Grundy counties, particularly around the towns of Joliet and Morris. These two counties probably lead in the production of sweet clover honey in Illinois.

Along the eastern border conditions vary greatly, and the soil is not suitable for obtaining the best flow of nectar from the clovers. In Kankakee County, along the Kankakee River there are great Spanish-needles swamps, as in northwest Indiana. Iroquois County contains many colonies of bees, and produces a large quantity of honey. As more than 400,000 acres of corn and oats are grown in Champaign County, much of it is poor bee territory. In Vermilion County similar conditions prevail. A beekeeper at Newman, Douglas County, reports that it is a poor locality for beekeeping. A good crop is obtained one season in three. There seem to be few who specialize in beekeeping in Coles County, but there are many small yards. A fair average is about one super to a hive. White clover is the main dependence, but it is a partial failure whenever the weather is too wet or too dry. Field aster is very common in the fall, but the bees get little nectar from it.

From the eastern border let us turn our attention to the Mississippi Valley, which forms the western boundary of Illinois. Here two crops may be gathered in a favorable season. At Hamilton, Hancock County, large crops are sometimes secured both in spring and fall. An immense acreage of fall flowers, Spanish needles, boneset, heartsease, and asters, is found on the bottom-lands along the Mississippi River.

"In the valley of the Mississippi River in Illinois," writes Frank C. Pellet, "conditions vary widely. In places where there is no lowland on the Illinois side, clover is the principal dependence; and when that fails there is no honey to sell. Where there is a wide valley there is an abundance of Spanish-needles, heartsease, and boneset, which insure a dependable fall honey flow. The same conditions prevail largely on the Missouri side of the great river."

Pike and Calhoun counties are bordered on one side by the Mississippi, and on the other by the Illinois River. A local beekeeper asserts that there is no better location in the state. The early-blooming honey plants are willows, maples, dandelion, and the great apple orchards. The summer plants are white clover, alsike clover, sweet clover, and basswood; and the fall flowers are heartsease, Spanish needles, and asters. The clovers bloom from May 15 to July 30, and, if the weather is right, yield heavily, but not, however, as well as in the northeastern section. In a dry fall an average surplus of 100 pounds has been obtained from heartsease. The honey from asters granulates very quickly. It is clear and strong, and is secured in large quantities.

In the central portion of the state the best locations are found along the Illinois River and the small streams. Most of the apiaries at Decatur, Macon County, are of small size, but there is one of 50 colonies and another of 100 colonies. A profitable crop is obtained three years in five, and once in 15 years there is an extra-good crop. While not more than 150 pounds per colony of bulk comb honey is usually obtained, 300 pounds or even 400 has been secured. In addition to the usual honey plants, sweet clover is being introduced to improve the land, and is making great changes. At Pekin, in the Illinois River Valley, the apiaries range from 50 to 250 colonies, and, except in dry seasons, a surplus of 50 to 75 pounds is obtained. "The Illinois River Valley," writes A. L. Kildow, State Bee Inspector, "I consider the best region for bees in the state. The border of the Illinois River from its source to its mouth consists of bluffs and bottom-land. The bluffs are used for pasture, which is plentiful in favorable years, and the bottom-land furnishes an abundance of spring and fall flowers." In the vicinity of Springfield, Sangamon County, the larger apiaries range from 25 to 120 colonies, and there are about 35 containing from one to ten colonies, which are run on the "let alone" plan.

In the southern portion of the state, or "fruit belt," an immense acreage is devoted to fruit-growing. A dense center for the production of apples is in Marion, Clay, Richland, Wayne, and Jefferson Counties, which contain over 2,000,000 apple trees. Another center is along the west boundary in Adams, Calhoun, and Pike counties. The value of the land per acre is much less in the southern part of the state than in the north-central part. White clover and alsike clover are said to thrive better and to yield more nectar in the northern part of the state than in the southern, as northward there is more snow, which protects the clover during the winter months, while in the south it often winter-kills. Clover is also injured by dry weather.

WISCONSIN

Total area, 56,066 square miles. The surface of Wisconsin is a broad rolling plain rising between the basins of Lake Superior and Lake Michigan and the Mississippi Valley. Its length from north to south is 300 miles and its breadth 250 miles. There is a range in elevation of over 1350 feet from the valleys along the Mississippi River and near Lake Michigan, which are about 585 feet above sea level, to the summit of Rib Hill (1940 feet) in the north-central part of the state. The divide between Lake Superior and the Mississippi River extends largely as a broad flat plain across the northern counties, about 1600 feet above sea level, from which there is a rapid descent of 1000 feet to the Lake on the north.

The Great Wisconsin glacier extended over the northern and eastern portions, but about one-fifth of the state in the southwestern section, or 10,000 square miles, remained free from the invasion of ice. The fertile soils of this area have been chiefly formed by the weathering of the underlying limestones. This section is rough, with steep hills and deep valleys, and contrasts noticeably with the rolling lake-dotted prairies and woods of the remainder of the state.

The glaciated area is covered by a coarse drift which varies in thickness from a few feet to 200 feet in depth. In northern and eastern Wisconsin there are more than 2000 small lakes formed by erosion and the heaping up of morainic material. There are also large areas of swamp land, where asters and other marsh flowers are abundant. As in Michigan, the southern half of the state was formerly covered with a hardwood forest of oak, maple, elm, basswood, birch, poplar, and hickory, while northward coniferous trees were most abundant.

As there are no pronounced physical features dividing Wisconsin into well-defined sections, the honey-plant regions, as in Illinois, are to a large extent artificial. The underlying rocks in their relation to the origin and composition of the soils and the distribution of the vegetation, especially of the clovers, probably offer as good a basis for such a division as any. Accepting the geological structure of the state as our guide, four fairly distinct honey-plant regions may be recognized: 1. A southern and eastern region, where the underlying rocks are Silurian limestone. 2. A western or St. Croix Valley region, where the underlying rock is magnesian limestone. 3. A central crescent-shaped region where the rock is Potsdam sandstone; 4. A great northern region where the very ancient rocks are granites, schists, gneisses, and igneous mineral-bearing beds which were once molten matter.

The southern and eastern region might well be called the alsike and white clover region, as these plants are here the chief sources of surplus honey. It is bounded on the north by a line running easterly from La Crosse on the Mississippi River, through the southern part of La Crosse and Monroe counties to the east boundary line of the last-named county, thence southeasterly to Merrimack on the west boundary line of Columbia County; thence northeasterly to Michigan, at a point about 10 miles north of Green Bay. The underlying rocks of this region are chiefly Silurian limestones belonging to the Niagara and Trenton formations; but in the south central portion there are large beds of sandstone. The soils are fertile silty loams in the southern portion of this region, and excellent heavy clay loams south of Green Bay, all of which are suitable for the growth of the cereals and clovers. In this section are located the greater part of the population, the larger cities and towns, the most valuable lands, and many of the most important industries. From 90 to 100 per cent. of the land is in farms, and three-fourths of the total value of crops in 1919 were contributed by the cereals and hay. Dairy farming is very generally pursued, and thousands of acres of the clovers are under cultivation, but most alsike clover is cut for hay so early that it is only moderately important as a honey plant. The surplus comes to a great extent from alsike and white clover, but sweet clover is becoming more important each year. Contrary to general opinion, red clover sometimes contributes an appreciable amount of surplus when the season is quite dry and the soil thin. Under these conditions the blossoms are smaller and several observing beekeepers have reported finding bees working extensively on red clover, while other clovers were being neglected. During the exceptionally dry year of 1922 in the southeastern part of the state several beekeepers reported that honeybees devoted their whole attention to the alfalfa fields.

The principal honey plants in the southern region in the order in which they bloom are maples, dandelion, fruit bloom, white clover, alsike clover, basswood, sweet clover, buckwheat, goldenrod, and aster. Dandelion is valuable for a stimulative flow, and a strong colony will sometimes store a super of honey from dandelion and fruit bloom. Sweet clover has not yet become a common field crop in any part of the state, but each year finds a few more advocates of it. It has, however, spread itself along the roadsides and waste places in the southern part of the state until the beekeepers have come to depend upon it to furnish a small surplus and keep the bees busy from the main flow until frost. This prevents a large consumption of stores and tends to prevent robbing.

Basswood has greatly decreased, but a second growth has sprung up to take the place of the larger trees to some extent. While these trees blossom fairly regularly, they do not yield the enormous amount of nectar that the original trees did, owing to the lack of the virgin forest conditions, of wind protection, and moisture-retaining leaf mulch. However, the beekeepers in almost every county in the state continue to consider it a possible source of a supplement to the clover crop; and occasionally it surprises even the most hopeful, as in 1923, when it yielded practically the only honey obtained in the southwestern part of the state and enough in all parts of the state to flavor the honey appreciably. Bottlers even called for some pure clover honey to tone down the strong basswood flavor. On the average it can be depended on about two years out of five to yield a fair crop.

White clover and alsike clover are both abundant in the pastures and along roadsides and fence rows. In the eastern part of the state there is a large section in Dodge, Washington, Manitowoc, and Calumet counties where two of the leading white clover seed producing sections of the country are found. To a somewhat less extent the same sections produce alsike clover seed. Both begin blooming early in June, and the main flow continues in the southern part of the state from two to four weeks, and often continues to yield an appreciable amount several weeks longer. The growing season of a narrow belt of land in the eastern part of the state is lengthened ten to twenty days by the influence of the waters of Lake Michigan. More than a million apple and cherry trees are under cultivation in this section. While a large majority of the beekeepers are located in the southern part of the state, they are frequently handicapped by the absence of a surplus-producing fall honey flow. However, that is made up to a certain extent by a better retail market than is found in the northern part of the state.

The central honey-plant region is a broad crescent-shaped band, sweeping round from the Michigan line through the center of Wisconsin to the northern part of Minnesota. The underlying rock is Potsdam sandstone. As the soils are sandy and relatively low in fertility, this is the poorest section of the state for beekeeping. The line bounding this region on the south is the same as the northern and western boundary line of the southern limestone region. Its north boundary line runs from the middle of Marinette County southwesterly to the middle of Portage and Wood counties, thence in a very irregular manner northwesterly to the southwest corner of Douglas County. The St. Croix Valley region in the west, where the underlying rocks are chiefly limestone, must be distinguished from this sandstone region. The soil of the central section is often thin, sterile, and, on account of its acidity, unsuitable for a good growth of the clovers. In Marquette and Adams counties the soils are mostly poor sands, and there are in both counties only a few bees. Along the bottom-lands of the Wisconsin River, extending clear to Prairie du Chien, where the soil is likewise sandy, considerable buckwheat is grown, yielding some honey. Many of the adjoining counties northeastward and northwestward have also sandy soils of low fertility. But in spots there are areas of white clover, as at Reedsburg in northern Sauk, Mauston in southern Juneau, and near Sparta in western Monroe. Of the 34,000 acres of buckwheat under cultivation in 1919 more than two-thirds are found in this region. Probably on account of climatic conditions buckwheat does not secrete nectar in Wisconsin as consistently as it does in the eastern states; but short crops of buckwheat honey are fairly common in the best buckwheat sections of the state.

In the counties in the northeastern arm of the central sandstone region beekeeping is only fairly successful. The counties of Marinette and Portage have comparatively few bees. In the northwestern arm, partly because of the great acreage of buckwheat and partly because of better soils, excellent results are obtained. Chippewa County, which has a large acreage of buckwheat and fertile silty soils in its southern portion, in 1919 produced a considerable amount of honey, as did Barron County also.

In the western part of the state, enclosed on the Wisconsin side by the sand-soil section, there is an excellent region for beekeeping, comprising the counties of St. Croix, Pierce, Buffalo, and portions of Dunn, Pepin, and Trempeleau counties. This is known as the Pierce County or St. Croix Valley region. The underlying rock is chiefly magnesian limestone, covered by fertile clay soils on which the clovers thrive. Along the Mississippi River there is a deposit of wind-blown loess from which have been formed brown silty soils; while easterly the productive soils extend over a portion of the adjoining counties. The main reliance for surplus honey is white clover and basswood, the more northern latitude being very favorable for the secretion of nectar by clover. In Buffalo County there are many small beekeepers near Mondovi and Alma, but the largest number of yards is in Pierce County.

The northern region is largely undeveloped, and in the ten most northern counties less than 15 per cent. of the land is in farms. There is a large area of swamps and lakes, and much land is covered by coniferous trees. The soils derived from the glacial drift are generally fertile, though there are several large tracts of sandy soils. The north-central portion of the state consists of a great mass of warped and twisted granite rocks, which belong to one of the oldest formations. In the northwest there is an immense area of igneous copper-bearing beds which have overflowed the older rocks. The southern boundary line of this region, where it is overlapped by the sandstone of the central region, is an irregular line running from a central point on the east line of Marinette County southwesterly to the center of Portage and Wood counties, thence northwesterly with many large curves and bends to the southwest corner of Douglas County.

In Iron, Vilas, Sawyer, Price, Oneida, Forest, Florence, Bayfield, Marathon, Langlade, and Rusk counties there are few bees. Wilson believes this may become good bee country when the territory has been developed. The principal honey plants of this region are white clover, alsike clover, raspberry, basswood, willow-herb, buckwheat, goldenrod, and aster. The climate of northern Wisconsin, with its warm days and cool nights, stimulates the clovers to a very rapid secretion of nectar. The bloom also often yields for a longer period than in the southern part of the state. Raspberry, though less abundant than formerly, may in normal seasons yield a profitable surplus, but it is practically always mixed with clover honey. Willow-herb abounds for a few years on the areas which have been cleared from forests, but is soon succeeded by raspberry bushes. In the fall an abundance of goldenrod, aster, and other late-blooming flowers may yield a surplus. However, years of short crops are known here as well as in other sections.

According to the Wisconsin Crop Reporting Service, in 1919 about 27 per cent. of the colonies of bees in the state were engaged in the production of comb honey, and 73 per cent. in the production of extracted honey. The average per colony of both kinds of honey was 54 pounds; of comb honey 34 pounds per colony; and of extracted honey 61 pounds per colony.

MINNESOTA

Total area, 84,682 square miles. The southern portion of the state is an undulating plain which was formerly heavily forested with a variety of hardwood trees; the western section is a level, treeless, fertile prairie; while northwest of Lake Superior there is a rugged, hilly area of granite rocks, containing rich deposits of iron ore. There are not far from 10,000 large and small lakes, which have a combined water surface of 3824 square miles. In north-central Minnesota from Lake Leech to the Red Lakes there is a flat, poorly drained area abounding in swamps, bogs, small streams, and lakes, with a sandy soil partially covered with pine and juniper, which is the poorest location in the state for beekeeping. All of its surface, except the extreme southeastern counties, is covered with a deep layer of glacial drift, from which have been derived by weathering its very fertile soils. For the production of honey it may be divided into three well-defined regions: A northeastern, poorly developed region; a southeastern or white-clover region, also known as the "Big Woods," and a western or prairie region.

The northeastern coniferous region, or the "cut-over lands," extends southward as far as Chisago County, and westward beyond the Red Lakes. Formerly a great belt of white pine reached from Lake Superior to the Red River Valley, north of which along the border there was a forest of dwarf pine and stunted juniper. Much of the pine has been ruthlessly lumbered and the stump land burned over. Then hundreds of acres of fireweed sprang up, which later was largely crowded out by raspberry, cherry, and other bush growths. The soils around Lake Superior are clay loams; but in the northeastern and north-central portions of the state are sandy and gravelly out-wash soils, which are suitable for growing buckwheat, potatoes, and the clovers. In the great counties of Cook, Lake, and Koochiching, which comprise more than four million acres of land, there are only a few farmers. Agricultural methods are primitive and bee culture is entirely neglected. In most of the counties of this region less than five per cent. of the land is in farms. Except in the vicinity of the flourishing city of Duluth and along the Iron Range from Hibbing to Ely, it is a thinly settled wilderness with few towns and roads (the main roads being cement), and the traveler may journey a hundred miles without meeting a human being.

Northern Minnesota is a most promising region for beekeeping, as there is no disease, little competition, and alsike clover grows luxuriantly in the cleared lands and secretes nectar most freely. Other surplus honey plants are fireweed, raspberry, white clover, goldenrod, and aster. A few years ago the purple-red fireweed was very abundant, but more recently it has been largely supplanted by shrubby vegetation. The season opens in June and closes in August. The nights are much cooler than the days, and the honey flow is so rapid that a colony may store 200 pounds of honey. The extreme north is uninhabited, and excessively cold in winter, but almost tropical in summer. Good bee pasture has been reported along the Rainy River, which connects the Lake of the Woods with Rainy Lake.

Although there are many excellent locations in this region, there are at present very few beekeepers. The counties are of immense size, and northward are not inhabited. Duluth on Lake Superior offers a ready market for all the honey produced in this portion of the state. In the great county of Beltrami, which includes the Red Lakes and much swampy and sandy soils, there are only a few small towns in its southern portion. The roads connecting the towns are excellent, being either cement or gravel.

The white-clover region occupies the southeastern portion of the state, beginning about 75 miles south of Lake Superior. A generation ago it was covered with a solid hardwood forest of sugar maple, basswood, oak, and elm, which has been largely lumbered and the land converted into productive farms. The underlying rocks are limestones and the dark silty soils are rich in lime. In this, the most populous and productive section of the state, are located the great cities of St. Paul and Minneapolis. The area of the counties in this region is very much smaller than in the northern part of the state, and does not exceed half a million acres each.

Most of the surplus honey comes from white and sweet clover, and from basswood, a complete failure of which in Fillmore County is said not to have been known in twenty years. Expert beekeepers may average not far from 100 pounds per colony, and 20 pounds have been stored by a single colony during a favorable day. White clover is much more reliable in this region than in Illinois or Iowa, and during five years one maximum crop, one failure and three fairly good yields may be expected. In 1918 white clover along the Mississippi River was a complete failure, while buckwheat yielded better than for years. A good basswood flow in some years is obtained in this region, but it is often unreliable. In the spring, brood-rearing is stimulated by willows, maples, and especially by fruit trees and dandelion, which gave a large flow in 1919; in the fall, the bees get their winter stores from goldenrod and aster. The honey flow from clover begins about the middle of June and ends the last of July. As in other localities, the rapidity of the flow is greatly influenced by the weather. Clover honey is very white, or sometimes tinged with yellow if mixed with basswood, and of the finest quality.

The southeastern counties greatly surpass all other parts of the state in bee culture, and are almost overfilled with beekeepers. White and sweet clover are very reliable, and nearly 100 pounds per colony have been obtained. In addition to the honey plants already mentioned, alsike clover, alfalfa, buckwheat, and heartsease are valuable. Nearly all of the counties bordering the eastern side of the Mississippi

River, with their fertile limestone soils, give excellent results. Farther northward beekeeping is pursued successfully on both sides of the Mississippi River, and many commercial apiaries are reported in Sherburne, Benton, and Morrison counties. On the western side of the river the average surplus is smaller and the yards contain fewer colonies. Good locations may also be found along the western banks of the scenic St. Croix River, which forms a part of the eastern boundary of the state.

The prairie region occupies the western portion of the state, extending in the south to half its width, but rapidly narrowing northward. It is an open, level, or gently rolling country on which the eastern hardwood forest has failed to encroach; but the streams have worn many small valleys crossed by belts of hardwood, as basswood, oak, box-elder, and cottonwood. From the high winds which sweep over this unbroken surface apiaries require protection by windbreaks of evergreens or board fences. There is very little land that is not arable, or that does not sell at a high price per acre. The soil over a large part of this area is a dark loam of great fertility and durability, varying into a limestone soil on the western slope of the Mississippi River, and to a clay loam in the valley of Red River, where the land, which is the bed of an ancient glacial lake known as Lake Agassiz, is very level.

The honey plants are buckwheat, aster, clover, alsike clover, and sweet clover, but the surplus comes chiefly from sweet clover which is extensively cultivated for seed and forage. In the southwest corner of Minnesota there is a much smaller number of beekeepers, and a much smaller amount of honey is produced than in the southeast corner. Northward conditions do not improve; Pope, Stevens, and Traverse counties have only a few bees. At Villard 25 colonies have averaged 50 pounds of honey from buckwheat alone. For several years in succession it may be too dry for the clovers. In the Red River Valley grain-farming is the chief agricultural pursuit and beekeeping receives little attention. But the acreage of sweet clover is increasing and the outlook for beekeeping improving.

IOWA

Total area, 55,475 square miles. Iowa lies in the prairie region, and is a vast rolling plain, 300 miles long by 200 miles wide, bordered on the east by the Mississippi and on the west by the Missouri River. With the exception of a small portion in the northeastern corner, the entire state was glaciated and is covered with a layer of glacial drift 15 to 200 feet deep, which is overlaid along the Missouri and Mississippi rivers by deep fertile loess soils. The larger rivers flow eastward, and have worn in their lower courses narrow valleys 200 to 300 feet in depth. The north-central part of the state is very level, but southern Iowa is hilly. The wooded area does not exceed 7000 square miles, and is confined chiefly to the river bottoms, where the common trees are willows, cottonwood, honey locust, ash, and elm. The land is well drained, and there are few swamps. The soils, which are dark clay loams and silt loams, are unsurpassed in fertility, and an immense area of cereals is planted annually. The severity of the winters renders it difficult to winter bees outdoors.

With scarcely an exception white clover is reported to be the principal source of honey throughout the state. In many localities no other honey plant yields a surplus. It may bloom from the first of June to the beginning of August, and yield an average of 100 pounds per colony; but it is not reliable more than two years in three in the eastern portion. It may winter-kill, or the season may be too dry, or too wet and cool. Occasionally there will be a year when clover blooms freely but yields very little nectar. "White clover," writes F. B. Paddock, "is certainly developing into a very uncertain plant for nectar secretion." At Buck Grove, Crawford County, in the west, during 12 years it has been killed out only once, and another season there was an extreme drought when it yielded for only 15 days. At Colo, in Story County, in the center of the state, during 23 years there have been four years (1903, 1911, 1917, and 1918) when white clover was nearly a total failure. During the other years the crop has been from fair to good. It is, therefore, desirable to select a location where, in addition to white clover, there is abundance of sweet clover and autumn-blooming honey plants, or not to depend entirely on beekeeping as a vocation.

The acreage of sweet clover has steadily increased westward until it is now found to some extent in every county in the state. There are two areas in which this valuable forage plant is abundant. One is in the extreme east, especially in Jackson County; the other is in the western part of the state along the Missouri River from Sioux City for more than 60 miles southward.

Along the streams there are three species of willow—the almond-leaved willow (*Salix amygdaloides*), the black willow (*S. nigra*), and the heart-shaped willow (*S. cordata*). There are two species of maples—the silver or soft maple, and the later-blooming black or hard maple; but the latter is not common westward. The dandelion is spreading rapidly, and is beginning to yield a surplus. It occurs in every county in the state. Heartsease (*Polygonum Persicaria*) and Pennsylvania smartweed (*P. pennsylvanicum*) are common weeds and often yield well in the fall. Other autumn honey plants are goldenrod and Spanish needles. Wild mustard (*Brassica arvensis*) is very abundant, especially in the northwest section of the state. Twice during the past thirty years the low pasture lands of Linn County have been a purple sea of vervain bloom (*Verbena stricta* and *V. hastata*), and two supers of honey have been filled from this source. Less important plants are milkweed, motherwort, buckbush, boneset, yellow sweet clover, mint, catnip, sunflower, along the Missouri River, artichoke in the northern part of the state, partridge-pea, rosinweed, Canada thistle, and aster. Of the early-blooming plants, false indigo (*Amorpha fruticosa*) is important in a few localities along the streams. Black locust is often planted for ornament. Buckwheat is reported to yield very little nectar. All cultivated fruits are common in the southern half of the state, where the climate is milder, but cherries and bush fruits thrive in nearly all sections. In recent years many young orchards have been planted in western Iowa, which are beginning to bear fruit.

Probably no other state is more uniformly adapted throughout its entire extent to the production of honey in moderate amount than Iowa. While good results may be obtained in every section, the smallest number of colonies is found in the northern and northwestern counties of Worth, Winnebago, Kossuth, Emmet, Dickinson, Osceola, Lyon, Sioux, and O'Brien. There are also only a few bees in the southwestern counties of Fremont, Mills, and Montgomery. The northern part of the state is colder and more hilly than the south, and the west is dryer than the east.

The most favorable location for beekeeping is the eastern section of the state along the Mississippi River and the streams which flow into it as far westward as they are timbered. Excellent opportunities may be found in the valleys of the Wapsipinicon, Red Cedar, Iowa, and Des Moines rivers. While farmers maintain only about a dozen colonies each, apiaries of 200 and 300 colonies are not unusual. Sweet clover is extensively cultivated and yields a surplus every year.

Many beekeepers are also located in the southeast corner, which has a milder climate and a greater rainfall than any other part of the state. After the flow from white clover is over there is a fall flow from heartsease, Spanish needles, and buckbush.

The western counties of Woodbury, Crawford, Harrison, and Pottawattamie, near the Missouri River, also furnish excellent records. In the region around Sioux City sweet clover covers thousands of acres, making a luxuriant growth in the rich loess soil bordering the river. It is an ideal crop for preventing the erosion of the steep hillsides and for enriching the land.

Iowa is far from being overstocked with bees; and, according to a conservative estimate, the crop might be ten times larger than it is to-day. S. W. Snyder states that at Center Point, Linn County, while two apiaries produced 20,000 pounds of honey in one season, probably not more than two-thirds of the nectar available was gathered. In many localities the bee pasturage will sustain from 75 to 100 colonies, and a return of $2000 from a single yard has not been uncommon. Nevertheless, few beekeepers depend entirely on beekeeping, as there are years when no surplus is produced, and the bees must be fed.

MISSOURI

Total area, 68,727 square miles, or 43,985,280 acres. Approximately 19,000,000 acres, found chiefly in the Ozark region and the southeast lowlands, are uncultivated, and of this area 5,000,000 acres are so rugged that they can probably never be improved for crops. Missouri, which is situated near the center of the Union, is bordered on the east by the Mississippi River, and is divided into two nearly equal parts by the Missouri River, which crosses the state from west to east. It is the meeting-place of four physiographic provinces which have a wide extension beyond its limits — the prairie region north of the Missouri River, the Great Plains region, the Ozark region, and the southeast lowlands.

The majority of the successful beekeepers, according to L. Haseman, of the Missouri College of Agriculture, are located within reach of either the Missouri or the Mississippi river, or one of their tributaries In the drained areas of southeast Missouri beekeeping is rapidly coming to the front. The larger part of the beekeepers in the state have but few colonies; but there are a number who operate from 200 to 700 colonies. There is a seasonal difference of about three weeks between south Missouri and north Missouri; but, aside from this and the local influence of certain honey plants, beekeeping conditions do not vary greatly in the river valleys.

"After an experience of thirty years in commercial beekeeping in Missouri," writes J. F. Diemer, of Liberty, "I am convinced that there are many good locations in this state, both along the larger rivers and the smaller streams. An ideal location is on the edge of the bottom-lands, where there is a chance for two honey crops — one from white clover, which grows on the upland; the other from Spanish needles, heartsease, and other flowers which grow on the bottom-lands. The best locations are found on the rivers Platte, Grand, Nodaway, Chariton, Osage, Meramec, Missouri, and along the Mississippi on the east border. White clover yields a surplus only about once in three or four years, and then it yields very abundantly; but it is very seldom that we fail to get a crop from fall flowers. One season, without white clover, we had a surplus of over fifty pounds, and, in addition, so large an amount was left in the hives that they entered the winter weighing about 80 pounds, or with more stores than were required. Most of the rivers mentioned are north of the Missouri River. The southern part of the state is not to be compared with the northern part for the production of honey. There are not many commercial beekeepers in Missouri; but most of those who are engaged in this industry are successful. My yards contain each from 50 to 100 colonies."

The most important honey plants are white clover, heartsease, and Spanish needles, which are widely distributed over the state, blue-vine (or shoestring vine), sweet clover, black locust, and aster. Of less value are willows, hard maple, dandelion, basswood, tupelo, catalpa, and goldenrod. Too often, says Haseman, the hot dry weather of July and August cuts short prematurely the crop from white clover, or it may, on the other hand, be lost as the result of excessive rain. Spanish needles, heartsease, sweet clover, and blue-vine are less affected by unfavorable weather conditions. Sweet clover is rapidly increasing, but basswood and the other native trees are fast disappearing; and in the southeast lowlands it is only a question of a few years before the timber will all have been cut and the land reclaimed. The dandelion in its western migration has become valuable for both pollen and nectar in Missouri, and blooms in sheltered places every month in the year. At Brunswick, Chariton County, in the cornfields on the bottom-lands of the Missouri River, blue-vine during the autumnal months may be found twining around every cornstalk. The vine is no larger than a baling wire, but it may reach a length of forty feet. (See Blue-vine.) Wild crab-apple, red-haw, and a large variety of shrubs help build up the colonies for the white clover flow. In the southwestern part of the state and along the Missouri River there are large areas of apple orchards, and many other fruit trees are under cultivation.

The eastern part of the prairie region, north of the Missouri River, is nearly level or gently rolling; but there are large areas which have been eroded by the streams. North Missouri is also called the Glacial Region, since during Pleistocene time it was covered by an ice sheet which deposited over most of its surface a layer of drift 10 to 200 feet in thickness. After the glacial age the wind spread over part of this region a fine fertile soil, known as loess, which reaches its greatest depth along the Missouri and Mississippi rivers. The loess soil covers not far from 1000 square miles, and is found in 46 counties. There is very little land which is not under cultivation, and hundreds of thousands of acres are devoted to the production of corn, wheat, oats, and hay.

As would be expected from the physical features of northern Missouri, beekeeping is only moderately successful in most of the counties, and in none of them does the amount of honey produced equal the large yields secured in the best part of the white clover belt. At Maywood, on the Fabius River, in the northeast corner, commercial apiaries contain from 50 to 75 colonies, but the former number is preferable. The honey plants are white clover, alsike clover, sweet clover, heartsease, basswood, Spanish needles, aster, and many minor plants. A good honey flow is usually obtained from either spring or fall flowers. In dry seasons the bloom on the hills is almost entirely neglected by the bees, while they visit in great numbers the flowers of the river bottoms.

In Montgomery County the apiaries are much smaller at present than formerly. A beekeeper at Rhineland describes local conditions as follows: "I have about 80 colonies, and seldom have an entire failure. In a good year I average 60 pounds. White clover is the main summer (June) source of nectar, but in dry seasons it burns out. I seldom fail to secure some honey in autumn." In Carroll County, near the center of the state, on the Missouri River, the yards seldom number more than 25 to 35 colonies. There are great areas of bottom-land along the river, 40 miles long by two to fourteen wide, which are covered with clover, heartsease, and Spanish needles. There are a few progressive beekeepers here who are fairly successful. Much alsike clover is sown, sweet clover grows wild in many places, and in the fall blue-vine springs up over a large acreage from which wheat has been harvested. The upland, as has already been explained, is not as good territory, and a large crop is harvested only about once in three years. On the lowlands there is, nearly every year, a fall flow of dark thin honey. Extracted honeys may granulate in 40 days, and become solid in three months.

In the northwestern portion of the state white clover is the most important honey plant; and when it fails only a small surplus is secured in locations away from the rivers. A beekeeper in Atchison County, who lives near the Missouri River, reports that he has secured from his best colonies for a period of 7 years 200 pounds of extracted honey annually. At St. Joseph, Buchanan County, the pasture lands are largely covered with white clover; and basswood, dandelion, black locust, sweet clover, alfalfa, and buckwheat are abundant; while on land under cultivation there are the usual fall flowers, such as heartsease and Spanish needles.

The Southwest Prairie, which is an eastern extension of the Great Plains, is a triangular-shaped area, about one-third of the size of the northern prairie, covering about 17 counties south of the Missouri River and west of a line running from Cooper County to Jasper County. It is the smoothest area in the state, the streams having broad, shallow flood plains with wide, gently sloping valleys and rounded divides. The soils, which are deep and fertile, are derived from shales, sandstones, and limestones.

The number of colonies of bees in many counties in this region has decreased greatly during the past 15 years. Dade and Polk counties contain the largest number of colonies, but Saline produces the largest amount of honey per colony. At Independence, in Jackson County, on the south side of the Missouri River, the larger apiaries range from 50 to 100 colonies; but the smaller yards are disappearing. This is not as good a location as it was 50 years ago, as there is to-day not much uncultivated land. Lafayette and Cass counties also contain good locations. At Harrisonville the apiaries range from 15 to 20 colonies, but there has been a heavy mortality among the bees during the past few years. The best location is the broad flood plain of the Missouri River; but back from the river on the higher land, where grain is largely grown, there is little space available for honey plants. White clover, moreover, fails often on account of dry weather.

The Ozark Plateau, so called from the Ozark Mountains, includes all that portion of the state south of the Missouri River, exclusive of the Southwest Prairie and the southeastern lowlands, or about two-thirds of the entire area of the state. The western boundary line extends from Cooper County to Jasper County. This region contains the highest land in the state (1800 feet in Iron County), and is much more deeply dissected than the prairies to the north. The valleys of the streams are deep and narrow, and the intervening land is hilly, or almost mountainous, much of it being unsuitable for agriculture. The roughest part of this region is found in Crawford, Washington, Iron, St. Francois, Madison, Wayne, Shannon, and Carter Counties, where much of the land is full of boulders, and fit only for forests or cattle ranges. All of the important soil-forming rocks of the Ozark region are limestone; but since the soils are the oldest in the state, and the land is hilly, most of the lime has been removed by leaching. It was formerly covered with a forest composed chiefly of short-leaved pine, white, black, red, and scrub oak, none of which are of value to the beekeeper. But much of this has been removed, and to-day tracts of virgin woodland more than 320 acres in extent are rare. White clover and sweet clover are the chief honey plants, and there is some prospect of sweet clover largely superseding other hay crops. Buckeye, papaw, redbud, buckbush, hackberry, basswood, sugar maple, and various wild flowers eke out the nectar supply.

Many of the counties in this region during the past fifteen years show an increase in the number of colonies of bees, while a few show a decrease. The average production of honey per colony is low; and counties in the white clover belt with the same number of colonies of bees secure much larger yields. Dry weather in

July and August often injures the honey flow. Many farmers keep a few colonies in nail-kegs and log gums, but honey production for market receives little attention. The bees are never fed in the fall, and receive no extra protection in winter. Modern hives are almost unknown; but five out of six seasons the beekeeper "gets enough to rob."

The southeast lowlands offer much greater inducements than the Ozark region, and contain more colonies of bees than the rugged country directly north of it. Commercial beekeeping is very successful in some places, though the honey crop is uncertain over most of this area. It is a broad plain, the northern extension of the Mississippi Lowlands, sloping gently to the south, and comprising about 3000 square miles. The surface drainage is poor, and there are large areas of swamps and morasses. The soil, which consists partly of sandy and partly of clay loams, was brought down by the Mississippi and the other rivers which flow through this area. Very little of the land is subject to overflow, and thousands of acres have been reclaimed by dredging ditches. It was formerly covered with a forest of cypress, ash, and gum, with a mixture of elm, hickory, oak, and catalpa. On the northwest it is bounded by a line of bluffs running from Cape Girardeau to Ripley County on the south state line.

The honey plants of this region are willows, red maple, redbud, persimmon, dandelion, fruit bloom, white clover in spring and early summer; and in the fall cotton, heartsease, Spanish needles, goldenrod, and aster. There is an immense acreage of cantaloupes and watermelons, which are shipped to market by the carload. They yield a delicious honey, which is so white that it is almost transparent. Cow-peas are the source of a honey with a beanlike flavor. The odor of the newly gathered nectar is very noticeable in the evening, but both flavor and odor largely disappear with the ripening of the honey. White clover is abundant on the hills and alsike clover is grown in places; but until sweet clover is more generally grown there will always be a dearth of honey during the summer. The fall flow is almost always sure in the better honey-producing areas.

Three-fourths of the bees in this region have been in box hives and there are many wild colonies in the trees; but better methods of beekeeping, with modern equipment, are being promoted in this region more than in any other part of the state.

NORTH DAKOTA

Total area, 70,837 square miles; length, 320 miles; breadth, 210 miles. North Dakota is a vast fertile prairie bounded by the Red River on the east, and crossed by the Missouri River in the west, which enters the state 65 miles south of the northwest corner, and crosses the south boundary line near its center. The area drained by the Missouri River comprises nearly one-half of the state, or 20,000,000 acres. One hundred square miles near the center of the north boundary are covered by the Turtle Mountains, the only mountains in the state. There are no large bodies of water. The forest land, found principally in the Turtle Mountains and along the Missouri and Little Missouri rivers, comprises only 500,000 acres, consisting of scattered growths of cottonwood, box elder, bur oak, elm, poplar, and cedar. There is an area of waste or bad lands along the Little Missouri River in the southwest. The rainfall over the whole state is sufficient for growing grain crops without irrigation. In the eastern half it ranges from 18 to 24 inches, and in the western half from 15 to 18 inches; but in the southwest corner it is only 12 inches. Four-fifths of the total value of the crops are contributed by the cereals, and the remainder largely by hay and forage. More than twelve million acres of wheat, oats, barley, and flax are grown; but wheat is by far the most important crop. Out of every 100 farms, 80 grow wheat. As the winters are severe, the acreage of apples, pears, and plums is small; but blackberries, raspberries, and currants are perfectly hardy. Alfalfa is not largely cultivated. The cultivated crops, except sweet clover, which is increasing very rapidly, offer a poor pasturage for bees.

North Dakota may be divided into three regions: The Red River Valley, a narrow belt of land in the extreme eastern portion of the state; west of this valley a glaciated region, a wide rolling upland occupying the larger part of the state; and a southwest unglaciated region, west of the Missouri River, where the soils are largely residual, or derived from the decomposition of the underlying rock. The eastern portion of the state and the valley of the Missouri River are the sections most suitable for beekeeping. The Bad Lands, in the southwest, are useful only for grazing, and only partially available for this purpose. In the central part of the state, along the Missouri River, bees are kept chiefly for pollination and for the production of honey for home use.

The Red River Valley, on the eastern border of the state, is a very level plain 25 to 75 miles wide, in which are situated the eastern parts of the counties of Pembina, Walsh, Grand Forks, Trail, Cass, and Richland. The soils, which are predominantly dark highly calcareous clay loams, were deposited by an ancient glacial lake which once filled the valley. The land is very largely occupied by grain fields; but sweet clover has been more widely grown during the past ten years, and is the hope of bee culture in North Dakota. Along the Sheyenne River sweet clover is the main honey plant. White clover is common and yields well. Basswood is abundant along the Red River, but it is a source of nectar only every other year. There is little buckwheat, and alfalfa is visited little by bees since sweet clover became abundant. Minor honey plants are dandelion, fruit bloom, wild mustard, Canada thistle, goldenrod, and aster; but around Fargo, according to R. L. Webster, bees pay little attention to these plants because sweet clover is so much more attractive. High winds often interfere with the flight of the bees if they are obliged to travel far. Success in this section depends largely on an abundance of sweet clover and on out-apiaries.

Rising 300 to 500 feet above the Red River Valley, a broad glaciated plain or plateau, occupying the larger part of the state, extends westward to the valley of the Missouri River, and in the northwest to the west boundary line, where it attains an altitude of 2700 feet. A mantle of drift derived chiefly from the limestones and granites of Manitoba covers its surface to a depth of one foot to several hundred feet. The soils are principally sandy calcareous loams, but glacial lake soils occur in portions of Bottineau, McHenry, Rolette, Pierce, and Ward counties, deposited by a smaller glacial lake south of Turtle Mountains. Bees located on the high prairies would starve unless there were fields of sweet clover and alfalfa planted near by. There is no shelter from the wind, and the chief crops are the cereals. In the western part of the state even solitary bees and bumblebees are rare. In Sargent County, in the southeast, farmers are beginning to plant sweet clover and alfalfa. With the increase of the sweet-clover acreage the outlook is promising.

The section of North Dakota southwest of the Missouri River was not covered by the great ice sheet. The soils are residual, or formed from the decomposition of the underlying rocks of shale and sandstone; but directly west of the Missouri River the soils consist partly of drift. Fertile alluvial soils are found in the flood plain of the river. Along the Little Missouri River the land is so broken by innumerable ridges, buttes, and ravines that it is unfit for agriculture, and is known as the "Bad Lands." In the valley of the Missouri River alfalfa is grown to a considerable extent, and sweet clover has become established on uncultivated land.

The number of acres in the state which can be irrigated is estimated at 1,540,000; and the number requiring drainage at 3,255,000. While by utilizing water the land will produce much larger crops, there will be no crop failures in North Dakota if good dry-farming methods are employed.

The future of beekeeping in North Dakota will depend very largely upon the cultivation of sweet clover. Where large acreages of this valuable fodder plant are grown, good crops of honey are obtained. Of the 44,800,000 acres in the state, 42,150,000 are arable; and, as the soils of the Red River Valley and of the section covered with glacial till are for the most part calcareous, sweet clover can be grown over the larger part of the state. The acreage of sweet clover in the Red River Valley is rapidly increasing, and in the entire state it is estimated that there were more than 400,000 acres in 1924. After the wheat has been harvested, sweet clover grows up and blooms profusely the following year. An average surplus of as much as 500 pounds per colony has been reported, and an average of 150 to 200 pounds per colony is not uncommon. Although the winters are very severe, winter losses are not great, as bees winter well in this climate in wholly underground cellars. According to R. L. Webster sweet clover does not yield nectar every year in North Dakota. The native honey plants are confined chiefly to the Red River Valley and the valleys of the James, Missouri, and other large rivers, and to the region of the Turtle Mountains. While fairly abundant in localities, they are for the most part of secondary importance. The number of species of seed plants in North Dakota is small — not much exceeding 900. Many eastern plants do not extend westward beyond the Red River.

As the honey flora of North Dakota is almost wholly unknown, a list of the more important species, compiled with the assistance of O. A. Stevens, of the Agricultural College, is given below:

Alfalfa (*Medicago sativa*).—Cultivated rather commonly in central and western parts of the state.

Aster (*Aster paniculatus, A. multiflorus,* and *A. laevis*).

Basswood (*Tilia americana*).—Abundant along streams in the eastern part.

Buffalo-berry (*Shepherdia argentea*).—Common, especially along the Missouri River and in the Bad Lands; blooms with the earliest willows, and is much frequented by honeybees.

Buckwheat (*Fagopyrum esculentum*).—Cultivated to a small extent.

Catnip (*Nepeta Cataria*).—Occurs locally.

Choke Cherry (*Prunus virginiana*).—Common through the state.

Clover, alsike (*Trifolium hybridum*).—Not common. White clover (*T. repens*) is common in the Red River Valley, but rare in most of the state. See Sweet Clover.

Dandelion (*Taraxicum officinalis*).—Very common in the eastern part.

Goldenrod (*Solidago canadensis, S. rigida,* and *S. serotina*).—Common.

Gooseberry (*Ribes gracile*).—Common in the eastern part. *R. setosum* is common in the western part.

Hawthorn (*Crataegus succulenta*).—Common throughout the state.

Maple, soft (*Acer saccharinum*).—Often planted in the eastern part.

Mint (*Mentha canadensis*).—Common throughout the state.

Mustard (*Brassica arvensis* and *B. juncea*).—Common. Other species of *Brassica* also occur.

Plum (*Prunus americana*).—Common.

Raspberry (*Rubus villosus*).—Abundant.

Smartweed (*Polygonum lapathifolium*).—The most common species.

Sweet clover (*Melilotus alba* and *M. officinalis*).—Commonly cultivated and wild, especially in the Missouri River Valley.

Willow-herb (*Epilobium angustifolium*).—Common.
Willows (*Salix interior and S. amygdaloides*).—Common along streams.

SOUTH DAKOTA

South Dakota extends from east to west 370 miles, and from north to south 207 miles, the total area being 77,615 square miles. Its surface is a high rolling plain or tableland, diversified by the narrow valleys of the streams. The Black Hills in the southwest cover an area of 5000 square miles, and vary in altitude from 3000 to 5000 feet. The timber land comprises about 2000 square miles in the Black Hills, and a few smaller tracts in the northwest and on the bottom-lands of the Missouri River. Mixed farming in the eastern portion of the state, and stock-raising in the western, are the principal industries. The winters are long and severe; but the air is so dry that there is little suffering from the cold. There is sufficient rainfall for growing crops in the eastern half of the state, but in the Black Hills irrigation is extensively practiced.

The region east of the Missouri River, which comprises about two-fifths of the state, was glaciated, and the brown fertile soil is derived from the glacial drift. An immense acreage is planted with wheat, corn, oats, and flax, four-fifths of the total value of the crops being contributed by the cereals. There are numerous lakes in the glaciated portion, and it is estimated that there are 300,000 acres requiring drainage, which are either constantly marshy or are periodically overflowed. West of the Missouri River the state was not glaciated, and the soils are residual, consisting of sandy and clay loams formed by the weathering of the underlying sandstones and shales. There is a large area of waste land in the Black Hills and in the Bad Lands.

The best locations for beekeeping in South Dakota are in the southeast corner along the bottom-lands of the Missouri River, and in the irrigated areas of the Black Hills section. In general the eastern portion of the state, with its greater rainfall, more abundant flora, and larger population, is naturally better adapted to bee culture than the western part. As in North Dakota, the future of bee culture in South Dakota is largely dependent upon the more general cultivation of sweet clover. No other honey plant is so reliable or yields so well. It is the source of the larger part of the surplus. "Where sweet clover is abundant in the east half of the state," says L. A. Syverud, state bee inspector, "beekeeping is profitable;" and he adds, "I tried out a location in the North part of last season (1924) with an average per colony that netted me over 200 pounds." East of the Missouri River alfalfa is uncertain and yields little or no nectar. The poorest locations for beekeeping are in the northwest corner and in the Bad Lands.

Reports from the northeast corner of South Dakota are brief and unsatisfactory, and would indicate that at present there are only small yards widely scattered.

In the counties near the center of the east border, as Brookings, Kingsbury, Miner, and Lake, beekeeping is pursued at present chiefly as a side line.

In Lincoln and Clay counties sweet clover reigns supreme as a honey plant. Introduced about 1890, it is steadily spreading northward, and is destined to be the main reliance of beekeeping in this state. "We have forgotten," writes a beekeeper of Vermilion, "that there is any other honey plant. Alfalfa a short distance west of us, and white clover 200 miles eastward in Minnesota, are good honey plants; but here we recognize only sweet clover. It is the most valuable of forage plants, and the greatest soil-redeemer and land-restorer known. Other plants may fail, but it is reliable every year." Less important honey plants are dandelion, white clover, catnip, fruit bloom, heartsease, goldenrod, and aster. At Scotland,

Bon Homme County, where sweet clover is making rapid headway, alfalfa yields very little nectar. It is, however, usually cut as soon as it blooms.

Many farms in the southeast section maintain with good results from one to 20 colonies of bees. At Vermilion, Clay County, 20 colonies netted in one year $17.50 per hive. There is considerable loss from foul brood and poor methods of wintering. Beekeeping in this section is a very young industry, many beekeepers stating that they have been in the business only one or a few years.

On the Missouri River, near the center of the state, there are only a few bees. The river flows for a distance of 547 miles in South Dakota, and the flood plains or bottom-lands are about two miles wide. Tracts of timber consisting chiefly of cottonwood, elm, cedar, butternut, and ash occur in the bends and on the islands in the river.

Only a small percentage of the tableland in the northwestern portion of the state at the beginning of this century was in farms. Formerly this constituted a vast open range where horses, cattle, and sheep grazed and roamed at will. With the rapid increase in population in recent years the number of farms has greatly increased, and the free range has become much restricted. There are very few colonies of bees reported from any part of this area. The strange formation in the central west of South Dakota, known as the Bad Lands, covers 2000 square miles between the Cheyenne and White rivers. They consist of a labyrinth of winding ravines and narrow ridges which in places widen into broad buttes and rounded domes, often surmounted with slender spires. The region is bounded by a high clay bluff. Very little of it can be used for farming.

Irrigation in South Dakota is confined almost wholly to the western counties in the vicinity of the Black Hills, where the irrigation projects, completed or under way, include 201,625 acres. The most important honey plant in this section is alfalfa, of which more than 40,000 acres are grown, much of which is under irrigation. White sweet clover is also of great value, growing abundantly on the banks of the ditches.

Spearfish Valley, in Lawrence County, is the seat of many prosperous farms, luxuriant gardens, thriving fruit groves, and apiaries that average 70 pounds per colony. At St. Onge six apiaries are reported to contain 1400 colonies, and to average about 100 pounds of honey per colony.

A most important reclamation project is that of the Belle Fourche Valley, in Butte and Meade counties. The total irrigable area is 100,000 acres, divided into more than 1000 farms. The soil is fertile, and free from alkali and stones. The natural growth in its wild state is sage brush, cactus, and wild wheat grass. "The Belle Fourche Valley," declares a beekeeper in this locality, "is the best beekeeping county, not only for this state, but for many other states." At Fruitdale there are many large apiaries which average 100 pounds of honey per colony. One producer of honey is reported to have shipped a carload. An apiarist who has 200 colonies writes that he winters his bees outdoors successfully in well-wrapped hives. At Nisland the apiaries range from fifty to several hundred colonies; a crop is obtained every year, and the surplus may reach 200 pounds per colony. "Beekeeping is as yet primitive in South Dakota, and it will be a long time before, in the Black Hills region, the thousands of pounds of honey which are going to waste will be saved by the bees."

NEBRASKA

Total area, 77,520 square miles. Nebraska is a great upland plain, 208 miles wide and 413 miles long, sloping gradually from an altitude of 5000 feet in the extreme western portion to about 1000 feet along the Missouri River, which forms its east boundary. A great number of broad shallow valleys, from a quarter of a mile to 20 miles wide, have been worn in its surface by the many streams. Between the valleys the land is gently undulating prairie. In the north-central and western portions there is a large area of sand hills and buttes; but there are no mountains. The average annual rainfall in the east is 30 inches, and in the west 15 inches. Irrigation is largely restricted to the western counties. The eastern and southern portions are highly productive, and are largely devoted to growing grain.

According to its surface features, altitude, and flora, the state may be divided into three regions: (1) The eastern or prairie region; (2) the sand-hill region; and (3) the foothill region, the last two regions belonging to the Great Plains. The best locations for beekeeping are along the Platte River, a broad shallow stream with nearly treeless banks, which flows through the whole length of the state from west to east; along the Republican River in the extreme southern part and along the Loup and Elkhorn rivers and their tributaries.

At Central City, Merrick County, there are a number of commercial apiaries. The surplus is stored chiefly from white clover, sweet clover, alfalfa, and heartsease. Farther westward in the alfalfa region there are numerous apiaries which will be described later.

The eastern or prairie region has an average width of 145 miles; but along the Platte River it extends westward to Dawson County. The land is level or gently rolling, and is diversified by the broad valleys of the many streams. On the east border the Missouri River winds in a tortuous course through swampy bottom-lands nine miles in width. Fertile loess soils, retentive of moisture, support a great variety of plants. The hardwood forests of Nebraska are unimportant, and are confined chiefly to a narrow belt along the Missouri River; but in the northwest corner of the state there are extensive stretches of pine woodland. Trees of interest to the beekeeper are basswood, hackberry, hawthorn, prickly ash, redbud, honey locust, Indian cherry, buckeye, choke cherry, and sumac. Of shrubs, gooseberries, snowberry or buckbush, and red raspberry are common, while woodbine, wild clematis, and poison ivy abound in the thickets.

In late summer the prairies are covered with tall grasses and bright-colored flowers, as blazing stars, verbenas, goldenrods, asters, five-fingers, thistles, and sunflowers. On waste land, on the broken prairie and around the towns and cities are patches of sunflowers, and the roads are bordered on each side in places with white and sweet clover.

The larger part of the territory suitable for beekeeping is included in the eastern or prairie region. The conditions are fairly good for apiaries of moderate size, and hundreds of farms maintain a few colonies. Alfalfa along the eastern border yields little nectar, as there is often too much rain; and immense areas are sown with grains which are of no value as sources of honey.

The surplus in eastern Nebraska comes almost exclusively from white clover, alsike clover, and sweet clover. In the fall heartsease and Spanish needles in localities also yield a surplus. The more important minor plants are willow, maple, box-elder, dandelion, fruit bloom, mustard, Rocky Mountain bee plant, mallows, false indigo, wild sunflowers, goldenrods, and asters. The cultivation of sweet clover is increasing rapidly, making some of the best bee territory in the United States, and when pastured it furnishes good bee range from July 1 until frost. A small surplus may also be obtained from black locust and sumac. Heartsease covers great stretches of land, and a surplus of 250 pounds per colony has been obtained from it, but it is of no value in a dry season. The orchards of apples and peaches are also becoming more numerous.

In northeastern Nebraska in the counties bordering on the Missouri River beekeeping was formerly only moderately successful, but in later years it is becoming more profitable on account of the introduction of sweet clover. From Omaha northward to the Dakota line may be found some of the best bee ranges of Nebraska.

Richardson County, in the southeast corner, has been called "the garden of the state;" but a beekeeper at Humboldt writes: "Twenty years ago beekeeping was a profitable industry in this county; but to-day you can buy colonies for less than the cost of new hives. During the last three or four years, partly as the result of dry weather and the ploughing up of alfalfa and sweet clover, and partly on account of foul brood, I have lost a large part of my colonies, and expect to give up beekeeping."

Reports from the counties in the eastern section away from the Missouri River vary greatly. White clover often suffers from drought; heartsease yields well in wet seasons; sweet clover is not abundant, but is increasing; and alfalfa yields little nectar.

The sand-hill section, which occupies the central portion of the state, has an average width of 175 miles. The sand hills, the chief characteristic of this region, vary in height from 15 to several hundred feet. They are usually conical, but vary greatly in shape, and often contain crater-like hollows called blow-outs, two or three hundred feet in diameter and 40 or 50 feet deep, formed by the wind. The vegetation is sparse, consisting chiefly of bunch grass with a few scattered shrubs and sunflowers. The sand hills occupy a broad belt in the central portion of the state, and are believed to owe their origin to the wind. Much of this region north of the Platte River is not suitable for cultivation.

Beekeeping in this section is almost wholly dependent on alfalfa and sweet clover. It is surprising, says J. H. Wagner, how rapidly sweet clover and alfalfa are being put in. The larger part of the 1,214,649 acres of alfalfa grown in the state is found in 28 counties in the south-central portion. Where it is grown without irrigation the yield varies greatly in different years. To secure an abundance of bloom there must be sufficient moisture in the ground by May first to produce a vigorous growth of the first crop, and the harvesting of each succeeding crop should be followed by ample rain. If the season is very dry the growth will be short and the bloom scanty. But during a very rainy season little nectar is gathered; for instance, in Custer County, in 1906, there were in June 12.73 inches of rain and in July 9.26 inches of rain, and only six pounds of surplus per hive was stored from the first and second crops. In a normal season the main supply comes from the second crop in July. The first crop blooms in June and the third in August. Large areas of alfalfa are grown in Custer County without irrigation.

In Lincoln County, through the center of which flows the Platte River, there is a large area of alfalfa under irrigation. Sweet clover and alfalfa furnish the surplus honey. They are reliable nearly every year, but the flow may be lessened either by very wet or very dry weather. Sweet clover yields for about six weeks. Alfalfa, which is the source of the main crop, lasts for about two months. It is cut three times in a season, and the honey flow from the second and third crops is better than from the first.

In Custer and Sherman counties, near the center of the state, there are a few bees. From sweet clover and alfalfa an average of 100 pounds is obtained near the center of the state; but 240 pounds per colony, spring count, has been secured. Alfalfa secretes nectar best in dry weather, and sweet clover always yields, but it is most reliable in seasons when there is an average rainfall. It is one of the best-paying crops, and each year a larger acreage is sown. There is a dearth of nectar in this locality until buckwheat blooms about June 15.

On the Republican River, where it crosses the south border, there are several apiaries. A full crop of 60 to 100 pounds per colony is usually obtained every other year, and a fair crop every year. Alfalfa is the best honey plant. For hay it is cut four times in a season in this locality. Sweet clover is also largely sown here, and is reliable. There is also an abundance of heartsease. Bees can be wintered on the summer stands in winter cases packed with leaves or chaff. Beekeeping is still in its infancy, but this region promises to make a good bee country.

The foot-hill region of Nebraska has an average width of 75 miles, and differs in surface features from the other portions of the state. It is characterized by steep ridges, deep ravines and canyons, and numerous isolated buttes. The buttes are conical hills with flat tops and an elevation of 60 to 1000 feet. A large area in the northwest corner of the state is occupied by "Bad Lands," where the streams have cut the clayey soil into deep canyons, forming pinnacles, bluffs, and chimneys. There is almost no vegetation, as there is great heat in summer and little rain. Sagebrush and greasewood are common; but in one tract, Rydberg says, "Not a green spot was to be found." The rolling or prairie land of the foothills is covered with grasses amid which many flowers bloom profusely, as lupines, ground plum, psoraleas, and asters. There are very few bees in this section, Scott's Bluff County, which has 100,000 acres of land under irrigation and 35,000 acres of alfalfa, containing the largest number. There are many thousand acres of irrigated alfalfa also in Dawes County.

A beekeeper at Oshkosh, on the Platte River, writes that, although sweet clover and alfalfa are abundant, at an altitude of almost 4000 feet the air is so dry that little nectar is secreted, and his experience with bees has been a failure. The honey consumed is shipped from Colorado. A much more favorable report comes from Lakeside, in the southern part of Sheridan County, where wild flowers build up the bees in spring, and over 400 acres of sweet clover and some alfalfa furnish the surplus.

J Howard Wagner, of Central City, Nebraska, after reviewing the foregoing, adds: "A great coming bee section of Nebraska is the territory lying between a line running from North Platte through Burwell, O'Neill, and northeastward to the Platte River and the Missouri River. There are hundreds of good bee locations as yet unoccupied. The South Platte offers opportunities, but is more subject to drouth. For years I did not think Nebraska was a good bee state, but I have changed my mind. As we beekeepers advance with the world the results will be greater. With the control of foul brood in sight, I believe Nebraska will take its place in the Union as one of the best honey-producing states."

KANSAS

Kansas is 410 miles long and 210 miles wide, and has a total area of 82,158 square miles. The eastern third of the state belongs to the Central Lowlands, and the western two-thirds to the Great Plains. Its surface is a vast undulating plain consisting of an endless succession of shallow valleys and broad level uplands, which, from an altitude of 4000 feet in the west, slopes gradually downward to 900 feet in the east. The Missouri River forms the northeast boundary. The north-central portion is crossed by the great valley of the Kansas River; and in the south-central part there is another broad valley formed by the Arkansas River. The northeastern counties, north of the Kansas River, are covered with glacial till, which is overlaid in turn by fertile loess soils. Southeastern Kansas has clay soils derived from shale, and the central portions of the state sandy loams derived from the underlying limestones and sandstones. Loess soils occur again in the northwestern, and limestone soils in the southwestern portion. The rainfall decreases from 42 inches in the southeastern counties to a little more than 15 inches on the Colorado line. In eastern Kansas it is sufficient for growing all crops; but the western portion of the state is semi-arid, and irrigation is necessary. Along the streams in the eastern part of the state there is a fringe of cottonwood, willow, oak, and a few other trees; but the western part is almost devoid of timber, and in the early history of the country the prairies were covered with buffalo grass. Corn and wheat combined represent nine-tenths of the total acreage of crops.

From the point of view of the beekeeper, Kansas may be divided, according to J. H. Merrill, into four regions: A northeastern rainbelt; a southeastern rainbelt; a central alfalfa belt 165 miles wide; and a semiarid alfalfa belt in the west 170 miles in width. Good alfalfa locations for beekeeping are found in the central part of the state, along the Arkansas River, and all streams west of Topeka. In the Arkansas River Valley the water is so near the surface that the roots of alfalfa are able to obtain sufficient moisture without irrigation. Beekeeping is developing very rapidly in the western part of the state, owing to the introduction of alfalfa and sweet clover. Large areas under irrigation are yielding heavy crops of alfalfa.

Northeastern and southeastern Kansas, or the "rainbelt," is about 75 miles in width, and includes 25 counties in three tiers. The leading crops are corn, wheat, forage, oats, and potatoes. In this region are located a large part of the bees in the state. Throughout the "rainbelt," alfalfa, of which there are some 240,000 acres, yields little or no nectar. A beekeeper who has resided in eastern Kansas for 25 years reports that he has never known of a pound of alfalfa honey produced in this section. Around Topeka it is only occasionally that bees work on the bloom. In eastern Kansas the number of colonies of bees in a county is not determined by the alfalfa acreage. The surplus comes chiefly from white clover, sweet clover, smartweed, and Spanish needles. Other species of some value are maple, dandelion, fruit bloom, catalpa, box-elder, wild mustard, horsemint in localities, buckbush, honey locust, basswood, sumac, goldenrod, aster, and wild sunflowers. The large number of colonies in the "rainbelt" is clearly not due to an excellent honey flora, but must be attributed partly to its dense population, which much exceeds that of any other section of the state. Many small apiaries are maintained by feeding, and others solely for the purpose of securing honey for home use.

In the northeastern or glaciated section of the state, north of the Kansas River and west of the Missouri River, the surplus is secured chiefly from white clover, sweet clover, and heartsease. In the early spring red maple and dandelion, and in the fall goldenrod and aster, furnish a large amount of pollen and nectar. There is a dense area of apple trees and other fruits, as peaches, pears, and cherries. White clover is more valuable in this section than in any other part of the state, but it does not yield every year. When weather conditions are favorable it is the source of four or more supers of section honey.

There is little commercial beekeeping in the central counties of the "rainbelt," and only a small amount of honey is produced. A beekeeper at Humboldt, Allen County, describes the general conditions in his locality as follows: "A crop is usually obtained three years in five, and is sold locally. The honey plants are white clover in wet seasons, fruit bloom, sweet clover, buckbush, broomweed, heartsease in some seasons, and Spanish needles. Asters are common along rivers and roads, but I have never seen a bee on goldenrod."

In the southeastern part of the state, or the southern portion of the "rainbelt," the soils are derived from the underlying shale, and are not suitable for the cultivation of the clovers. The best results are obtained in Wilson and Montgomery counties, where there are 30,000 acres of alfalfa under cultivation. It is reported to yield nectar in favorable seasons. There is also some white sweet clover. Cherokee County, in the southeast corner, is a much poorer location.

West of the eastern rainbelt is the central section of the state, extending 165 miles from Topeka, or the west line of Shawnee County, to the west line of the tier of counties containing Russell and Barton counties. The elevation is 1000 to 2000 feet. The rainfall varies from 20 to 30 inches. It is in this section that the majority of beekeepers are located on account of the large amount of alfalfa and sweet clover. Alfalfa does not yield equally well in all the territory. It yields the most nectar in the valleys of the rivers and smaller streams, where immense crops of forage are harvested from three to four cuttings. On the higher ground sweet clover is being grown to some extent. According to a beekeeper who has a good alfalfa location on the bottom-land of a western river, alfalfa will yield the entire season if water can be reached at a depth not exceeding 10 feet. On high ground alfalfa yields only after showers. The best yields appear to come from localities where alfalfa is grown for seed. If there has been an abundant rainfall in March there will be sufficient water in the soil to provide for the crop during the rest of the season; and if it is very hot in July and August, without much rain, the bloom will yield nectar. But alfalfa is not as good a honey plant in central Kansas as it is in the irrigated sections of the western part of the state, and of Colorado, Idaho, California, Nevada, Arizona, and New Mexico. On the uplands in the central section it usually yields a little nectar from all cuttings, except perhaps the first, which comes between June 15 and July 15, while fields left for seed furnish a slow honey flow for a longer time.

In central Kansas the only honey plant of great importance besides alfalfa is sweet clover. On the uplands the land is so largely occupied by the immense acreage of corn and wheat that the native nectar-yielding plants are not abundant. But sweet clover will grow on soil unsuitable for alfalfa, and in a few years greatly improves its fertility. The acreage under cultivation in Kansas is rapidly increasing; and, according to the Kansas State Board of Agriculture, 48,891 acres were grown in 92 counties in 1923.

In Washington, Republic, and Jewell counties, along the north border of the state there is a large acreage of alfalfa. Probably more than half of the farmers keep a few bees and produce honey for home use. At times alfalfa yields nectar well in these counties. Horsemint and catnip are helpful; but sweet clover is the coming plant in this part of Kansas, where it yields an abundance of nectar.

The central counties of the central belt report relatively few colonies of bees, although the area of alfalfa ranges from 20,000 to 40,000 acres. The yields from alfalfa and sweet clover run about 75 pounds per colony.

The southern portion of the central belt, especially in the valley of the Arkansas River, offers many excellent opportunities for beekeeping. Outside wintering is the rule in Kansas, as frequently there are warm days which permit the bees to obtain a flight. Brood-rearing begins in February, when the elm and maple furnish pollen. The surplus is gathered from alfalfa, of which there are 47,000 acres, sweet clover, and heartsease. The other honey plants serve only to stimulate brood-rearing. Sweet clover grows well on the uplands.

The western third of Kansas, in portions of which alfalfa is at its best, is a broad rolling plain with a semiarid climate, where in most seasons irrigation is required for maturing the crops. The soil is more sandy than in the eastern part of the state, and was covered formerly with bunch or buffalo grass. Large streams may disappear in the sand, as White Woman Creek, which is lost in the sand in Scott County. Practically all the water for irrigation purposes that can be utilized in western Kansas is furnished by three streams, the Cimarron River, the South Fork of the Republican River, and the Arkansas River. Of the 47,000 acres under irrigation in 1919, more than 97 per cent. were along the Arkansas River and its tributaries. But losses from evaporation, seepage, and diversions for irrigation in Colorado, now exhaust the summer flow of this river.

The enormous supply of underflow or under-ground water should not be overlooked. In the stream valleys this supply of water is available at moderate depths by pumping, and more than one-fourth of the land irrigated in the state is supplied from this source. In the upper Arkansas Valley from the Colorado line to Dodge City it has been estimated that 100,000 acres could be irrigated by pumping. But on the high plains the ground water occurs at such great depths that the cost of pumping is too great to permit of a large use of water from wells. There are thousands of acres in western Kansas for which water is not available for irrigation purposes.

In western Kansas in the region of the foothills of the Rockeies the surplus of honey comes chiefly from alfalfa and sweet clover. The Rocky Mountian bee plant is common, but there are no early-blooming honey plants. Some of the largest commercial apiaries of the state are located in the Arkansas Valley.

"While alfalfa," says Merrill, "is the principal honey plant of Kansas, sweet clover, because it yields honey all over the state and because its acreage is increasing at a rapid rate, will soon take the front rank. Western Kansas, especially in the Arkansas Valley, is more certain to produce results in honey from alfalfa and sweet clover than any other part of the state."

OKLAHOMA

Total area, 70,057 square miles. A triangular mountainous area in eastern Oklahoma belongs to the Ozark Plateau, or uplift, which has been described under Missouri. The central section consists of broad rolling plains and prairies; and the western portion of the state, which has an altitude of over 2000 feet, lies in the Great Plains. The Red River forms the southern boundary line. The Canadian River crosses the state from west to east near its center, and the Arkansas River traverses the northeastern portion. The normal annual rainfall along the eastern boundary is 40 inches, which decreases to 11 inches in the western part of the state. Cereals contribute one-half of the total value of the crops; cotton one-third, and hay, forage, and potatoes most of the remainder. The area of alfalfa is over 350,000 acres.

The line bounding the eastern mountainous region on the west extends from the northeast corner of the state to the northwest corner of Marshall County on the Red River. The northern part is occupied by the Ozark Hills and the southern by the Ouachita Mountains, between which lies the rugged valley of the Arkansas River. Among the mountains only the more level portions are suitable for agriculture. Much of the land, especially along the river bottoms, is densely wooded.

The largest number of colonies of bees in the eastern, or Ozark region, is found in the counties bordering the Arkansas River, as LeFlore, Haskell, Sequoyah, Cherokee, and Muskogee. At Westville, Adair County, a little farther north, bees are kept chiefly for pollinating apple orchards. The rainfall in eastern Oklahoma is so large that alfalfa secretes very little nectar. In the northeast corner there is considerable white clover. The acreage of sweet clover is steadily increasing, and will doubtless become the most important source of honey in the eastern counties. There is relatively little cotton planted in the mountains. Other honey plants are willows, maple, fruit-bloom, redbud, sumac, heartsease, and goldenrod.

West of the mountainous region, or Ozark Uplift, the Prairie Plains, which have an elevation ranging from 800 to 2000 feet above sea level, occupy the whole central portion of the state. Westward this region gradually merges into the Great

Plains, a treeless expanse covered with bunch or buffalo grass, where formerly cattle-raising was the chief industry. In south-central Oklahoma there are two small mountain ranges — the Arbuckle Mountains and the Wichita Mountains. The best locations for beekeeping are along the larger streams, as the Canadian and Washita Rivers, where the bottom-lands are a half-mile or more in width. The largest number of colonies of bees is found in Kay, Paine, Lincoln, Oklahoma Pottawatomie, Garvin, Pontotoc, and Johnston counties. The apiaries are usually small, and are maintained mostly by farmers to obtain a home supply of honey. At Stillwater, Payne County, according to W. R. Wright, the honey plants and their approximate dates of blooming in normal seasons, are as follows: White maple, February 2-9; apricot, peach, plum, and pear, March 25-30; water willow and redbud, April 1-7; black locust, April 19 to 29; raspberry and blackberry, April 12-16; white clover, April 21; alfalfa, May 3; persimmon, May 19; catalpa and yellow sweet clover, May 18; white sweet clover, May 25; sumac, June 10; horsemint, June 19; sunflower, July 2; chittimwood, July 20; cotton, July 26; heartsease, August 1; cowpeas, September 1. Alfalfa, cotton, and sweet clover are the three most abundant honey plants, but the two former are in many localities of very little value.

There are more than 350,000 acres of alfalfa grown in Oklahoma; but the area in the eastern section is relatively small. The largest part of the acreage is confined to the north-central and western counties. Beginning on the north border in Alfalfa, Grant, and Kay counties, the alfalfa belt extends southward to the center of the state, including Oklahoma and Canadian counties, whence it extends to the west border and southwesterly to the Red River. The secretion of nectar by alfalfa in Oklahoma is influenced by the same conditions as in Kansas. The largest yield is obtained near the rivers and in dry hot weather, when there is sufficient moisture in the ground. In Alfalfa County, on the north border, in which there are 17,000 acres of alfalfa, a beekeeper reports that much seed is raised, and four full crops of blooming alfalfa are cut during the season, assuring a continuous flow of nectar. In Cleveland County there are also three or four successive crops of alfalfa, and a more or less continuous honey flow from spring until October.

In 1919, 2,732,900 acres of cotton were cultivated in Oklahoma, cotton outranking corn, both in acreage and value. Temperature and rainfall permit of its cultivation in every part of the state, except along the north border and in the northwest corner. A dense area occurs near the center of the state, and another in the southwest counties. The humid conditions required to stimulate the secretion of nectar occur only occasionally, and it is, in consequence, a very unreliable honey plant. It is, however, frequently reported as furnishing more or less surplus, the honey flow lasting in Love County from July 20 to September 30. The honey is white, of heavy body, and excellent flavor.

At Stillwater, Payne County, there are no strictly commercial apiaries; but in a radius of five miles there are 50 small yards, and a surplus of 60 pounds may be obtained every other year. In Alfalfa County, so called because of its large acreage of alfalfa, there are so few bees that much nectar is reported to be lost. There is a moderate honey flow for many weeks, which does not end until the first frost late in October. At Oklahoma, Oklahoma County, there are several apiaries. A surplus of 40 to 60 pounds is secured nearly every season. A beekeeper of this city states that in his opinion more than half of the surplus honey is secured from alfalfa.

The western section of the state, or the Great Plains area, is a rolling, treeless expanse, producing chiefly bunch or buffalo grass. In the northwestern counties. cactus, yucca, and sage brush are familiar plants. The land rises in altitude from 2000 feet to 5000 feet in Cimarron County in the Panhandle. The soils are sandy loams, and the climate is semiarid with an average annual precipitation of 17 inches. In the northwest the salt plains, perfectly level tracts covered with snow-white crystals, are wholly devoid of vegetation. This section of the state lies west of the 97th meridian, or west of Alfalfa and Canadian counties and the Wichita Mountains in the south. With the exception of a few small apiaries there are no bees in the extreme northwest. A dense area of kaffir corn and milo maize furnish an abundant supply of pollen.

In Washita County there are probably 250 farmer beekeepers. With few exceptions they neglect their bees, and are ignorant of the methods of modern bee culture. In some instances colonies are located five miles from a dependable supply of water. A surplus of from 75 to 90 pounds has usually been gathered from alfalfa, cotton, goldenrod, and broomweed. Broomweed is common only after very wet springs and summers. Cotton yields best on hot, moist mornings, but it is not dependable. Alfalfa secretes nectar most freely on very hot days following a good rain.

In the southwest corner of Oklahoma there are nearly half a million acres of cotton, and a large area of alfalfa. Muskmelons are also extensively grown. Notwithstanding this great area of honey plants, relatively few bees are reported from the southwestern counties, or from the counties along the Red River. Conditions for outdoor wintering in central Oklahoma are excellent, and the hives are seldom moved from their summer stands. Severe cold weather does not often last longer than two or three days, and usually it is warm enough on one or two days nearly every week for the bees to obtain a flight. The critical period is in the spring from early in March to the middle of May. During the larger portion of this time the weather is cool and the winds are high, and the bees are able to gather very little nectar. Unless there is an abundance of winter stores or feeding is practiced, many colonies dwindle down to a mere nucleus or die of starvation. The summers are long, and usually favorable for gathering nectar. With three or four successive crops of alfalfa and many other nectar-bearing flowers, there is a light continuous flow from May until October. But in Oklahoma, as in Kansas, the future of beekeeping will be largely dependent on the increase of the acreage of sweet clover.

MONTANA

Total area, 146,572 miles. In land area Montana ranks third among the states, only Texas and California being larger. It is three and one-half times the size of Ohio. The eastern three-fifths belong to the Great Plains, and the western two-fifths are occupied by the Rocky Mountains with their spurs and outlying ranges. Montana is a semi-arid state with an average precipitation of 13 inches in the eastern portion, and of 20 inches among the mountains. Throughout a large part of the state the rainfall is sufficient in most sections for maturing grain crops without irrigation. The irrigated acreage is about 1,679,000 acres, 75 per cent. of which is in the valleys of the mountains. Of the 92,998,400 acres comprising the land surface of the state, it is estimated that 6,000,000 acres can be irrigated from all sources, including streams, reservoirs, and wells. In addition there are 10,000,000 acres of bench lands which are suitable for dry farming. Probably 50,000,000 acres can be utilized only for grazing. The cereals are very largely grown by dry-farming methods; but three-fourths or more of the acreage of clover and alfalfa are irrigated. At least 50 per cent. of the acreage of small fruits and orchard trees is also irrigated.

The eastern portion, or the Great Plains, rises from an elevation of 2000 feet along the eastern border to approximately 4500 feet in the foothills of the mountains. Its surface is a vast upland plain with a gentle slope toward the east, broken only by the valleys of the Missouri and Yellowstone rivers and their tributaries and widely distributed buttes and low hills. Cottonwoods, willows, wild plum, service berry, choke cherry, buffalo berries, and in some sections ash and Canadian poplar are common on the clay loams of the river valleys; but the vast sandy plains are treeless, and produce little except scattered clumps of bunch grass and patches of cactus and sagebrush.

In the eastern portion of the state commercial beekeeping is largely restricted to the irrigated land along the Yellowstone River and its southern tributaries; but promising locations may be found along the Musselshell River and the Milk River. The native honey flora is of little importance except for stimulating brood-rearing. Several species of willows furnish the first nectar and pollen, while thornbush, choke cherry, service berry and cultivated apple and cherry trees are helpful during the spring season. Dandelion is unusually abundant along the Yellowstone Valley, and often yields a small surplus which blends well with the later honey, although it crystallizes early. Then come the mustards (*Brassica campestris* and *B. arvensis*), common in grainfields. White clover and alsike clover are found only in very limited areas, and furnish little or no surplus. Alfalfa covers a larger area than any other plant; but sweet clover is reported to yield a larger surplus. Bees often show a marked preference for the bloom of sweet clover to that of alfalfa. While both the white and the yellow sweet clover are annually becoming more common in some parts of the state, in other localities they are decreasing, due to more intensive cultivation, especially west of Billings. At Miles City, on the Yellowstone River, and in Tongue River Valley, there is a profusion of herbaceous wild flowers, which, although of little value to bees, are an endless source of pleasure to the beekeeper. "Between my home and the river on the flats among the cottonwoods," writes a beekeeper of Miles City, "there are countless little delicate flowers of every hue and design. It is glorious."

In eastern Montana the only honey plants which are commercially important are alfalfa and sweet clover, which are restricted chiefly to the irrigated areas. Beekeeping, both present and prospective, is, therefore, largely dependent on irrigation.

In the northeast corner of the state there is a nearly square area containing 25,000 square miles, through the center of which flows the Missouri River. The principal crops are the cereals. There are few bees in this region, even in the valleys; and on the open plains they would speedily die of starvation. At Frazer, Valley County, there are two apiaries which together contain 20 colonies. From sweet clover and alfalfa 100 pounds of surplus per colony are obtained in a good year. Settlers are reported as beginning to manifest greater interest in bee culture. At Sidney, Richland County, on the Yellowstone River, there is an apiary of 100 colonies and another containing a much larger number. There are also some ten small yards in this vicinity. At Savage, in the same county, there are three apiaries which together number 600 colonies, besides a number of smaller yards. The bee territory is very limited, and is confined to the bottom-land and irrigated land along the river. Alfalfa is restricted to the irrigated land; but sweet clover grows also on the islands, and on the banks of the rivers and irrigating ditches, and will also grow on dry land, and in some places is now used for pasture. The bench lands back of the rivers are not so suitable for bee culture.

The counties along the north border and in the central portion of the state are Hill, Blaine, Phillips, Chouteau, Cascade, Fergus, and Musselshell. They have a combined area of 17,000,000 acres, and are in the "dry farm" section of the state. At Chinook, in the Milk River Valley, beekeeping at present receives little attention; but sweet clover and alfalfa are yearly becoming more abundant, and the valley promises to become a good location in the future. In Musselshell County, at Roundup and Lavina on the Musselshell River, and at Flatwillow Creek, there are several apiaries which range from 10 to 30 colonies, and one which contains about 100. In this region drought may greatly delay the blooming of alfalfa and sweet clover, and injure the prospects of the beekeeper. Winter losses are a serious problem, and often run from 10 to 75 per cent.

The portion of the Great Plains which is best adapted to bee culture is the irrigated land along the Yellowstone River and its southern tributaries. But from Big Timber, Sweet Grass County, to Glendive in the eastern part of the state, the Yellowstone Valley is narrow, and apiaries can be placed to advantage only at special points. In the counties of Custer, Rosebud, Treasure, Big Horn, Yellowstone, Stillwater, and Sweet Grass there are 218,000 acres under irrigation and more than 87,000 acres of alfalfa. At Miles City, Custer County, an apiary containing 200 colonies produced 7000 pounds of honey, a fair average per colony of extracted honey being 75 pounds. At Forsyth, Rosebud County, an apiary of 150 colonies secures a fair crop nearly every year, chiefly from sweet clover, as alfalfa is cut so early that it affords little nectar. Near the center of this county, at Ashland on the Tongue River, the population is small, and there are only a few yards which average about 20 colonies. The country is covered with a carpet of wild flowers throughout the summer. Wild plum, cherry, thorn-apple, dogwood, and different kinds of cactus are common. A beekeeper writes: "I have traveled the Tongue River Valley from Miles City nearly to Big Horn Mountains in Wyoming, and I know that sweet clover is extending over all the valley, and that more alfalfa is sown every year. I have seen 1500 acres of alfalfa and sweet clover under irrigation with only wild bees to gather the nectar."

Yellowstone County contains more colonies of bees, and produces a larger quantity of honey, than any other county in the state. There are 240,000 acres of irrigated land and 23,000 acres of alfalfa. In 1920 there were at Billings, on the Yellowstone River, three companies which operated 300, 350, and 1700 colonies respectively, divided into apiaries of 100 to 150 colonies, located 5 to 20 miles apart. In a good year a surplus of 100 pounds of extracted honey per colony was obtained from sweet clover and alfalfa; but alfalfa was usually cut as it was beginning to bloom. According to R. A. Cooley, State Entomologist, the number of colonies of bees within 100 miles of Billings is probably about 5000. A beekeeper who moved 450 colonies to Fromberg five years ago has had fair crops every year until 1920, when little surplus was gathered. Another beekeeper reports that for 11 years his colonies stored an average of seven to eight cases of comb honey annually, but that in 1920 only 1½ cases per colony were obtained. Eight years ago this territory was heavily stocked with bees; but the smallest yards have been largely destroyed by American foul brood, and the larger apiaries have suffered severely from this disease. Losses of 10 to 75 per cent. every year from wintering are reported by several of the larger producers.

At Hardin, Big Horn County, there are in the valley 14 apiaries, together containing 1700 colonies, and several smaller apiaries have been started. A crop is obtained every year which has averaged 100 to 150 pounds per colony, and in an exceptional year as high as 300 pounds per colony. The sources of surplus here as elsewhere are sweet clover and alfalfa, of which there are 12,000 acres. At Big Timber, Sweet Grass County, on the Yellowstone River, there are five apiaries which together have a total of 510 colonies. There are also 20 small yards which would average not far from 15 colonies each. The surplus ranges from 60 to 150 pounds. In some seasons 20 pounds of dandelion honey are secured in May. "We always," writes a beekeeper, "have a fair flow from sweet clover which grows along the irrigation ditches; but our winter losses are heavy — in 1920 about 40 per cent., largely the result of not providing any protection."

The western two-fifths of the state are occupied by a series of valleys and mountain ranges belonging to the Rocky Mountains. The winds from the Pacific Ocean reach the northwestern portion of Montana, modifying the climate and bringing considerable rain, which is supplemented by irrigation throughout this section. Alfalfa, clover, and timothy are the principal hay crops. Sweet clover and white clover are used extensively for pasture. Both the mountains and valleys are densely forested with various species of pine, fir, and cedar, and hundreds of acres of apples, pears, peaches, plums, and cherries, thrive in the fertile valleys of the Kootenai, Bitterroot, Missoula, and Flathead rivers. At the foot of the Bitterroot Mountains the river of Clark's Fork takes its rise and flows westerly to join the Columbia River. The climate of this country resembles that of the Pacific coast.

The counties in northwestern Montana have few acres of alfalfa, and few colonies of bees. Near Helena in the west-central portion of the state the apiaries range from 10 to 100 colonies, and the surplus comes mainly from sweet clover and alfalfa. The honey flow continues for about two months from June to September. In a part of this region the elevation is too high for the best results. At Corvallis, Ravalli County, on the west border, there is a large number of apiaries of good size. There are also a great number of small apiaries which range from 5 to 20 colonies. A beekeeper writes: "I obtain a crop every year, and average about 125 pounds of extracted honey. I came here nine years ago with nothing, and to-day have a fine house and a modern honey-house in which everything is handled by steam and electricity. I have made it pay." The honey plants in Ravalli County are white sweet clover, yellow sweet clover, alfalfa, mustard, white clover, alsike clover, fruit bloom, fireweed (*Epilobium angustifolium*), buckbush (*Symphoricarpus occidentalis*), and the Rocky Mountain honey plant (*Cleome serrulata*). Dandelion makes a remarkable growth here, and in spring carpets the ground with yellow. Strong colonies will store from 6 to 10 pounds a day from this source. There are in this county 94,000 acres under irrigation.

The southwestern portion of the state requires irrigation. Three large valleys are the Jefferson, Madison, and Gallatin, in which rise the streams which unite to form the Missouri River. As the result of the dryer climate the valleys are treeless and covered with grasses, but the mountain-sides are timbered. Gallatin Valley is 28 miles long, and about 14 miles wide, and is surrounded by mountain ranges which, southward, rise into sharp-pointed snow-covered peaks. The soils are loams mixed with sand and gravel. The mercury at times falls as low as 50 degrees below zero. Frosts always occur in May and September, so that the growing season is comparatively short. Light frosts also sometimes occur in the summer months. A beekeeper writes from Jeffers, Madison County: "I have the only apiary of importance. It contains 50 colonies, and produces a little surplus every year, averaging 30 to 40 pounds per colony. Alfalfa is the chief honey plant. There is only a little sweet clover. Our winters are so long that, when spring comes, the colonies have dwindled down to only two or three frames. It is not until August that they are strong enough to store any honey in the supers, and the season closes about the 20th of the month. There are a few farms which have one or two colonies."

In the extreme southwest corner is the great county of Beaverhead, containing 3,000,000 acres. A beekeeper at Dillon writes: "I do not know of a single apiary in this county, which is larger than the state of Connecticut. The lowest altitude is above 5000 feet, and the summer season is short. Hay and livestock are the chief products, although there are many thousand acres of alfalfa. Bees should do well in the valleys, but at best the season is short. So far this spring I have not seen one honeybee."

WYOMING

Wyoming is 355 miles in length and 276 miles in width, and has a total area of 97,890 square miles. The eastern section, east of the Big Horn Range in the north and the Laramie Range in the south, forms a part of the Great Plains. The western fourth of the state is covered by the Northern Rocky Mountains, and the southwestern region is indented by three fingerlike projections of the Southern Rocky Montains. The southern and central regions, intervening between the two mountain systems, is a gently rolling country, barren of trees, known as the Wyoming Basin. It extends 250 miles north and south and an equal distance east and west. In the northwest corner of the state is the Yellowstone National Park.

The forests, in which the prevailing trees are pine and spruce, are confined chiefly to the Rocky Mountains and the Big Horn Mountains. While large areas of the treeless plains produce nutritious grasses, westward there are many sections in which the vegetation consists chiefly of sagebrush and saltbush. The presence of sagebrush is always a reliable indication of a good sandy loam; but greasewood (*Sarcobatus vermiculatus*) indicates a poor alkaline soil. Big Horn County and Carbon County are sagebrush counties; but on the low flat alkaline land of southwestern Sweetwater County greasewood is abundant. The eastern counties of Crook, Campbell, Weston, Niobrara, Converse, Goshen, Platte, Laramie, Albany, Natrona, Johnson, and Sheridan are largely covered with native grasses; but there are areas in which sagebrush and even greasewood predominate. The mean elevation of the state is 6000 feet; but in the Rocky Mountains, where the winters are severe and the snowfall is heavy, an altitude of 14,000 feet is reached. Of 62,645,120 acres of land in Wyoming, about one-sixth is included in the forest reserves; 6,000,000 acres are estimated to be irrigable, and 4,000,000 are suitable for dry farming. The remaining acres are classified as grazing lands.

Wyoming is a semi-arid state with an annual rainfall ranging from 10 to 20 inches. In 1920 the area irrigated was 1,207,982 acres, and the area included in enterprises was 2,564,688 acres. The whole eastern portion of the state is drained by the tributaries of the Missouri River; the southwestern region drains into the Colorado River, and the area along the central and western border into the Columbia River. Practically the entire summer flow of the streams is now utilized, and the future of irrigation in Wyoming will depend on the construction of large reservoirs for the storing of the winter flow and for its gradual distribution.

Commercial beekeeping is entirely dependent on sweet clover and alfalfa, which grow only on irrigated land. These areas are mere garden spots compared to the great region of dry desert land which produces little besides sagebrush, saltbush, and cactus. "A colony of bees would starve on a million acres of such a range." No one has attempted to keep bees in the mountains, as the snowfall is heavier, the winters colder, and the seasons shorter than at lower altitudes. While there are many wild flowers it is doubtful if they would yield a surplus. American foul brood is prevalent in many localities. The loss of bees during the winter from extreme cold weather is often serious, and may run as high as 50 per cent.

There is usually an excellent honey flow from mid-June to September 10 from sweet clover and alfalfa. According to T. S. Parsons, Specialist in Crops, sweet clover is the principal honey plant in the state, producing more honey than all other sources combined. It is very common along the canals and lateral ditches, and in 1919 there were 330,000 acres. White clover is rather common in the northern part of the state. There is also a variety of wild flowers which are helpful in maintaining the strength of the colonies, as willows, dandelion, wild fruit bloom, goldenrod, and aster. In Big Horn County there is a large area of alfalfa and a greater number of colonies of bees than can be found in any other county in the state. Second in importance is the region around Landers in Fremont County, which has the largest irrigation project in Wyoming. Sheridan County, east of the Big Horn Range, and Goshen and Platte counties in the southeast, also produce a large surplus of honey.

East of the Big Horn and Laramie Ranges the eastern portion of the state, or the Great Plains, slopes gently from an altitude of 6000 feet at the foot of the mountains to 4000 feet along the east border. This area is without trees, but is largely covered with native grasses, sagebrush, and in places with greasewood. It is largely used for grazing purposes.

The counties in the northeast corner have many streams, but the late summer flow of water is so small that it is necessary to rely largely on dry-farming methods. Beekeeping in this section has been almost entirely neglected, as there is only a small acreage under irrigation. Sheridan County, on the north border, east of the Big Horn Mountains, is a much better location. In 1919 it produced over 29,000 pounds of honey, chiefly from 20,000 acres of alfalfa. There were under irrigation 68,000 acres, and 108,000 were included in enterprises. Under a great storage system Sheridan County will ultimately reclaim all irrigable land within its borders. There are seven or eight apiaries in this county which contain from 30 to 150 colonies of bees, but not more than three exceed 100 colonies. A crop of 50 to 100 pounds per colony is obtained every year from sweet clover and alfalfa."

In the southeast corner of the state Goshen and Platte Counties reported 122,000 acres under irrigation and 48,000 acres of alfalfa. The irrigated areas are located chiefly along the North Platte River and its southern tributary, the Laramie River. Douglas, in the southeastern part of Converse County; Wheatland, in Platte County, and Torrington, in Goshen County, are in the heart of these areas. Near Douglas there are apiaries which contain one hundred colonies. At Wheatland, from which a carload of honey is usually shipped yearly, several apiarists operate from 500 to 800 colonies, but the average is nearer fifty. There is almost an entire absence of native honey plants, and the surplus is gathered wholly from successive crops of alfalfa from June 20 to September 10, the best flow coming in the hottest weather when there are occasional showers. A few years ago from Old Fort Laramie to the state line of Nebraska there were great areas of irrigated alfalfa and sweet clover by the roadsides and on the banks of the ditches, with only a few small yards of bees to gather the nectar. The altitude of the valley is 4200 feet, and the climate is nearly similar to that of Colorado. Sweet clover grows well wherever it is permitted to establish itself. The extreme cold, both early in the fall and late in the spring of the winter of 1919 and 1920, proved disastrous to beekeepers in this section. The bee inspector of Goshen County writes: "At the beginning of the winter I had 280 colonies, and lost 110." At Rock River there are some 10 apiaries which range from 50 to 200 colonies. The average surplus has been 60 pounds of comb honey and 100 extracted, except in 1919.

The Big Horn Basin, which includes the counties of Big Horn, Park, Washakie, and Hot Springs, has a larger number of colonies of bees, and produces more pounds of honey, than any other section of the state. In Big Horn and Shoshone counties the irrigation enterprises are situated chiefly on the Big Horn River and its tributary, the Shoshone River. As the result of agricultural development many new towns have grown up, as Worland, Byron, Cowley, and Lovell. In 1919 Big Horn County reported 4800 colonies of bees, 512,000 pounds of honey, 52,000 acres of alfalfa, and 109,000 acres under irrigation. In the vicinity of Basin, the county seat, more than twenty thousand acres have been reclaimed by the Big Horn Canal, which is about 35 miles long. Sweet clover grows wild along the canal ditches, and yields the larger part of the surplus honey. Every 8 or 10 miles there is an apiary, ranging from 50 to 150 colonies, which secures from sweet clover and alfalfa an average surplus per colony of 125 pounds of extracted honey and 60 pounds of comb honey. A beekeeper at Basin writes: "Four or five years ago, before this locality was overstocked, my average surplus was 400 pounds of extracted honey per colony, fall count; now it is 90 pounds per colony, spring count. Besides the hives of bees there are a large number of swarms in the cottonwood trees which form a narrow fringe along the river — 34 were counted in a distance of one mile. Many acres are planted with sugar beets, which yield no nectar, and every acre of beets supplants an acre of alfalfa. There are four commercial apiaries which average about 150 colonies each."

At Hyattville, Big Horn County, there are 10 apiaries which are estimated to contain over 2000 colonies. In a good year a surplus of 200 to 300 pounds of honey is obtained, but the flow may be closed by an early frost, or delayed by a wet spring. The valley here is less than a mile in width, and north and south there are "bad lands" for 20 miles. Three apiaries number 200, 400, and 500 colonies respectively. At Worland and Powell equally large apiaries are reported. In Park County seventy-seven thousand acres of irrigable land have been reclaimed with water diverted from the Shoshone River.

Another important center for beekeeping is at Landers, Fremont County, in the west-central portion of the state, which has produced as high as 1,000,000 pounds of extracted honey in a season, and where there are 31,000 acres of alfalfa and 115,000 acres under irrigation. In the Landers Valley dandelion is so abundant that it affords a valuable early crop. White sweet clover is the most important honey plant. Alfalfa is of much less value. The production of honey in this locality is largely confined to three companies which operate from 500 to 1000 colonies each. The bee-man here is forced to encounter serious difficulties. The winters are long and cold, and nearly all bees in Fremont County are wintered in cellars. The summers are limited to June, July, and August. Bad spring weather is often disastrous to whole apiaries, and at times there are killing frosts in June. Sweet clover grows almost wholly on the banks of the irrigating-ditches and in low wet places. Alfalfa is usually cut twice, and is in bloom from 10 days to three weeks. Ultimately 300,000 acres in one area will be reclaimed in this county in the ceded portion of the Shoshone Indian Reservation. The soil and climate are very favorable to agricultural purposes, but in the northern part of the county there is an area of "bad lands."

There are no bees in Sweetwater County in the southwest; no fruit trees, and only a small acreage of alfalfa. The soil in the southwestern portion of the county is strongly alkaline, and covered with greasewood (*Sarcobatus vermiculatus*). The great Continental Divide passes through eastern Sweetwater County, west of which there is a belt of land where water is so scarce, and there is so little rain, that it is used as a winter range for sheep. In the northern part of the county there is an alkali and greasewood flat. But in the northwestern portion the Big Sandy River project provides for the irrigation of more than 10,000 acres.

The western section of Wyoming is occupied by the Rocky Mountains, and has an elevation of 6000 or more feet. So severe are the winters, so heavy are the snows, and frosts are so liable to occur in May and September, that much of Lincoln and Uinta counties, on the west border, are unsuitable for beekeeping. The crops consist chiefly of cereals, hay, and forage. At present there are no commercial apiaries. The streams flow westward, and are tributaries of either the Columbia or Colorado rivers, with the exception of Bear River, which finds an outlet in Great Salt Lake. There are 271,000 acres under irrigation, and 22,000 acres of alfalfa in the two counties. Green River rises in the high mountains where there is a heavy snowfall, and carries a large volume of water, much of which is unused. The valleys through which it flows are undeveloped, and lack transportation facilities. The course of Snake River is through high mountain valleys. But the southern part of Uinta County is either sagebrush country or rolling grassy prairie; and the Upper Green River country is a continuous meadow.

In the northwestern corner of the state is the Yellowstone National Park, notable for its hot springs and geysers and its magnificent natural scenery. Its area is 3575 square miles, all of which is at an altitude of more than 6000 feet. There is a great variety of wild flowers, but the forests consist mainly of Douglas spruce and yellow pine. The underlying rock is volcanic. Frosts may occur throughout the summer, and snow begins to fall in September.

COLORADO

Total area, 103,948 square miles. Colorado is 290 miles wide and 380 miles long, and ranks seventh in area among the states. It is crossed near the center by the Rocky Mountains, a series of lofty parallel ranges containing 180 peaks which exceed 12,000 feet in height. Two-fifths of the territory east of the mountains are level or rolling, and belong to the Great Plains. West of the Rocky Mountains is the Colorado Plateau, or Western Slope, the surface of which is broken by isolated mountains and small ranges, and in its central portion is deeply dissected by canyons. In a more general way Colorado may be divided into two sections: an eastern broad level expanse, and a western mountainous area. Nearly all the cities and towns are located in the foothills and mountains.

The Great Plains, which occupy the eastern two-fifths of the state, have an essentially arid climate with a rainfall of 11 to 15 inches per annum. The land rises from an altitude of 3000 feet along the eastern border to an elevation of 7000 feet in the foothills, where this section is terminated by the bold outline of the front range of the Rocky Mountains. Stock-raising and dry farming are the principal industries. While dry farming without irrigation will in some years yield large profits, in other years it is a complete failure. If, however, from one to ten acres can be irrigated from wells or small reservoirs the dry farms will secure large crops in wet years and escape serious loss in years of drought. The nights are dry and without dew, and in summer the heat is seldom oppressive, although the days are clear and cloudless. Light snowfalls are common in winter, but the snow seldom remains on the ground for more than a day or two at a time. The soils of the river valleys are alluvial, and support belts of cottonwoods and willows, but otherwise the Plains are treeless. The uplands are covered with sandy loams, which are residual, or formed by the weathering of the underlying rock.

In 1919 there were in Colorado 3,348,000 acres under irrigation, of which 1,300,000 acres were in the valleys of the South Platte and Arkansas rivers. It is chiefly in these two river valleys that beekeeping is an important industry in the Great Plains. Rising in the region of perpetual snows, they descend as mountain torrents in steep narrow channels to the Plains, which they cross in broad sandy valleys. In the east-central portion of the state, as in Kit Carson, Cheyenne, and Lincoln counties, there is practically no honey flora and no bees.

The South Platte River flows through the northeastern portion of the Great Plains. On the immense area under irrigation there are thousands of acres of alfalfa and sweet clover, which furnish most of the surplus honey. A beekeeper in

Sedgwick County says he has never had a failure, and that his average per colony has been 125 pounds of honey, and in 1918 it was 200 pounds per colony. Alfalfa and sweet clover grow throughout the valley, sweet clover being abundant on the bottom-lands and islands of the river. In the spring the bees get much pollen from the cottonwoods and willows, and in some seasons store a surplus in the fall from asters. In Morgan County apiaries of specialists range from 45 to 500 colonies, and average, with rare exceptions, 80 pounds of surplus. More than one hundred thousand acres are under irrigation. In Weld County, which has 383,000 acres under irrigation, there are many bees. Commercial apiaries may be counted by the dozen. In some seasons the average production per colony has been 60 pounds of comb honey and 80 pounds of extracted honey. A beekeeper at Greeley with 500 colonies in nine yards, has secured as high as 20,000 pounds of honey, but this is above the average, as some seasons are less favorable. According to Frank Rauchfuss, the seasons of 1923 and 1924 were lean ones for beekeepers on the eastern slope of the Rocky Mountains, due to light rainfall and a small amount of snow in the mountains, which reduced the supply of irrigation water. In the six counties of Sedgwick, Logan, Washington, Morgan, Weld, and Adams there are 650,000 irrigated acres and 166,000 acres of alfalfa.

Larimer, Boulder, and Jefferson counties lie partly in the Great Plains and partly in the mountains; but it is only in the eastern portion of this region that beekeeping is an important industry. In Larimer County 200 apiaries reported, in 1916, 5500 colonies of bees, and an average per colony of 44 pounds of comb honey and 60 pounds of extracted honey. (Eighth Annual Report of the State Entomologist of Colorado.) Winter losses are at times heavy. In 1920 a beekeeper reported that 40 per cent. of his colonies perished. In Boulder County, in 280 apiaries containing 8500 colonies of bees, there has been secured an average per colony of two cases of comb honey and 60 pounds of extracted honey. There are a few apiarists in this section who have operated as high as 2000 colonies in apiaries of about 200 colonies each. No extra packing or protection is given the bees in winter; and as almost every day is clear, most of the colonies survive, although much weakened. At Arvada, Jefferson County, in a radius of three miles, there are several apiaries which average about 70 pounds of honey per colony. Dandelion is abundant, and in warm seasons strong colonies store a surplus. In 20 years there have been two complete failures — in 1902 and 1910. American foul brood is prevalent, and in some locations the smoke from the smelters is highly injurious.

After flowing through the mountains for 100 miles the Arkansas River emerges from the Grand Cañon near Cañon City, and crosses the Plains through the counties of Pueblo, Otero, Bent, and Prowers. There are in the valley of this river about 300,000 irrigated acres and 140,000 acres of alfalfa. Its many tributaries in the Plains are dry channels during eight months of the year, but the snows in the mountains, melting during the months of May and June, provide a supply of water during the irrigation season. Some very large yields in Prowers County have been reported. In Bent County 4000 colonies of bees in 140 apiaries have reported an average per colony of 30 pounds of comb honey and 45 pounds of extracted honey. Near Las Animas, on the Arkansas River, there are numerous apiaries. There has been only one failure in 15 years. The honey plants are alfalfa, sweet clover, and cleome. From Otero County about five carloads of honey have been shipped in a single year. At Rocky Ford four specialists are reported to operate over 3000 colonies. The location has rather too many colonies for the pasturage, as alfalfa is often cut so early as to be of little value.

From Campo, Baca County, in the southeast corner, a beekeeper writes: "There are no bees, so far as I know, in this county, which up to five years ago remained largely unsettled."

The center of the state is occupied by the Rocky Mountains, a series of lofty ranges running nearly north and south, in which there are many peaks rising to the height of 12,000 to 14,000 feet. West of the Eastern or Front Range are four extensive valleys or parks which have a combined area of 13,000 square miles. They are known as the North, Middle, South, and San Luis parks, and were formerly the basins of great lakes. The soils of the park lands are fertile, and covered with abundant vegetation. There is little irrigation in the mountains except in the San Luis Valley, and the area of alfalfa is small. Many counties report no bees, and others only a few hundred colonies. Adverse conditions occur in Fremont County, where orchardists spray their fruit trees before the petals fall, and American foul brood is common. Near Cañon City and Florence there are a few bees, and most of the apiaries are small. In Custer County, at Westcliffe, the altitude is 8000 feet, and, while the natural meadows are covered with wild flowers, very few bees work on them. Many ranchmen have a few colonies which are much neglected. Dandelion and aster at times yield a surplus; but here, as on the plains, alfalfa and sweet clover are the chief sources of nectar.

In the southern portion of the Rocky Mountain region there are 500,000 acres taking water from the Rio Grande and its tributaries. Rising in the mountains of southern Colorado this river flows through the San Luis Valley into New Mexico. The surface of the valley, which has an area of 9400 square miles and an elevation of 7500 feet, is nearly flat, as it was formerly the bed of a great inland lake. The principal crops are alfalfa, cereals, potatoes, and Canada field peas. Beekeeping in this valley is in an undeveloped condition, but there are probably many localities in which it would be fairly successful. At Moses and at Hooper there are a few bees. During the past four years there has been a fair flow in July and August. The honey plants are alfalfa, sweet clover, and Rocky Mountain bee plant (*Cleome serrulata*). At Hooper, in 1919, one colony stored 200 pounds in August; and the entire yard averaged over 200 pounds, spring count, and increased 150 per cent.

The Western Slope, or Colorado Plateau, comprises the counties situated on the western boundary line of the state, but the level portion of Moffat County belongs to the Wyoming Basin. Distributed irregularly over this section are numerous isolated mountains and short mountain ranges, elevated treeless plains, and mesas, or grassy tablelands bounded by steep scarps or walls. Along the Gunnison, Uncompaghre, and other streams are deep canyons which extend for miles. The rainfall is so small that agriculture can not be carried on without irrigation. For the production of honey the Western Slope may be divided into three regions: the northwestern region, the Colorado River Basin, or the Grand Junction and Montrose region, and the southwestern region, or the San Juan River Basin.

In the northwestern corner of the state, in Moffat County, there are 17,000 acres under irrigation, chiefly along the Little Snake River. On both sides of the river are comparatively level arid lands. Yampa River flows in a part of its course through a deep cañyon, south of which there is a ridged tableland. Only a few bees are reported in this county. The adjoining county of Routt is largely included in the Park Range National Forest. Rio Blanco County has 28,000 acres under irrigation on White River, and 15,000 acres of alfalfa under cultivation. At Meeker, on White River, there is an abundance of white clover, and in May and June many wild flowers. Nearly every one has a few colonies. A beekeeper writes: "I know of no other locality better than this with fewer bees. The possibilities are great." But this is an isolated locality, rather limited, and the honey produced here must be hauled a considerable distance over the mountains to reach a railroad, as there is practically no local sale for it.

The entire valley of Grand River from Glenwood Springs to Grand Junction is well adapted to bee culture. In Garfield County there are 73,000 acres under irrigation along Grand River, and there are 31,000 acres of alfalfa. An average production per colony of 34 pounds of comb honey and 20 pounds of extracted has been reported. From Newcastle a beekeeper writes: "My experience with 400 colonies in my present location has been profitable, as the weather conditions are most favorable, and the climate ideal. All land above 7000 feet is not suitable." From Rifle another apiarist writes: "I have been in the business for 15 years, and handle about 400 colonies with fair results. There are some 15 commercial apiaries in this locality."

In Mesa County also there is a large acreage of alfalfa and many commercial apiaries. Sweet clover grows only along ditches. Foul brood and spraying of fruit bloom with poisonous solutions have destroyed the bees in the orchard districts. Instead of two or three sprayings there are now in some orchards seven or eight, and greater loss is caused by the poison on the cover crops than by that on the bloom. Most of the beekeepers have in consequence moved away from the orchards. This will prove in the end injurious to the fruit-growers, as there will not be sufficient bees present to pollinate the flowers properly. Alfalfa is often cut as soon as it begins to bloom. In Delta County, along the Gunnison and Uncompaghre rivers there are 30,000 acres of irrigated alfalfa and 7000 colonies of bees. An average surplus per colony in 150 apiaries was two cases of comb and 40 pounds of extracted honey.

In Montrose County, which adjoins Mesa and Delta counties on the south, the irrigated area along the Uncompaghre River comprises about 94,000 acres. The area of alfalfa in 1919 was 34,000 acres and a few years previously the county bee inspector reported 7000 colonies in 75 apiaries. Four thousand cases of comb honey have been produced in a season besides some thousands of pounds of extracted honey. In no counties of Colorado have better results been obtained than in Mesa, Delta, and Montrose counties, but beekeepers in these counties feel that they are becoming crowded and are opposed to having others come into this territory.

In the counties of San Miguel, Dolores, and Montezuma, in the southwest corner, there are few colonies of bees and a small acreage of alfalfa. In Dolores County there is only a little over 1000 acres under irrigation. In Montezuma County, in the extreme southwest, conditions are much better; and along the small rivers there are 44,000 acres under irrigation and 22,000 acres of alfalfa. A beekeeper at Mancos writes: "Commercial apiaries contain from 30 to 300 colonies. I have 90 colonies, and obtain a good crop every year." Another beekeeper, at Cortez, writes: "After an experience of 10 years I can say that honey production here is a very reliable and a fairly remunerative employment. The most serious drawbacks are foul brood, which requires constant attention, our isolation, and high freight rates. Our seasons are very certain, although always late." In La Plata County, which adjoins Montezuma County, there are 63,000 acres under irrigation, and 26,000 acres of alfalfa. Many apiaries report an average surplus of 50 pounds per colony. The percentage of bees in box hives in many counties is very small; but in La Plata it is 15 per cent.

A list of the honey plants of Colorado, by C. P. Gillette, State Entomologist, was published in the Fifteenth Annual Report of the Agricultural Experiment Station, 1902. The following plants yield only pollen: American elm (*Ulmus americana*): wind-flower (*Anemone patens*, var. *nutalliana*); Lombardy poplar (*Populus dilatata*); white ash (*Fraxinus americana*); cottonwood (*Populus balsamifera*); sedge (*Carex. sp.*); western ragweed (*Ambrosia psilostachya*); poppy (*Argemone platy-*

ceras). The poplars yield large quantities of pollen, which is eagerly gathered by bees.

The more important nectar-yielding plants are as follows: Willow (*Salix amygdaloides*), box elder (*Acer Negundo*), Plum (*Prunus americana*), dandelion (*Taraxicum officinale*), apple (*Pyrus Malus*), mountain maple (*Acer glabrum*), rattleweed or loco (*Astragalus caryocarpus*), wild gooseberry (*Ribes irriguum*), choke cherry (*Prunus virginiana*), figwort (*Scrophularia marylandica*), white clover (*Trifolium repens*), alfalfa (*Medicago sativa*), yellow sweet clover (*Melilotus officinalis*), raspberry (*Rubus deliciosa*), salmon berry (*Rubus parviflorus*), black locust (*Robinia Pseudo-Acacia*), Red gaura (*Gaura coccinea*), purple gilia (*Gilia pungens*), honey locust (*Gleditsia triacanthos*), deer's tongue (*Frasera speciosa*), Rocky Mountain Bee Plant (*Cleome serrulata*), sunflower (*Helianthus annuus*), white sweet clover (*Melilotus alba*), red mallow (*Malvastrum coccineum*), rabbit brush (*Chrysothamnus pumilus*), oreocarya (*Oreocarya virgata*), rosinweed (*Grindelia squarrosa*).

The wild bloom is of much value in building up the colonies in spring and should be familiar to every beekeeper. The Rocky Mountain bee plant is the source of only a small amount of surplus to-day, but in localities it is still important both in the eastern and western counties. In the foothills no shrub yields more nectar than the choke cherry. Rattleweed, raspberry, and mallow are freely visited by honeybees. White sweet clover is rapidly becoming more common, as it will thrive on land not adapted to growing alfalfa. It is believed that sweet clover will play an important part in the development of dry farming. There is a wealth of wild flowers in Colorado, many of which are to some extent visited by honeybees, but as a whole they are of little value to bee culture.

NEW MEXICO

Total area, 122,634 square miles. The eastern part of the state belongs to the Great Plains; the north-central portion is occupied by the Rocky Mountains; and the western and southern portions are desert plains diversified by short mountain ranges with many intervening valleys, mesas, and lava beds. Of the 70,000,000 acres, approximately 2,000,000 acres are irrigable. Agriculture and beekeeping are mainly restricted to the valleys of the Rio Grande, Pecos, Canadian, and San Juan rivers and their tributaries. The soils in the river valleys are fertile sandy and heavy loams, which have been derived from the rocks of the neighboring hills. In the desert plains the soils are composed largely of gravel and sand. The higher mountain ranges and plateaus are forested with pine, fir, cedar, and oak. On the arid plains northward sagebrush and greasewood are abundant, and southward cactus and yucca. The only trees common along the rivers in the valleys are willows and cottonwood.

In 1920, 538,000 acres were under irrigation, and 66,000 acres were available for settlement. Irrigation is practiced mainly in four sections of the state. In the north-central portion, in the Rocky Mountain region, in the counties of Colfax, Mora, Taos, Rio Arriba and San Miguel, there are 206,000 acres irrigated from the Canadian and Rio Grande rivers and their tributaries. The annual rainfall in this section is from 18 to 20 inches. In the southeast section, in Chaves and Eddy counties, 93,000 acres are irrigated from the Pecos River and its tributaries. In the Pecos Valley the annual rainfall is 15 inches. At times the river carries large floods, and at other times the water is mostly lost in the sands and by evaporation. In the south-central portion of the state in the Rio Grande Valley, in Dona Ana and Sierra Counties, there are 40,000 acres under irrigation. The average annual rainfall is less than six inches. On the high plateaus in the western part of New

Mexico there is little water for irrigation, and the land is mostly barren of vegetation for a large part of the year. In the southwest corner the Gila River, and underground water pumped from wells, provide irrigation for 20,000 acres. Beekeeping is confined almost entirely at present to the irrigated areas of the Pecos, Rio Grande, and San Juan Rivers.

In New Mexico the Great Plains comprise the larger part of the twelve western counties. The southeastern region is a western extension of the well-known Staked Plains of Texas. It is a generally level country, covered with coarse grasses, and chiefly utilized for stock-raising. There is no opportunity for irrigation, and no honey flora which would support bees.

Beekeeping in the southeastern part of the state is restricted largely to the valley of the Pecos River. In Chaves and Eddy counties there are 93,000 acres of irrigated land and 31,000 acres of alfalfa. A large acreage of fruit trees, cotton, and melons also succeeds well on the fertile alluvial soil. The altitude of the land is 3000 feet, and the temperature ranges from zero to 110 degrees F. Near Carlsbad, Eddy County, there are several commercial apiaries which obtain an average surplus of 75 pounds of honey. Besides alfalfa, mesquite and catsclaw are important. At Loving the average surplus secured by commercial apiaries was formerly about 100 pounds of honey per colony; but during the Great War alfalfa was partly replaced by cotton, and at present the average is about 60 pounds per colony. A beekeeper at Hagerman, Chaves County, reports that in that locality there are numerous commercial apiaries which range from 50 to 150 colonies. The average surplus is from 60 to 75 pounds of honey, and it is obtained every year. The area adapted to bee culture does not extend north of Roswell. From the Hondo River, a western tributary of the Pecos River, some 8000 acres of fertile soil are successfully irrigated, and the local community is one of the most prosperous in New Mexico. Besides alfalfa, black willow, sweet clover, catsclaw, mesquite, thistle, and goldenrod are important honey plants.

The winter, which is very short and mild in the Pecos Valley, seems like summer to a northern man. Bees are universally wintered on their summer stands without extra protection. But at an altitude of 3500 feet the nights are cool throughout the year, and it is difficult to get strong colonies for the opening of the alfalfa flow, about May 15. A period of cool weather often causes the colonies to dwindle badly. There is little trouble from swarming; but there is much difficulty in keeping the colonies supplied with queens. A continued honey flow of about 90 days enables the careful beekeeper to harvest 100 pounds of extracted honey per colony nearly every year. The great acreage of apple bloom is not of much benefit to the bees. The weather is too cool, and there is danger from poisoning, as some orchardists spray their trees when in bloom. Thus colonies must be built up during the first part of alfalfa bloom, which involves the loss of a whole month.

The field around Roswell is well stocked with bees; but the area of alfalfa is constantly extending. For 60 miles in length and 10 miles in breadth this valley is like a beautiful garden. Great fields of alfalfa and immense apple orchards, truck gardening, and poultry, all yield profitable returns. Crop failures are almost unknown to the Pecos Valley beekeepers; but it is no place for a careless beekeeper. The honey is excellent, but it is light-amber-colored.

In the northern part of the state the western portions of Colfax, Mora, and San Miguel counties extend into the Rocky Mountains, where large reservoirs have been constructed providing for the irrigation of more than 140,000 acres. In Colfax and Mora counties there are about 16,000 acres of alfalfa At East Las Vegas, San Miguel County, there are no apiaries containing more than 50 colonies, and the average size is about ten. A crop of 75 to 100 pounds per colony is obtained eight years in ten. Colonies must be watched closely, as many hives during the winter lose their queens. In the mountainous counties of Taos and Rio Arriba, on the north border, there are only a few colonies of bees, and the acreage of alfalfa is small, although there is a large area of irrigated land.

The Rio Grande River has its source in the Rocky Mountains in the southern part of Colorado, and flows southward through the central part of New Mexico. The valley is described by Dr. E. F. Phillips as follows: "The Rio Grande Valley from El Paso northward to Albuquerque is a typical irrigated alfalfa section. Mesquite occurs in this valley as far north as Belen. Beekeepers formerly depended on mesquite and other desert plants, but now pay attention only to alfalfa and sweet clover. The valley narrows northward so that there are few beekeepers north of Socorro, and I found only one commercial producer at Albuquerque. The honey is amber-colored. There are great apple orchards, but the beekeepers utilize the nectar only for brood-rearing. Beekeeping in New Mexico is crude, and little care is given to the colonies. There are a number of big producers near Las Cruces, but they have much to learn before they can produce maximum crops. I assume the mesquite is unattractive to them because of the universal difficulty of getting the colonies strong enough to gather the nectar."

Dona Ana County, in the southern portion of the Grande River Valley, has 32,000 acres under irrigation and 14,000 acres of alfalfa under cultivation. Near Engle there has been constructed a dam forming the largest artificial lake in the world. This gigantic reservoir will irrigate 180,000 acres in New Mexico, Texas, and Mexico. Apiaries range from 80 to 150 colonies. In the vicinity of La Mesa colonies average from 20 to 60 pounds, according to the season. In some seasons the mesquite flow is fine, but it is often a failure. Out of four cuttings of alfalfa, two may yield a surplus, but more often only one. There is only a little sweet clover along the ditches. The farmers in the valley are planting cucumbers and cantaloupes extensively, which are a great benefit, and will greatly improve the business. There are big bee-men there who operate 1500 colonies who do not average 20 pounds per colony, due to negligence.

The principal honey plants of the Grande River Valley are alfalfa, sweet clover, and mesquite; willows which in some seasons yield for three weeks; cachinella (*Berthelotia sericea*), a shrub about three feet tall, with pink flowers and silky leaves, which yields a light honey; tornillo, or screw bean (*Strombocarpa odorata*), a common large shrub in the upper valleys in the southern part of the state, the pods of which contain a large amount of sugar; catsclaw (*Acacia Greggii*); and sunflower and goldenrod, which yield a yellow honey of poor quality. The best crops are obtained when there have been a wet winter and a fair amount of rain in early spring. High winds or cold weather during the honey flow are very unfavorable.

From Socorro County a beekeeper at Socorro on the Grande River writes: "I have about all the bees in this country in out-apiaries, which range from 200 colonies down to 50." At Belen, on the Rio Grande River, Valencia County, T. L. Gunter reports that he operates 1650 colonies and covers all the available range. During eight years he has had three good crops, two half crops, and three failures. In a good year his average per colony from alfalfa and sweet clover is 80 to 100 pounds per hive. In Bernalillo County, farther northward, there are about 4000 acres of alfalfa. An immense reservoir has been built for irrigating the Bluewater and San Mateo valleys, which comprise about 21,000 acres. In Sante Fe County beekeeping has received very little attention.

Midway between the Pecos Valley and the Rio Grande Valley is the county of Otero. A beekeeper at Tulerosa writes: "There are 400 colonies here which obtain every year a surplus of 150 pounds per colony from alfalfa, sweet clover, and mesquite. I have lived here 20 years, and have kept bees the entire time. Our annual rainfall is only five inches, and the country is a desert except in the high mountains, where the seasons are short. I have kept from 50 to 300 colonies—all Italians. Our bee-range at Tularosa is three miles long and one and a half miles wide." Westward of Tularosa is the noted "Journey of Death," a desert 30 to 40 miles wide. A remarkable feature of this region is the "White Sands," an area of 300 square miles covered with dunes of almost pure granular gypsum. In Tertiary times an ancient lake, known as Lake Otero, occupied an area of nearly 2000 square miles.

The third important river valley in New Mexico for the production of honey is that of the San Juan River in the northwest corner. There are here nearly 1,000,000 acres of excellent red sandy loam lying mostly on sloping mesas, with numerous arroyos providing ample drainage, which can be brought under irrigation. Late frosts in spring seldom occur, so that this region is particularly well suited to apple orchards. According to a report from Kirtland there are probably 50 beekeepers in this county who ship annually three cars of comb honey besides some extracted honey. The larger apiaries contain about 100 colonies, but there is one at Farmington with 150 colonies and another at Kirtland with 200 colonies. A failure occurs only about once in seven years, and the surplus varies from 50 to 300 pounds per colony. The alfalfa weevil and grass-hoppers often curtail the production to a considerable extent. The early honey is nearly white, but the main crop is light amber. Besides alfalfa the most important honey plants are sweet clover, which yields on an average for three weeks; Rocky Mountain bee plant, which in some seasons yields for two months; wild sunflower; rabbit-brush and milkweed.

Along the western border in the desert valleys, on the great mesas and lava beds, and among the numerous mountain ranges there is little alfalfa and very few colonies of bees. From Tohatchi, McKinley County, a beekeeper writes that he brought there a few colonies in the spring, fed nearly all that year, and finally obtained a little surplus which was used for feeding the following year. There are a few colonies in Grant County in the southwest corner, but no commercial apiaries.

IDAHO

Total area, 83,888 square miles. The northern and east-central portions of the state consist of a series of mountain ranges belonging to the Rocky Mountain System. In this section are found the largest forests of white pine in North America, besides many thousand acres of spruce, cedar, juniper, and hemlock. Large areas in the timbered lands are suitable for grazing. The land adapted to agriculture, which lies along the streams and in the open prairie country, comprises about 6,000,000 acres, and is devoted chiefly to growing the cereals, wheat, oats, and barley. As the rainfall in the northern section of Idaho is sufficient for maturing crops, it is known as the humid belt in distinction from the southern or arid belt, where, over an area of 65,000 square miles, agriculture can not be carried on without irrigation. The southern portion of the state, with the exception of a small area in the southeastern corner, is occupied by the Snake River Valley, which along the west border extends northward as far as Clearwater River. The surface of the valley is generally level, and is covered chiefly with different kinds of sagebrush. The soil, derived mostly from volcanic rock and ash, responds quickly to irrigation and produces large crops of alfalfa, sweet clover, corn, and potatoes. In 1920 there were under irrigation 2,488,000 acres, and under cultivation 650,000 acres of alfalfa.

There are four localities in Idaho which are well adapted to commercial beekeeping: The extreme northern portion of the state, sometimes called the

"Panhandle," with Sandpoint as a center; the Boise Valley in the southwest; the vicinity of Twin Falls in the south; and the Idaho Falls section in the southeast.

In the northern mountainous region of Idaho the winters are long, the cold severe, and the snowfall heavy, while the summers are short and cool. Alfalfa is of little importance, and the beekeeper must depend chiefly on the native honey plants and the clovers. The most important sources of nectar are willows, dandelion, white clover, alsike clover, raspberry, buckbush or snowberry, willow-herb, and aster. On areas which have been cleared of forest and burned over the raspberry and willow-herb or fireweed flourish, and offer excellent opportunities to beekeepers who use modern methods. A surplus is also stored from snowberry. At Bonner's Ferry, on the Kootenai River, in Bonner County, at the extreme northern end of the state, very little land is under cultivation; and the best locations are near the bottom-lands of the river, which are overflowed every spring. The honey is a mixture gathered from fireweed, clover, and buckbush. It is light in color, thick in body, and has an excellent flavor. The flow lasts from June to the middle of August, and a surplus of 120 pounds per colony, spring count, is often obtained. In 1919 about 10 carloads of honey were produced in the northern part of the state.

A little farther south there are only small apiaries, as the conditions are not favorable for beekeeping. The winters are severe and the losses heavy.

The mid-western portion of the state is a rugged mountainous region with many narrow valleys, where there are small areas of arable land. In Shoshone, Clearwater, and Selway counties there are only a few colonies of bees and no alfalfa. The land lies at an elevation of 5000 feet, and is thinly populated. Likewise in the mountainous counties of Lemhi and Custer, which cover an area of 6,000,000 acres, beekeeping has received very little attention. But in Big Lost River Valley, Butte County, north of the Snake River Desert, there are a few farmers who have a few colonies of bees. At Arco there is one commercial apiary. The seasons are very short, and the first crop of alfalfa does not bloom until after the fourth of July. It is cut as soon as it blooms. A second crop is harvested in September, and a little later heavy frosts end the flow.

The southern portion of the state, through which flows the Snake River, is an elevated plateau completely enclosed by mountain ranges. The Snake River rises in the vicinity of Yellowstone Park, flows southwesterly to Twin Falls, where there is a vertical drop of 187 feet; and five miles farther west, at Shoshone Falls, it again plunges downward 210 feet; thence it runs nearly west across the state, turning northward in Canyon County and forming the state line as far north as Lewiston. Within the bend of the Snake River is the Snake River Desert, which in comparatively recent times was covered by an immense overflow of lava. The climate is arid and the vegetation consists principally of sagebrush, which, it has been estimated, covers 21,000 square miles in southern Idaho. The soils are sandy or ashlike silty loams, which are wonderfully fertile under irrigation. Of the irrigated land 89 per cent. lies in the valley of the Snake River, where alfalfa and sweet clover grow profusely and offer an extensive pasturage to bees. A number of native desert plants play an important part in building up the colonies in the spring. A species of sunflower furnishes pollen and a little nectar; and there are also many other minor honey plants, but their value is very imperfectly known even to the beekeepers of this section.

In the Boise Valley there is the largest area of irrigated land in southwestern Idaho. At Nampa, the center of the valley, the rainfall is less than an inch during the months of June, July, and August. The winters are open, with occasional light snowfalls. Spring begins early in April, and three cuttings of alfalfa are obtained in a season. Between Weiser on the west boundary and Mountain Home in Elmore

County, a distance of 110 miles, the Idaho-Oregon Honey Producers' Association reports 166 members, who, in 1918, produced forty carloads of honey.

At Meridian near Boise City commercial apiaries range from 100 colonies to 150; but several members of the association operate from 1000 to 1500 colonies in yards containing about 100 colonies. Ninety per cent. of the honey comes from alfalfa; but willows, dandelion, fruit bloom, white clover, alsike clover, and black locust all contribute to the crop. There has been only one complete failure in eighteen years. The honey is white or very light amber, and granulates quickly after it is extracted. A beekeeper at Nampa estimates that there are sixty specialists within a radius of twenty miles, while many farmers are starting small yards. In Canyon, Ada, and Elmore counties, in 1919, there were 352,000 acres under irrigation and 92,000 acres of alfalfa. According to the First Annual Report of the Idaho State Department of Agriculture these three counties contained 9000 colonies of bees and produced 475,000 pounds of honey.

The very large county of Owyhee, in the southwestern corner of the state, has an area of 5,048,000 acres, but only a small population. The surface of this section is largely a broken lava plateau, 5000 feet above the sea, covered with sagebrush. In the southwestern valleys the winters are mild, and the days intensely hot in summer. Apples, pears, and other fruits are successfully grown. A beekeeper at Grand View writes: "The valley here is ten miles long and two miles wide. The land has recently been brought under irrigation, and there is not more than one acre in twenty on which alfalfa is not grown. There is as yet very little sweet clover, and only a few trees, mostly willows, which bloom in June. There are three honey flows here, the second of which is the most reliable. The others give some honey nearly every year, but rarely a full crop. My best average per colony during five years was 75 pounds of comb honey, but usually I get about 60 pounds. The disadvantages in this locality are high winds in spring and fall; too hot summers because of the great area of desert land not far away; practically no spring pasturage; and the cutting of all the hay and forage in about ten days. In the Boise Valley the disadvantages are American foul brood and too many colonies of bees; but the honey flow is much longer. A few miles out in the desert, bees would quickly perish." In Twin Falls County, between the south boundary line and the Snake River on the north, there are many colonies of bees. The apiaries range from 25 to 200 colonies, but very light crops are obtained in some years. It is the reclamation of 261,000 acres of land and the cultivation of a great acreage of alfalfa that renders this locality of interest to beekeepers. The land has been largely taken up by settlers, and the town of Twin Falls has grown rapidly in population and wealth. In 1916 there was in this region a heavy production of honeydew excreted by the clover aphis, which during the following winter caused an enormous loss of bees. Good locations may also be found in the adjoining county of Cassia. There are comparatively few bees in this county, although there is a large acreage of alfalfa. Many of the smaller beekeepers lost their colonies during severe winters.

In the Idaho Falls region, in the counties of Bingham, Bonneville, Jefferson, and Madison, there are nearly half a million acres under irrigation and a hundred thousand acres of alfalfa. Beekeepers operate from 75 to 1500 colonies of bees, and in a good season produce from 80 to 160 pounds of honey per colony. According to the First Annual Report of the Idaho State Department of Agriculture there were, in 1919, 9258 colonies of bees in Bonneville County. American foul brood has been prevalent, and has greatly reduced the number of pounds of surplus obtained. Bees build up in the spring on willows, dandelion, and fruit bloom for the main flow, which comes from alfalfa and sweet clover. In this locality it is the first crop of alfalfa which yields the largest amount of nectar. Sweet clover grows profusely along the ditches and in the irrigated fields, and is yearly becoming more valuable as a source of honey.

UTAH

Total area, 84,990 square miles. The northeastern section of Utah is traversed north and south by the Wasatch Range, and east and west by the massive Uintah Range, two high mountain ranges forming a part of the Rocky Mountains. The eastern portion of the state is occupied by a series of very high plateaus, in which the rivers run in canyons, ranging from a few hundred to several thousand feet in depth. Widely distributed over this area are small detached groups of mountains. The land is chiefly used for grazing, but there are many small valleys which can be irrigated. The plateau section is bounded on the west by a line extending from St. George on the Virgin River, through Nephi, near the center of the state, to a spur of the Wasatch Mountains west of Cache Valley on the north border. The western portion of Utah, which is separated from the plateaus by a steep slope, and also the larger part of the state of Nevada, belong to the Great Basin, a desert region which has no outlet to the sea. The broad, nearly level desert valleys are interrupted by numerous steep mountain ranges running north and south, and produce little vegetation except sagebrush. The streams of this region either sink in the sands or collect in lakes in the lowest part of the valley basins. The largest of these deserts is Great Salt Lake Desert, to the northeast of which is Great Salt Lake.

Most of the cities and towns are located in the border land between the plateau region and the desert region. This belt of land, which includes the southern extension of the Wasatch Mountains, begins with Cache County, on the north border, and extends to the southwest corner of the state. The larger portion of the irrigated area, and consequently of the alfalfa acreage, lies in this central tract, where the waters from the higher levels are brought down to the sandy loams around the margin of the Great Basin region. In the north-central portion of the state the rainfall is about 15 inches, which is sufficient for growing grain crops on the higher levels without irrigation; but throughout the remainder of the state the annual precipitation ranges from 5 to 10 inches. There are great desert areas which are waste land, or which can be utilized only for grazing.

The best locations for beekeeping in Utah are in the Uintah Basin, in the northeast corner, south of the Uintah Mountains; Emery County in the east-central part; and the mountainous agricultural belt extending through the center of the state. The native honey plants are so few that beekeeping is almost wholly dependent on alfalfa and sweet clover. Other species worthy of mention are willows, fruit bloom, service-berry (*Amelanchier elliptica*), choke cherry (*Prunus melanocarpa*), hawthorn (*Crataegus rivularis*), dogwood (*Cornus stolonifera*), mountain mahogany (*Cercocarpus parvifolius*), Rocky Mountain bee-plant, wild currant, mustard, black locust (cultivated), and basswood.

The Uintah Basin in the northeastern corner of the state is an excellent location for the production of honey; but as the nearest railroad is at Price, 120 miles distant from Vernal in the center of the Basin, it costs one cent a pound to haul out freight. Alfalfa is grown extensively for seed, and in 1923 six million pounds were shipped from this county. There is thus an immense pasturage for bees, and irrigation and warm days and cool nights always ensure a large honey flow and the highest quality of honey. One of the largest producers in the Basin gave the following figures as fairly representative of the crops that can be secured. In 1920, 2245 cans of 60 pounds each were produced by 630 colonies. This was an average of 213 pounds per colony. In 1921, 1500 cans were produced by 650 colonies, an average of 138 pounds per hive. In 1922 the crop was much smaller, the average per hive for 750 colonies being 78 pounds, or 985 cans. In 1923, 700 colonies produced 1623 cans, or an average per hive of 139 pounds. The average per colony for five years was 144 pounds. While willows, dandelion, service berry, choke cherry, and thornbushes are common in the canyons and are very helpful for brood-rearing in the spring, the colonies can often be built up to advantage in Utah County and later moved to the Basin for the harvest. Besides alfalfa, sweet clover and alsike clover are important.

Emery County in the east central part of the state is bounded on the east by Green River, and crossed by the San Rafael River and its tributaries, which supply water for the irrigation of 91,000 acres. In the production of honey Emery County ranks in importance with the Uintah Basin.

In the rest of the plateau region, especially in Grand County, on the east border, and San Juan County, in the southeast corner, the land is largely arid and contains only a small acreage of alfalfa. But there are many small valleys which can be irrigated to some extent.

Washington County, where irrigation is most extensively practiced, that the major-

It is in the central mountainous belt of land extending from Cache County to ity of beekeepers in Utah are located. At Hyrum, Cache County, 10 commercial apiaries contain from 100 to 150 colonies. One of the most successful beekeepers in that section reports that in 10 years he has had three good crops, three fair crops, three poor crops, and one failure. Serious losses have been sustained here during the winter, and some years ago one of the largest honey producers in the United States lost 2120 colonies, having only 80 left. Another apiarist lost 1100 colonies. Alfalfa at Hyrum yields nectar in August so abundantly that a hive on scales often shows a gain of 11 pounds in one day. Weber County, south of Cache County, is an equally good location; and at North Ogden a beekeeper operates 20 apiaries which contain not far from 3000 colonies.

In the region surrounding Salt Lake City, honey producers are few, owing to the smelters located near the city. In Salt Lake City limits there are about 100 colonies, and in the Murray and Mill Creek districts adjoining 400 more. The valley of Salt Lake City would be a paradise for bees if it were not for the smelters located a few miles distant. By seeking out little coves near the mountains this disadvantage is overcome to some extent; and by careful attention and the use of modern methods it is possible to secure 100 to 150 pounds of surplus of extracted honey per colony in localities where alfalfa and sweet clover thrive.

At Provo, on Utah Lake, is located one of the largest beekeepers in Utah, who has several thousand colonies of bees at various points in Utah and Idaho. In Utah County there are thousands of acres of alfalfa, and white sweet clover is also the source of a large amount of honey. Dandelion, fruit bloom, and locust are abundant. Colonies can with advantage be built up here in the spring, and moved later to the Uintah Basin for the harvest.

At Nephi, near the center of the state, there are many apiaries. During recent years beekeeping has suffered from the alfalfa butterfly, grasshoppers, and occasionally from the Tachinid fly, *Archytas amalis*. A beekeeper at Mellington, Carbon County, writes that in the Price River Valley, 60 miles in length, there are about 50 apiaries which average 100 colonies. "I have been here 12 years, and have built up 12 apiaries. I obtain a crop every year from white sweet clover and alfalfa; but the amount varies, as occasionally there is not enough snow in the mountains to furnish sufficient water for irrigation." At Price specialists are reported to operate 700 or more colonies.

San Pete County is near the center of the state, and a beekeeper who has 225 colonies at Manti writes that he obtains from 90 to 120 pounds of honey per colony every year. Dandelion, although it seldom yields a surplus, is very valuable for building up colonies in the spring. Alfalfa yields the larger part of the surplus. It gives the best results on high, well-drained irrigated land. Sweet clover yields a continuous flow from July 15 to August 31. The Rocky Mountain bee-plant (*Cleome serrulata*) is very valuable, and is the source of a white, well-flavored honey. It grows on waste areas and barren lands which can not be cultivated for want of irrigation. A railroad passes through the central towns and cities—an advantage which neither the Uintah Basin nor Emery County possesses.

At Deseret, Millard County, a specialist operates 1200 colonies along the Sevier River. The Sevier River formerly terminated in Sevier Lake; but its waters have been so largly used for irrigation that the lake has disappeared. The county has many thousand acres of irrigated alfalfa. West of the Sevier River is the Sevier Desert. In Sevier County the number of colonies of bees, the surplus of honey, and the acreage of alfalfa is about the same as in San Pete County, which adjoins it on the north. Beaver is the next town of importance southwestward. There are several specialists who operate from 100 to 1000 colonies in this locality, and they seldom fail to secure a crop, although it is small in some seasons. There is a large acreage under irrigation. At Parowan, Iron County, there are several beekeepers who average about 50 colonies, but they are rapidly enlarging their yards. They never fail to obtain a surplus; but frosts and an insufficient supply of water diminish it in some seasons. An abundance of snow on the mountains, and a hot dry summer, are ideal conditions for a honey flow. Dandelion yields a surplus three years in four. Sweet clover and alfalfa remain in bloom until the last of September; and, after frost, rabbitbrush maintains brood-rearing until the beginning of freezing weather, about October 25. The farmer beekeepers have been driven out of the business by American foul brood, which has dstroyed many colonies of bees. In Washington County, in the southwest corner, there is less alfalfa, but otherwise conditions are very similar to those in the counties northward.

The western portion of the state consists of broad desert valleys between steep mountain ranges running north and south. The valley basins have no outlet for draining, and the streams which rise in the moutains either disappear in the sands or terminiate in lakes. The largest of these lakes is Great Salt Lake, which in former geological times had a much greater area than at present. To this ancient lake has been given the name of Lake Bonneville. Southwest of Great Salt Lake is the arid barren area, formerly known as the Great American Desert, which is about 125 miles long from north to south and 50 miles wide. The poet Bryant, 70 years ago, described it as a vast desert-plain which, as far as the eye could penetrate, was of a snowy whiteness and resembled a scene of wintery frosts and icy desolation. "Not a shrub or object of any kind rose above the surface for the eye to rest upon. The smooth hard plain was covered in wavy lines with a white incrustation. Beyond this we crossed what appeared to have been the beds of several small lakes, the waters of which have evaporated, thickly incrusted with salt, and separated from each other by small mound-shaped elevations of white sandy or ashy earth, so imponderous that it had been driven by the action of the wind into these heaps which are constantly changing their positions and shapes." Over thousands of square miles of these arid lands sagebrush is almost the only form of vegetation.

ARIZONA

Total area, 113,956 square miles. Arizona is divided into two natural provinces by a line running diagonally from a point ten miles east of the confluence of the Virgin River with the Colorado River, along the Grand Wash Cliffs, midway between Jerome and Flagstaff, whence it follows the southern rim of the Mogollon Mountains to the southeast corner of Apache County, where the Black River crosses the east border of the state. The northern province belongs to the Colorado Plateau, and consists of many relatively level plateaus, with an average elevation of 7000 feet above sea level, from the surface of which rise abruptly mountain ranges, isolated peaks, flat-topped hills or buttes, and mesas. The plateaus are separated by scarps and canyons, and differ greatly in elevation and climate. Through the northwestern portion of Arizona the Colorado River has cut the most stupendous gorge in the world — the Grand Canyon of the Colorado, which is more than a mile in depth. Over hundreds of square miles there have been worn in the horizontally stratified rocks a complex series of gorges and chasms, which, according to the tints of the different strata, are colored red, yellow, purple, and brown. The climate is temperate, with cold winters, during which there are severe frosts and snow falls occasionally. The summer season is similar to that of Kansas. The annual rainfall varies in different portions of northern Arizona from 10 to 20 inches, and agriculture is largely dependent on irrigation. The plateaus are covered with bunch grasses, and the mountains are timbered with one of the finest pine forests in the western states.

Southern Arizona forms a part of the Basin and Range Province, and consists of broad desert plains crossed by many low, short mountain ranges running north and south. A large part of this area is destitute of rivers and lakes, and the hot, dry, sandy plains are nearly treeless. More than 30,000 square miles, or a little more than one-fourth of the territory of Arizona, lying chiefly in the southwestern third of the state, will probably always remain waste desert land. The average elevation of southern Arizona is 2000 feet above sea level; but at Yuma the altitude is not more than 350 feet. The winters are mild, and a succession of crops can be grown throughout the year. The annual rainfall ranges from 5 to 10 inches in the east to 3 inches in the west.

The northern or plateau section of Arizona is divided into the four counties of Apache, Navajo, Coconino, and Mohave, which have a combined area of 36,116,-480 acres; but the portion of Mohave County lying between the Colorado River and Bill Williams Fork belongs to the southern province. The chief industry over this vast area is sheep and cattle grazing. Dry farming to be successful must be supplemented by irrigation at critical times. In all this vast region, covering 45,000 square miles, there are only a few hundred colonies of bees and a few hundred acres of alfalfa. As the Little Colorado River is lost in a deep canyon throughout its lower course, only its watershed in the eastern part of the state can be used for irrigation. Two or three crops of alfalfa are secured, and corn, winter wheat, deciduous fruits, and vegetables are successfully grown. In northwestern Arizona the water of the Colorado River can not be utilized for irrigation, as the deep canyons begin in eastern Utah and extend to the lower alluvial lands in the western part of the state. Of the plateaus north of Colorado River, the Kanab is the loftiest, reaching an elevation of more than 8000 feet. Its surface is covered partly with forests and partly with grassy parks on which bloom many beautiful flowers. But in the absence of any extended system of irrigation, except along the upper course of the Little Colorado, the outlook for agriculture and beekeeping in the larger part of northern Arizona is not promising.

Commercial beekeeping is confined almost entirely to the desert valleys of the southern half of the state, where large areas are under irrigation. Yavapai County is situated near the center of the state, and depends for irrigation on the Upper Verde River and its tributaries. This stream is a northern branch of the Salt River, and supplies water for the irrigation of 11,000 acres. The farms are small, and are situated in the narrow creek valleys, where the numerous ditches are usually owned by individuals. The climate is very favorable to the growth of deciduous fruits, apples, pears, and peaches, and to beekeeping. There are several large apiaries, as well as many smaller ones, in this part of the state.

Maricopa County, which, in 1871, was formed by a division of Yavapai County, is far in advance of any other county in Arigona, both in agricultural development and bee culture. The State Inspector of Apiaries inspected 34,710 colonies in the state in 1918, of which 21,319 were in Maricopa County; 4398 colonies in Yuma County; 2951 colonies in Cochise County; 2484 colonies in Pinal County; 1176 colonies in Yavapai County; 557 colonies in Pima County; and 256 colonies in Greenlee County.

One of the most famous irrigation systems in the world is the Salt River project, which has been constructed by the United States Reclamation Service. Eighty miles east of Phoenix, in a narrow canyon, the Roosevelt Dam rises to a height of 284 feet above the bed of the river, forming a beautiful lake 28 miles in length, surrounded by hills and mountains. The waters of Salt River and Tonto Creek are stored in the lake, and at the proper time are permitted to flow through the sluice-gates, taking up in their course down the valley the water of Verde River. At the diversion dam, at Granite Reef, the three streams are turned into great canals, and serve to irrigate 216,000 acres. The total cost of this system was about $10,000,000.

The altitude of Salt River Valley is 110 feet above sea level, and the average annual temperature is 69 degrees F. During 40 years the average annual rainfall has been about 8 inches, and the average number of clear days 232, and of rainy days 37. The soils are sand and gravel loams, loess loams similar to those found along the Mississippi River, and heavy clay loams. The loess soils are from 100 to 500 feet deep. The land is well adapted to irrigation, as it is almost level, with a gradient of seven feet to the mile. There are no rocks.

Alfalfa may be cut five or six times for hay, or, if preferred, profitable crops of seed may be grown. During the last years of the Great War alfalfa was partially replaced by Egyptian or long-staple cotton, used in the manufacture of automobile tires. The supply of a suitable grade of cotton for this industry having become wholly inadequate, in the spring of 1918 one of the large American tire companies bought several thousand acres of land in the Salt River Valley, and seeded them with Egyptian cotton. In 1918 about 40,000 bales of cotton of superior quality were harvested; in 1919, 90,000 bales were grown; and in 1920 growers expected 100,000 bales. The average yield was one-half a bale per acre; but on the older cultivated land one bale per acre was not unusual.

Cotton does not yield as much nectar per acre as alfalfa; and in localities where it has largely supplanted alfalfa beekeepers are not securing as large a surplus as formerly. The nectar is secreted much more freely by glands on the underside of the leaves than by the floral nectaries. Cotton honey is white, and has a mild flavor, or is almost without flavor, like sugar syrup. It may be expected that the larger acreage of cotton will, to some extent, compensate for the decrease in the alfalfa area. Cotton, moreover, remains in bloom from July until late in the season, while farmers cut the alfalfa for hay at a much earlier stage than formerly, not allowing the plants to come into full bloom. The alfalfa butterfly (*Colias eurytheme*) has so increased in numbers since 1895 that the honey flow, which once continued into September, is now cut short in July. In 1920 the price of cotton dropped so low that many acres, it was reported, in the Colorado River Valley lands could not profitably be picked. It is probable, therefore, that many farmers will return again to growing alfalfa.

In the vicinity of Phoenix there are 75 yards which range from 1 to 619 colonies. Ten yards report 240 or more colonies; and in five apiaries there are 305, 380, 425, 539, and 860 colonies respectively. A beekeeper writes that he has had as many as 1600 colonies, and at present has 10 apiaries. There are large apiaries located at Tempe, Glendale, Mesa, and Chandler. At Glendale two beekeepers operate 1600 colonies. A beekeeper at Cashion began beekeeping with 16 colonies, and in five years increased to 500.

West of Phoenix is the Buckeye Valley, which is about 20 miles long and 10 miles wide. The irrigation system is under private control, and consists of dams which divert the Gila River and distribute it over some 50,000 acres of land. The valley is noted for its fat cattle, grain, and alfalfa seed. A beekeeper at Buckeye writes that he has 1200 colonies, and that three other beekeepers operate 1200, 600, and 400 colonies respectively. The land was formerly devoted to growing alfalfa seed; but alfalfa has in some cases been supplanted by cotton, which here yields little nectar, and as a result nearly 2000 colonies have been moved out of the valley. From the same locality another beekeeper writes: "I have 19 apiaries ranging from 75 to 150 colonies. Good alfalfa and mesquite locations will support 300 colonies." The sources of honey are willow, March 15 to April 15; mesquite (*Prosopis velutina*), a bountiful yielder, May 15 to June 15, followed by a second blooming period two weeks later; and alfalfa, from May 25 to July 25 or later. The surplus comes chiefly from the second crop of alfalfa. Sweet clover starts well, but can not endure the extreme heat.

At Sentinel, Maricopa County, near the west border line, beekeeping is dependent on the native honey plants. A beekeeper writes: "I have about 500 colonies in six yards. The honey plants are mesquite, catsclaw, willow, and arrowweed. The first two species are the most important. A record crop does not come more than once or twice in ten years. A fair average per colony is 60 to 75 pounds, which is stored during the last three weeks of May and the first week in June. Losses range from 12 to 15 per cent. annually. This locality is a natural wild range, and there are no cultivated crops, so that there is no food for the bees in late summer, and it is then my loss is greatest." After the flow from mesquite is over, about June 10, it gets very hot and dry on the desert.

Pinal and Graham counties depend on the Gila River for irrigation. The valley of the Lower Gila River from Florence to Yuma closely resembles in climate and soil the Salt River Valley; but the flow of water is so uncertain that there has thus far been little development of agriculture in this region except under the Florence Canal. The river is not infrequently dry at Florence, sometimes for several months at a time, as from March to July, 1899. At Yuma the channel of the river has been known to remain dry for a year; but in some seasons it may discharge into the Colorado River a great volume of water. Without storage, and with such fluctuations in the water supply, the land remains largely uncultivated. The Upper Gila, near the eastern boundary, affords a supply of water adequate to irrigate about 25,000 acres between Duncan, on the east border of Arizona, and San Carlos in Graham County. This county has an elevation of over 2500 feet, and the climate is too cold for citrus fruits and eucalyptus trees; but the hardier vegetables grow throughout the winter.

In Graham County alfalfa hay is the principal crop, and four or five cuttings are harvested annually. In January, 1919, there were at Safford apiaries containing 224, 250, 340, and 517 colonies of bees. At Bryce there is an apiary containing 152 colonies; at Pima one with 196 colonies, and at Thatcher one with 68 colonies. A beekeeper writes from Solomonville: "During 12 years the crop, which ranges from 30 to 60 pounds of extracted honey, has never failed. The honey plants are alfalfa, catsclaw, and mesquite." In Pinal County several apiaries are located near

Florence, and there are also a few beekeepers at Winkleman, on the east border. At Casa Grande, in the desert, mesquite and catsclaw are the chief dependence of the beekeeper. If the spring is late, not much honey is gathered. Apiaries range from 25 to 100 colonies.

In Cochise County in the southeast corner of the state large apiaries near Benson contain 558, 505, and 87 colonies of bees. At St. David, a few miles distant, there are 10 yards ranging from 20 to 90 colonies. At Bisbee there is a yard with 105 colonies. A large area is irrigated from artesian wells, of which there are over 200 ranging from 125 to 800 feet on depth. At Bowie there is an apiary containing 315 colonies. A beekeeper writes from Dos Cabezos that in this locality there are four apiaries containing 700 colonies. There has been only one failure in 12 years. The average surplus is 120 pounds, gathered principally from mesquite and catsclaw. In 1920 rain checked the flow from mesquite.

In Pima County the Santa Cruz River supplies irrigation water for about 16,000 acres. This stream flows northward through the broad valley of Tucson toward the Gila River, but sinks in the sand before reaching it. In dry years it is insignificant in size, but it is subject to great floods in rainy seasons. Fifty miles north of Tucson a reservoir has been constructed. Near Tucson there are several apiaries which contain about 100 colonies each. From Wrightstown J. B. Douglas writes: "There are seven apiaries in this locality which contain a total of 1000 colonies. There has been but one failure in 14 years. The average in a good year is 150 pounds of extracted honey. There are no yards run for comb honey. The honey plants are mesquite and catsclaw and other desert plants. We have a long spring with a little honey coming in all the time. The main flow begins about May 20."

In the southwest corner of Arizona, in Yuma County, the rainfall is less than five inches annually. It is largely a region of hot, dry, nearly lifeless, sandy or gravelly plains lying between steep, low mountain ranges. The scanty vegetation consists of cresote bushes, mesquite, octillo, joint-pine, cactus, and yucca, with salt bushes and salt grass in the alkaline areas. The soils range from heavy adobe to light sands, the lighter soils predominating. In the vicinity of Yuma there are 49,800 acres under irrigation and 128,000 acres included in enterprises. The alluvial lands of this county are destined to become one of the richest agricultural regions in the state. The winters are mild, and the summers nearly rainless. The main crop is alfalfa, of which seven cuttings are harvested. On the bench or mesa land citrus fruits can be planted. Date palms also thrive in this section, but it is not adapted to tropical fruits like the pineapple and banana.

There are many opportunities for commercial beekeeping in Yuma County, as the area under irrigation can be greatly enlarged. Twenty-five apiaries reported more than 4000 colonies of bees, and three apiaries at Yuma numbered 614, 1050, and 1150 colonies. At Somerton there are yards containing 60, 145, and 200 colonies respectively. From Wellton a beekeeper writes: "I have 140 colonies, and have never averaged less than 60 pounds of extracted honey per colony. Our surplus is gathered from alfalfa and mesquite; and during the last two years cotton has been of value. The bees build up in the spring from willow and arrow-weed. Extracting begins about April 20, and lasts until June 15. Very little honey is gathered after August. There is room here for thousands of colonies." At Palomas, in the eastern part of Yuma County, there is an apiary of 65 colonies which stores about a can of honey per colony from mesquite. In 1919 this yard yielded a profit of $1200.

The honey flora of Arizona consists chiefly of cultivated crops, as alfalfa, cotton, fruit trees, melons, and the desert plants, which have, however, decreased greatly because of overgrazing by sheep and cattle, and the cutting of the trees for wood.

Arrow-weed (*Pluchea sericea*).—A shrubby desert plant which grows on lands liable to overflow. The pink flowers yield a light-amber honey of good quality.

Arizona Mesquite (*Prosopis velutina*).—Grows in the hot dry valleys of Arizona, southern California, and Sonora. The largest of the mesquites attain a height of 50 feet, though often they are not over 15 feet tall. Blooms from April to June 15, and a second time two weeks later Honey light amber, excellent flavor, crystallizes quickly after extracting.

Catsclaw (*Acacia Greggii*).—Western Texas to southern California. May to June. Honey white, excellent.

Arizona Acacia (*Acacia constricta*).—A spiny shrub, three feet tall, with bright-yellow flowers in globular heads. Dry hills. June.

Tornillo (*Strombocarpa odorata*).—Screw bean. A small tree found from western Texas to California. Yellow flowers in May. A common large shrub in the river valleys. Pods spirally twisted. (Synonym, *Prosopis pubescens*).

Palo Verde (*Cercidium torreyanum*).—Desert regions of Arizona, California, and Sonora. A small tree with green bark and yellow flowers in April. The leaves appear in April or May. Yields less nectar than mesquite; the honey is yellow, heavy, and of good flavor. (Synonym, *Parkinsonia torreyana*).

Creosote-bush (*Covillea tridentata*).—A spreading evergreen shrub with small bright yellow flowers. Dry plains and mesas. Near Tucson it blooms for three months.

Water Motor (*Baccharis glutinosa*).—Bottom willow. A common shrub in the river valleys, preferring land that is sometimes inundated; honey white, excellent; a large surplus is often obtained near Phoenix.

Willow (*Salix Wrightii*).—Along water-courses. Blooms in March. Also western willow, *S. lasiandra.*

Phacelia (*Phacelia tanacetifolia*).—Widely scattered over the slopes.

Gilia (*Gilia floccosa*).—Common everywhere in southern Arizona.

Prickly Pear (*Opuntia Bigelovii*).—There are 15 or more species near Tucson. Thirty species are listed for New Mexico. *O. Bigelovii* has a wide range on the southern deserts.

Near Phoenix, on the deserts, two species of plaintain (*Plantago ignota* and *P. fastigiata*), known locally as Indian wheat, are very abundant, and are eaten by cattle. They are of value to bees for pollen, but yield no nectar.

The average moisture of the honey in Arizona is 16.85 per cent., which is much less than that of eastern honeys.

NEVADA

Total area, 110,690 square miles. Nevada and western Utah form the larger part of the Great Basin, a vast tableland 4000 to 5000 feet above the level of the sea, lying between the Rocky Mountains and the Sierra Nevada Range. The surface of this region is not a plain, but is broken by many isolated parallel mountain ranges running north and south, between which are valleys 5 to 20 miles in width. At an elevation above 6000 feet the mountains are heavily forested with pine, juniper, and mountain mahogany. The valleys are bare of trees except for a few willows and poplars; but sagebrush is so abundant that it gives a grayish tinge to the landscape, whence Nevada is known as the "Sagebrush State." Greasewood and saltbush are also common; but in the southern valleys cactus and yucca replace the sagebrush. The Black Rock Desert, in northwestern Nevada, is barren of vegetation, due to the strongly alkaline soil; and in the southern part around Las

Vegas there are also great areas with a loose alkaline soil, which has hardened to a thin fragile crust on the surface. Much of the desert land, which is unsuitable for farming, can be utilized for grazing.

The Great Basin has no outlet to the sea, and the rivers which rise in the mountains finally sink in the sands or terminate in small lakes. Humboldt River, after flowing 300 miles, enters Humboldt Lake, and a number of other streams over 100 miles in length disappear in the sands, spreading out in great flats known as "sinks." The climate of Nevada is characterized by extreme aridity; and as the average annual rainfall in the northern part of the state is less than 15 inches, and in the southern part less than 5 inches, it is insufficient for growing crops without irrigation. The two principal areas under irrigation are the plains watered by the Carson, Truckee, and Walker rivers and the valley of the Humboldt River, reclaiming large areas in Washoe, Douglas, Lyon, Churchill, and Pershing counties. In the mountain valleys there are many places which can be irrigated to some extent by the intermittent mountain streams. The total area under irrigation is 561,000 acres. Several large irrigation projects are either under consideration or in a formative stage. Congress has already appropriated $500,000 for preliminary work on the Spanish Springs project. This project, when completed, will bring some 40,000 acres of land under irrigation in western Nevada.

The eastern counties of Eureka, White Pine, Lincoln, and Nye, which have an area of 26,700,000 acres, or more than one-third of the area of the entire state, reported very few colonies of bees and a relatively small acreage of alfalfa, although White Pine County reports 10,000 acres in alfalfa. All of these counties are large producers of live stock. The southern part of Nye County is occupied by the Ralston Desert.

At Las Vegas, Overton, and Bunkerville, in Clark County, in the extreme southern part of the state, beekeeping receives more attention, although the irrigated area and the alfalfa acreage are small. In the valley of Las Vegas there are four apiaries, the largest containing 70 colonies. The surplus is gathered from alfalfa, sweet clover, catsclaw, mesquite, and huajilla. During a period of seven years alfalfa, sweet clover, and catsclaw have proved reliable; but in 1919 mesquite failed on account of dry windy weather at the time of blooming. All cultivated crops are dependent on irrigation, as the climate is very hot and dry, the temperature running up to 129 degrees F., and the average rainfall is only four inches. Water is obtained from artesian wells 100 to 1000 feet in depth. From five to seven cuttings of alfalfa are made in a single season, but farming is as yet in its infancy. The little valley at Overton, also in Clark County, is better adapted to the raising of bees and the building up of colonies than for the production of honey. The apiaries seldom contain more than 55 or 60 colonies, as 75 colonies usually overstock the range. If the weather is hot and windy, little surplus is stored; but if there is rain and mild weather there is always a fairly good crop of light-amber honey. There are large areas of mesquite, which, with the mild winters and early springs, should be very valuable for an early honey flow. At Bunkerville, still farther north in Clark County, conditions are much the same as at Overton. An entire failure is unknown; but if it is very hot and windy only half a crop is obtained.

The Humboldt River rises in the northeastern part of the state, and after flowing westward and southwestward 300 miles empties into Humboldt Lake in the northern part of Churchill County. In the valley of this river, in 1919, there were in Humboldt County 207,000 acres under irrigation, and 27,000 acres of alfalfa. In Elko County 183,000 acres were irrigated, but there were only 8000 acres of alfalfa. Beekeeping does not appear to have received much attention, as only a few hundred colonies of bees have been reported in the two counties. In the southern part of the state the climate is semi-tropical, but in the northern part it is more nearly temperate.

While the valley of the Humboldt River is perhaps the most productive part of the state, as it produces more than one-half of the hay and forage crops, and has also a large acreage of cereals, another very fertile region is found near Pyramid Lake in Washoe County. Near Reno the apiaries contain from 100 to 150 colonies, and in some instances a larger number. Alfalfa and sweet clover are the only important honey plants, although dandelion, fruit bloom, and white clover deserve mention. Alfalfa honey in this locality is nearly water-white. Lyon County, in the western part of the state, although one of the smaller counties, containing a little less than a million acres of land, is the leading county in the production of honey. The soil, when reclaimed, is well adapted to the production of hay and forage crops, cereals, vegetables, and deciduous fruits. The water used for irrigation is taken from the Carson River, which rises in the Sierra Nevada and empties into Carson Lake. Within a radius of 35 miles of Yerington there are some 50 commercial apiaries which average from 50 to 100 pounds of surplus, according to the season and the strength of the colonies. Good locations are also found in Douglas County, west of Lyon County. In Mineral County irrigation is dependent on the Walker River, which rises in the Sierra Nevada and flows into Walker Lake. At Schurz, at the north end of the lake, there are apiaries which contain from 25 to 100 colonies. In none of the alfalfa-growing states does alfalfa secrete nectar more freely than in Nevada. Other honey plants of importance besides alfalfa and sweet clover, are willows, yellow cleome, and milkweed. All of the irrigated tracts of western and northern Nevada are suitable for beekeeping.

WASHINGTON

Total area, 69,180 square miles, of which 3114 square miles are water surface. Washington is divided by the Cascade Range, which extends north and south across the state, into two unequal parts, which differ widely in physical features and climate. The western section consists of the broad basin of Puget Sound Valley flanked on the west by the Coast Range and on the east by the Cascade Range. Puget Sound is about 100 miles in length; but it was formerly a great arm of the sea, extending southward into the Willamette Valley of Oregon; and high tides still overflow thousands of acres of land which it is possible to reclaim by dyking. Among some 55 species of deciduous-leaved trees and shrubs are broad-leaved maples, vine-maples, willows, Ceanothus, raspberries, blackberries, currants, cornels, bearberry, huckleberry, thornbush, and snowberry, all of which yield a little nectar.

The chief natural feature of the section east of the Cascade Range is the valley of the Columbia River. The Okanogan Highlands lie north of the river, and are forested to a great extent with an open growth of coniferous trees. West of the river are the eastern foothills, a rugged region which affords good grazing for cattle. East of the river is the Columbia Plateau, which a vast overflow of lava has covered with horizontal beds of black basaltic rock. It is an arid treeless region, producing little except sagebrush.

The following list of honey plants in the section west of the Cascades has been compiled from data furnished by H. A. Scullen:

Fireweed, or willow-herb (*Epilobium angustifolium*), is the source of most of the surplus produced by commercial yards. Considered from the standpoint of acreage it is equal in abundance to any two other plants in the state. But at present it supports a much smaller number of colonies than alfalfa, partly because it often grows in localities difficult of access, and partly because beekeepers do not fully realize its possibilities. It remains at its best for only four or five years at the longest. There is probably not a county in the state in which a few plants can not be found; but it is found chiefly in burnt-over areas, where the coniferous trees have been lumbered. The heaviest bloom is during July and August, but the date at which it begins to bloom varies considerably with the elevation. It ranges from sea level to the upper timber line, and is a very reliable yielder of a heavy white honey of mild flavor.

White clover (*Trifolium repens*) is the most abundant of the clovers, and is on the increase in the sections from which the timber has been cut west of the Cascades, but it probably secretes nectar less freely than in the Mississippi Valley. It is of very little importance in the irrigated areas. It begins blooming in May and continues into July. In British Columbia it seldom yields much nectar before July. Alsike clover has been planted on the lowlands of the west side, but it is nowhere of great value.

Two species of maple are important honey plants. Broadleaf, or Oregon maple (*Acer macrophyllum*) grows mainly west of the Cascades below 3500 feet; blooms in April and May, and is valuable for both nectar and pollen. Vine-maple (*A. circinatum*) blooms a little later than Oregon maple and is more abundant in some localities. The honey has a fine flavor, and is colored with a faint pinkish tinge. Dwarf maple (*A. glabrum*) is abundant in all wooded portions of the eastern part of the state.

Other honey plants of importance in western Washington are: Madroña (*Arbutus Menziesii*). Common except in dense forests; abundant around Puget Sound, yields well. Snowberry or buckbush (*Symphoricarpus racemosus*) and other species. Abundant on the west side and in the northeast corner. Yields a surplus in northern Idaho, and probably also in Washington. Cascara sagrada (*Rhamnus purshiana*). Abundant on the west side. A good honey plant. Dandelion (*Taraxicum officinale*). Abundant, except in arid regions. Cat's ear or California dandelion (*Hypochaeris radicata*). Abundant on west side. Honey amber-colored. Willows. Many species common along streams. Goldenrod. Abundant except in dry sections. Scullen states, however, that he has never seen a bee working on the bloom; but at Cloverdale, B. C., bees gather nectar from at least one species of goldenrod. Two species of barberry, Oregon grape (*Berberis nervosa*), and *B. aquifolium* are generally distributed in the northwest forests. Both bloom in March and April. Huckleberry. Very abundant in Mason County and in other localities. Other honey plants are sunflowers, milkweed, Phacelia, horehound, catnip, sumac, mustard, locust, wild currant, and fruit bloom.

While the apiaries in the western section of Washington are smaller than those east of the Cascades, the surplus per colony when well managed is reported to average larger. There are two honey flows in this region, one from clover beginning June 7, and a second from willow-herb beginning July 5. Climatic conditions are such that special manipulation is necessary in order that colonies may be at full strength for either flow. The same colony can not gather the maximum of honey from both flows. Except for 6000 acres in Clallam County there is practically no irrigation in this section, and only a few acres of alfalfa. The counties bordering on the Pacific Ocean contain the smallest number of colonies of bees, and the tier of counties adjoining the Cascades the largest number.

A large part of the northwest corner of the state, known as the Olympic Peninsula, is covered by the Olympic National Forest. This region is rugged and wild, and almost impenetrable in consequence of its broken surface, fallen timber, and, at high altitudes, glaciers and snowfields. On the western slopes the rainfall is from 100 to 150 inches annually, coming between November and May, while the summers are almost arid. Only a few bees are found here.

The occurrence of sugar on the foliage of the Douglas fir (*Pseudotsuga Douglasii*) was known to the Indians long before America was discovered by the white man. The sugar-yielding firs are confined chiefly to the dry belt of British Columbia, between the parallels of latitude 50 and 51 degrees and the meridians of longitude 121 and 122 degrees. The sugar is formed during the summer droughts on trees standing on gentle slopes facing north and east, and occurs on the leaves and branches in white masses one-fourth inch to two inches in diameter. According to Davidson and Teit, fir sugar is a natural exudation from the tips of the needles, which, as the result of evaporation, becomes a white solid. In leaves of the Douglas fir exposed to sunlight a larger quantity of carbohydrates is formed during the day than can be stored or carried away to the growing tissues, as happens in the case of trees growing in the dense forest. In the dry, hot atmosphere transpiration ceases, and the leaves become gorged with water which is forced out through their tips. A succession of dry sunshiny days is required to produce the sugar, which does not occur every year. Fir sugar very probably makes a poor winter food for bees, as its composition is very different from floral honey. Among other constituents it contains nearly 50 per cent. of the rare trisaccharide, *melezitose.*

At Port Townsend, at the entrance to Puget Sound, there are many small apiaries with from three to five colonies, but no specialists in beekeeping. The average surplus is not over 25 pounds, as there are many fogs during the honey flow, which comes mainly from white clover and fireweed. Vine-maple, dandelion, and berry bloom are valuable in stimulating brood-rearing. All things considered, this is a poor locality for bees.

The counties of Washington bordering on the Pacific Ocean are only moderately well adapted to beekeeping. However, back a distance from the coast some of the best fireweed sections are found. Better conditions prevail in the tier of counties east of Puget Sound adjoining the Cascade Range. In the five counties of Whatcom, Skagit, Snohomish, King, and Pierce the apiaries are usually small, but not infrequently contain 30, 40, 75, or even 100 colonies. An excellent grade of honey is gathered from the wild flowers, which never fail to yield every year, although the secretion and amount of surplus stored are affected largely by the weather. Winter losses are not heavy.

Two small counties, Cowlitz and Clarke, both of which have the Columbia River for their southern boundary, have in a single year produced large amounts of honey. For early brood-rearing there is a succession of bloom, beginning with the willows in February. In some springs a surplus is stored from vine-maple in May. White clover begins to bloom early in June, but the country is developed so little that it is not abundant enough to yield a surplus. Fireweed, which blooms from July 1 until frost, is the main source of surplus. The beekeeper in this section must contend against poor roads and both kinds of foul brood.

East Washington, or the section east of the Cascades, is crossed from north to south by the Columbia River, which, after describing a broad curve known as the Big Bend, unites with the Snake River near the south state line. The valley of the river is semi-arid, covered with sagebrush, and has long dry summers. During July and August the temperature is high, but the dry atmosphere lessens the effect of the oppressive heat. Most of the irrigated land is in proximity to tributaries of the Columbia River, which rise in the mountains and have a rapid fall to their outlets. Along the Columbia and Snake rivers the fall is so slight that the water must be lifted by pumping-plants.

The Okanogan Highlands are covered with an open forest of conifers in which the trees are much smaller than on the coast. The rolling hills produce a luxuriant growth of grasses, and many cattle have been raised in this country. The principal streams are the Okanogan and Methow rivers, which supply water for the irrigation of 36,000 acres. As the land is reclaimed it is seeded to alfalfa or planted with orchards. At Winthrop the Methow Valley is 30 miles long, and about two thousand acres are watered from Beaver and Frazer creeks. There are a few apiaries in the valley, and good yields are gatherd annually from alfalfa, sweet clover, white clover, and wild flowers. "I realize better profits from my bees," writes a beekeeper, "than from anything else on my farm."

In the mountainous counties of Stevens and Pend Oreille, in the northeast corner of the state, mining is the principal industry. Colville Valley is noted for its good farms and fine apples; but there are only a few acres of alfalfa, and beekeeping has received little attention.

Within the Big Bend of the Columbia River is the great, arid, treeless Columbia Plateau, from which the moisture-laden winds of the Pacific Ocean are cut off by the Cascade Range. In the northern part of these sagebrush plains various bunch grasses occur, but it is noteworthy that the buffalo grass so common on the Great Plains is absent. The soil of eastern Douglas, southern Lincoln, and northern Adams counties is strongly alkaline; but there are numerous small lakes and streams, which can be utilized for irrigating small areas by pumping. The chief crops are the cereals, and except for a few scattered colonies there is no opportunity for bee culture.

Spokane and Whitman counties on the east border produce one-third the total wheat crop of the state. Spokane Valley is 34 miles in length. The soil is a rich, black volcanic ash mixed with small gravel. The average annual rainfall is 18 inches. The acreage of alfalfa is small, but there are large orchards. The summers are hot and the winters cold. There are only a few bees.

The southeastern corner of the state, south of the Snake River, is noted for the fertility of its soil; but the Snake River flows in a deep canyon, and can not be used for irrigation except on bars and low benches along its course. In Walla Walla Valley, where there is a large acreage under irrigation, alfalfa is an important crop and is frequently cut four times in a season. Black locust has been planted extensively and yields a surplus of very white honey. This territory is believed to offer exceptional opportunities for beekeeping in the future.

The largest areas under irrigation in Washington are located on the eastern foothills and slopes of the Cascade Range in the counties of Chelan, Kittitas, Yakima, and Benton. In Chelan County there are twenty thousand acres under ditch from the Wenatchee River, a rapid mountain stream. The valley is a great fruit-growing center, and around Wenatchee there are 10,000 acres in orchards. The small number of colonies of bees in this county is probably due in part to careless methods of spraying the fruit trees.

The Yakima Valley, the best beekeeping territory in the state, extends through Kittitas, Yakima, and Benton counties, joining the valley of the Columbia River near the southern boundary of the state. It is a semi-arid district producing little except sagebrush, as the average annual rainfall is less than ten inches. The summers are clear and warm, and in winter the temperature seldom falls below zero. The Yakima River, fed by the melting snows in the mountains, generally affords an abundant supply of water. The total irrigable area in the basin of the river and its tributaries approximates 600,000 acres. At Ellensburg, Kittitas County, there are 22 commercial apiaries, which contain a total of about 2000 colonies of bees. The small yards exceed one hundred, and average about twenty colonies. About 40 pounds per colony are gathered from alfalfa.

"Alfalfa," according to H. A. Scullen, "is the leading honey plant in the state from the standpoint of honey production. The irrigated sections of the Yakima

Valley alone support nearly 30,000 colonies, most of which are in the hands of commercial beekeepers. Alfalfa is the leading cover crop for the orchard district, as well as the leading hay crop. If it were not for the effect of the poisonous spray on the bees, this valley would support possibly a third more colonies." In the Yakima Valley the bulk of the surplus comes from the second crop in July. Very little surplus is secured from either the first or third crop.

Many farmers in Yakima County have from 10 to 200 colonies, while specialists operate from 800 to 1200 colonies, with yields averaging around 100 pounds per colony. At Sunnyside the crop is gathered from sweet clover and alfalfa; but white clover, alsike clover, vetch, willow, locust, dandelion, and fruit bloom are minor honey plants of value for keeping up the strength of the colonies. The honey flow lasts from July 1 to about August 20. At Harwood there are nearly a dozen commercial apiaries which range from 50 to 100 colonies. The crop, which is gathered from alfalfa and sweet clover, is never a failure. Obstacles to beekeeping in Yakima County are cold winds in spring, a long interval between the close of fruit bloom and the beginning of alfalfa bloom, and the loss of bees from the spraying of fruit trees.

OREGON

Total area, 95,607 square miles. Like Washington, Oregon is divided by the Cascade Range into a western rain belt and an eastern arid belt. Western Oregon has a mild climate, heavy and incessant rains, especially near the coast, and is largely covered with a dense coniferous forest. Eastern Oregon is a broken tableland with an altitude of about 5000 feet, arid or semi-arid, with a rainfall ranging from 10 to 20 inches. Much of the southern portion of this section is a treeless desert producing little vegetation except sagebrush. In western Oregon beekeeping is restricted to a narrow coastal strip west of the Cascade Range and to the Willamette and the Lower Columbia valleys. In the semi-arid belt honey production is confined to those areas where large fields of alfalfa are grown under irrigation. The two regions differ so widely in climate, soils, vegetation, agriculture, population, and industries that they must be described practically as though they were different states.

Western Oregon comprises about one-third of the state. Ten miles inland from the sea rise the lower mountains of the Coast Range. The Umpqua and Rogue rivers break through this ridge in deep rugged canyons, forming narrow fertile valleys. East of the Coast Mountains lies the Willamette Valley, which is about 200 miles long by 30 miles wide, and has an area of over 6000 square miles. South of the valley, in Douglas, Josephine, and Jackson counties, the coast and Cascade Ranges converge to form a mountainous region which has a less humid climate than is found elsewhere in western Oregon.

The country lying between the Coast Range and the Pacific Ocean is very narrow and rugged. There are only a few large apiaries, and the average production per colony is small. The mountains are densely forested with giant firs, cedars, spruces, coast hemlocks, and pines; and at the lower elevations vine-maple and broad-leaved maple are common. Vine-maple and fireweed are the two most important sources of nectar. The valleys have an equable temperature without extremes, and bees often fly every week during the winter. They build up quickly in the spring and begin swarming in April and May.

In the extreme northwest corner, Clatsop County, bordered on the west by the ocean and on the north by the Columbia River, is an open dairy country where white clover grows well. This is the best fireweed county in the state. Farther east, and south of the Columbia River, there is a strip of sand dunes a mile in width. Sea fogs are common, and greatly reduce the length of the average working day of the honeybee. There is a rainfall of nearly 100 inches. The only month in which dry weather prevails is August. About the 10th of March the salmonberry begins to bloom, and yields a small surplus. The honey is slightly red, thin, and very sweet. It is closely followed by vine-maple, which yields a heavy honey of medium flavor with a pinkish tinge. The two honeys are usually blended. Vine-maple is in bloom during the last of April and the first of May. During the last of May the wild blackberry yields a dark inferior honey, which would be unsalable were it not mixed with honey from white clover. In July and August willow-herb, or fireweed, furnishes the choicest honey. The California dandelion is the source of yellow slightly bitter honey.

A beekeeper at Myrtle Point, on the Umpqua River, Coos County, writes that it is usually cold and rainy in April and May, when vine-maple blooms. This town is only 20 miles from the ocean, and the north winds in summer are cool, and the nights are often very cool. The apiaries are usually small and located on farms. The largest yard in the county contains 100 colonies. Vine-maple and willow bloom from April 20 to May 10. Apple bloom comes a little later; cascara sagrada from May 10 to June 15; poison oak from May 15 to June 15; and fireweed in July and August. Dandelion, buckbush, wild gooseberry, and huckleberry are less important honey plants.

The Willamette Valley, east of the Coast Range, is the most thickly populated portion of the state. The land is generally level and the soil fertile, supporting a rank growth of grasses, many groves of maple, ash, and poplar, and a great variety of shrubs. There is, however, a shortage of lime in the soil, and lime has to be brought from southern Oregon to supply the deficiency. The higher elevations of the mountain ranges on both sides of the valley are crowned with a dense fir and pine forest. The whole valley was once occupied by an arm of the sea which extended southward through Puget Sound. There are only a few commercial apiaries in the valley, and they seldom contain a large number of colonies. Because of excessive rains in the spring, and severe droughts in late summer, special management is necessary to secure a surplus. Alsike clover, red clover, and vetch are the leading honey plants. While the common vetch is good, the Bulgarian is more dependable, according to H. A. Scullen, Specialist in Bee Culture. It is better adapted to this soil and climate and is less subject to attacks by aphides. Both the vetches are extensively grown for seed, and have extra-floral nectaries.

In the northern part of the valley, Columbia County is bounded on the north and east by the Columbia River. There are several large apiaries and a few small ones. From Ranier, on the Columbia River, the report is less favorable. The farms are small, and most of the settlers work in the logging camps. There is a heavy rainfall in winter. The springs are cold and windy, and it is very dry in late summer. The larger part of the colonies in this location have died, apparently from bad management or brood diseases. But at Oak Point, in the state of Washington, 21 miles from Ranier, the German settlers are said to be very successful. Locally at Portland, on the Willamette River, Multnomah County, the surplus is derived from white clover only. Many, however, after the flow from white clover is over, ship their bees to the mountains, where they secure a good crop from fireweed. There are no large apiaries in this vicinity.

The larger part of the Willamette Valley is included in the counties of Clackamas, Marion, Linn, Washington, Yamhill, Benton, and Lane. The counties of Washington, Marion, and Linn are, according to Scullen, the most promising for honey. In Clackamas County in a good year the average surplus per colony is about 50 pounds, but very large crops are never obtained. A beekeeper writes from Woodburn: "I consider this a good location for an average yield every year, but know of no large yields. The past six years we have had trouble with foul brood." At Silverton, in the northern part of Marion County, the average surplus does not much exceed 25 pounds of honey, but single colonies have secured a much larger surplus. In every half dozen years there is likely to be one failure. The honey plants are maple, willows, evergreen, blackberry, fruit bloom, and white clover. The evergreen blackberry escaped from cultivation many years ago, and is now common throughout the Willamette Valley. It yields an amber honey of good body and aromatic flavor. Extracted honey is produced chiefly in this locality. Some years ago there were three apiaries in the adjacent mountains in a fireweed region, but they have disappeared. At Jefferson a good crop is obtained nearly every year. Beekeeping in Linn County is essentially the same as in Marion County.

The southern part of the Willamette Valley is occupied by Lane County, in the mountains of which are the sources of the Willamette River, which flows northward through the valley to the Columbia. Bees are kept mainly for pollinating the fruit trees in this county. There are thousands of acres of fireweed in the sections from which the forest has been lumbered, which make the finest kind of bee pasture. In the foothills there are openings for many more beekeepers.

A heavy rainfall in the spring is the rule in the Willamette Valley. In some years, according to E. J. Ladd, but little or no honey is secured from hundreds of acres of fruit trees in full bloom; but if the weather is favorable the strongest colonies gather a small surplus from this source. Bees fly many days in January. In February they bring in pollen from the willows, and from mustard, which is very abundant. In March dandelion stimulates brood-rearing. In April swarms are common, and the colonies are booming in May. Soft maple and vine-maple yield well in April, and fruit trees bloom in April and May. In June come white clover and alsike clover, and in July late berry bloom. Settled clear weather may be expected after July 4, and rain is rare until September. Clover dries up, and the bees must depend on berry bloom and late flowers. The surplus is stored in ten days to three weeks, and it is the alert beekeeper who succeeds. Some rarely fail; others seldom get much surplus. The average per colony is about 25 pounds. No fall crop is dependable here, and the colonies lose rather than gain in weight. Excellent crops are in some seasons obtained by moving the bees to the mountains, where whole slopes are pink with fireweed. The lumbered districts with their great areas of this honey plant offer opportunities of great promise to strong colonies where the rainfall is 50 inches or more.

South of the Willamette Valley the Cascade and Coast Ranges converge to form a mountainous region occupied by Douglas, Josephine, and Jackson counties, where most of the beekeeping is carried on as a side line. The crop is not always certain and is frequently small, which in part is due to negligence. At Jacksonville, Jackson County, the honey plants are manzanita, chamise (*Adenostoma fasciculatum*), madroña, poison ivy, sweet clover, white clover, and vetch. There are few commercial apiaries in this district that are getting good results.

Eastern Oregon, which comprises about two-thirds of the state, is a semi-arid region, where agriculture is mainly dependent on irrigation, and alfalfa and sweet clover are almost the sole reliance of the beekeeper. In the northern portion there are many streams from which water is diverted for reclaiming the fertile land of their lovely valleys, but the Snake River, which forms a large part of the eastern boundary of the state, flows through a deep canyon with precipitous black basaltic walls 2000 to 5000 feet in height—an impassable barrier. In the northeast the Blue Mountains modify the climate and offer many rich valleys and tablelands. Near the close of the Tertiary age a tremendous lava flow covered the southern portion of eastern Oregon with volcanic rocks to a depth of two thousand feet. It also extended over eastern Washington, southern Idaho and northeastern California, covering a total area of 200,000 square miles. The whole southeastern section is an arid sterile volcanic plateau on which the principal vegetation is sagebrush. The streams are few and small, and often disappear in alkaline sinks.

Probably in no part of Oregon is commercial beekeeping more successful than in Umatilla County. Twenty years ago the land was almost a desert, producing nothing but sagebrush, cactus, and a little bunch grass in the spring. The average annual rainfall is about 9 inches, and without irrigation no crops can be raised. The climate is very mild, and fruits and vegetables mature here two weeks earlier than in other parts of the state. East of the junction of the Umatilla River with the Columbia River thousands of acres are irrigated from a storage reservoir, which receives the flood waters of the Umatilla River. From this point up the river to Pendleton numerous ditches divert water directly from the river, and provide for the irrigation of more than 10,000 acres.

At Hermiston, six miles south of Umatilla, the valley is large and nearly level or gently rolling. There are many bees here. "During 12 years," writes a beekeeper, "I have never known a failure, but some years are far better than others." About one-fourth of the farms have one or more colonies. The best honey flow comes from the second crop of alfalfa.

Few large apiaries are reported in Union and Wallowa counties east of the Blue Mountains, in the northeast corner of the state. In Union County alfalfa and sweet clover are almost the sole dependence of the beekeeper, but dandelion tides over lean intervals.

The southern half of East Oregon is a great volcanic tableland, which has a rainfall of only 10 inches and is largely covered with sagebrush. The population is small, there are no railroads, and only a small portion of the land is under cultivation. Near its center is the Great Sandy Desert where the soil and lakes are alkaline. The small rivers disappear in the sands or marshy areas; and in the valleys there are in the wet season lakes which dry up later, leaving snow-white incrustations. Many small mountains, buttes, and ridges are scattered over this whole section. The soils are volcanic ash or the sedimentary deposits of lakes.

The larger part of this tableland is divided into the four counties of Malheur, Harney, Lake, and Klamath. There are more than 400,000 acres under irrigation from the larger lakes and the rivers which flow into them; but the principal occupations at present are stock-raising and the growing of hay. In all this vast territory there are only a few bees, except in Malheur County, in the northeastern corner, where there are a large number, especially around Nyssa and Ontario. There are also a number of small yards. During ten years a beekeeper here has had three failures and one severe winter loss of about 75 per cent.; but the average winter loss is only about 10 per cent. Most of the honey is obtained from the first and second crops of alfalfa. Doubtless in future years, with the increase of the area of alfalfa, beekeeping will become an important industry in the other counties.

The alfalfa weevil, says Scullen, has been a serious pest in Malheur County during recent years. This is being held in check, and the future looks brighter.

CALIFORNIA

Total area, 158,207 square miles. California ranks second in size among the states of the Union, being surpassed in area only by Texas. The state is commonly divided by beekeepers into northern and southern California, the dividing line being the Tehachapi Mountains in Kern County, which run nearly east and west. Its chief physical feature is the Great Central Valley lying between the Coast and the Sierra Nevada ranges, which converge at each end of the valley into a single range. The Siskiyou Mountains and Mount Shasta stand at the north end of the valley, and at the south end it is closed by the Tehachapi Range. The counties north of the valley are rugged and mountainous, and are largely covered with forest. Southern California, which comprises about one-third of the state, consists of a series of mountain ranges running mostly parallel with the coast, east of which there is a desert area which lies partly below sea level. Nearly seven-eighths of the surface of the state are covered by mountains.

The surface, soils, and agriculture of California are extremely diversified; and, owing to the nearness of the ocean and the prevailing westerly winds, there is a very wide range in temperature and rainfall. In the northern portion the soils are derived from lavas and volcanic ash; in the southern part they are sandy loams, and on the overflow lands of the rivers clay loams. The year is divided according to the rainfall into a dry and a wet season. The greatest rainfall comes in the winter months, and the months of June, July, and August are almost rainless over the greater part of the state. The amount of rain varies greatly in different years and in different sections of the state. At San Francisco the annual rainfall ranges from 7 to 49 inches. In the northwest corner of the state it is 60 or more inches, while in the Imperial Valley, one of the dryest regions in the United States, it is from 1.5 to 2.5 inches. In the Death and Saline valleys there is practically no rain. The winters are much milder than on the Atlantic Coast, and there are many localities where there are 200 clear days in a year.

Of the 99,617,280 acres of land surface, 20,000,000 acres are waste land, bare mountain ledges and deserts; 60,000,000 acres are in forest and mineral lands; and only a part of the remaining 20,000,000 acres is arable and suitable for agriculture. The land requiring irrigation is confined to the interior valleys and the southern coast. From San Francisco northward along the coast the heavy rains and summer fogs furnish sufficient moisture for the needs of agriculture. Existing enterprises are capable of irrigating about 4,219,000 acres, while 7,800,000 acres are included in projects. Nine-tenths of this land are watered from streams, and one-tenth is served by pumping-plants. Alfalfa and orchard and tropical fruits are the crops most extensively irrigated.

THE GREAT CENTRAL VALLEY

The Great Central Valley is 450 miles in length and 40 miles in width, so that the level area contains 18,000 square miles. The only drainage outlet of the valley is the Golden Gate, near San Francisco, which in its narrowest part is only a mile wide. The northern portion is drained by the Sacramento, and the southern by the San Joaquin River, the two rivers uniting before entering into San Francisco Bay. In the delta region of the two rivers, especially in Suisun Bay, there are 50,000 acres largely covered with tule (*Scirpus lacustris,* var. *occidentalis*), a nectarless, wind-pollinated, grasslike plant belonging to the sedge family. The valley of the Sacramento River is only about half the width of that of the San Joaquin; but, owing to the greater rainfall, it has a much greater water supply. Under the present system of irrigation the area of irrigable land in the Sacramento Valley is 630,000 acres; in the San Joaquin Valley, 2,150,000 acres. The San Francisco Bay region, where there are 5000 acres of irrigable land, is not included in the above figures.

The chief source of surplus honey in the Central Valley is alfalfa. A large quantity of amber-colored honey is stored from the second and third crops; but the first crop is of very little value, being cut before the blossoms appear. The first pollen and nectar gathered in some sections are from the Eucalyptus groves. There are nearly 2000 acres of these trees in the Berkeley Hills behind Richmond, and there is another large center between Newark and Arden. Willows may yield a surplus in Sacramento, Fresno, and Tulare counties; and in mid-summer and autumn a dark honey-dew is often gathered from the foliage. In the San Francisco Bay region southward there is a dense area of pears, plums, and prunes; but the largest area of prunes is in Santa Clara County. The almond-producing counties are Butte, Sutter, Yolo, Sacramento, and San Joqauin. Fiddleneck, or sheep's tansy (*Phacelia tanacetifolia*), blooms for about six weeks, beginning in April, and yields an amber-colored honey having an aromatic flavor. Alfilerilla, or filaree (*Erodium cicutarium*), is common everywhere on the dry plains and barren hillsides, and blooms for several months. Bur clover (*Medicago hispida*) is also very common, and is excellent to stimulate brood-rearing after a wet winter. With the passing of the spring flowers there comes a barren interval of several weeks until the second crop of alfalfa blooms about the first of June. In the Sacramento Valley, Yolo County has the largest acreage; but Tehama, Butte, and Sacramento counties also have large acreages. In the San Joaquin Valley, Stanislaus, Merced, Fresno, Tulare, and King counties have many thousand acres of alfalfa.

Carpet grass (*Lippia nodiflora*) is very abundant on the banks of the lower Sacramento and San Joaquin rivers. It blooms from May until frost. The larger part of the surplus in Sutter County formerly came from carpet grass, but much of the basin land has been reclaimed. The blooming of blue curls (*Trichostema lanceolatum*) in August marks the advent of the fall honey plants. It yields a surplus in Fresno County, and is abundant on the plains. Jackass clover or stinkweed (*Wislizenia refracta*) grows well in an alkaline soil. It is spreading widely over the poor lands of the San Joaquin Valley, and blooms throughout the autumn months. On the alkaline plains of the upper San Joaquin, spikeweed (*Centromedia pungens*) covers tens of thousands of acres; and in Fresno County spikeweed honey is produced by the carload. It is also abundant on the alkaline land of Solano County. This plant often forms thickets four or five feet tall. In late fall, after the spring vegetation has crumbled to dust, tarweed (*Hemizonia virgata*) may extend for miles, a single plant sometimes bearing 3000 flowers. It is common in both river valleys and on hills. It is a very reliable source of honey at Elk Grove, Sacramento County. Another species of tarweed blooms in the spring. Star thistle in the Sacramento Valley yields a heavy, white, mild honey, known commercially as Shasta honey. It is very abundant in grain fields, blooming from June until frost. The valley is carpeted with a great variety of grasses and beautiful flowers, such as lupines, Gilias, Godetias, poppies, buttercups, and clovers, few of which add much to the surplus.

The Sacramento River Valley is a fair location for beekeeping, and some counties have a large number of colonies of bees, especially Butte and Tehama. In Lake County fruit-growers employ 300 or more colonies for the pollination of prune and pear orchards. Commercial beekeeping in the past has received very little attention in Glenn, Yuba, Amador, Napa, and Solano counties; but Glenn and Yuba now have much alfalfa and star thistle, and bee culture is increasing in importance. At Willows, in Glenn County, there are numerous apiaries, the number varying greatly in different years. Migratory beekeeping is practiced, and the crop comes chiefly from alfalfa and star thistle. A beekeeper reports that from 1912 to 1920 he has had four years in which a surplus of 100 or more pounds per colony was secured. During three years it averaged only 45 pounds per colony. At Chico, in Butte County, there are numerous apiaries varying in size from 50 to 1500 colonies, which depend on alfalfa, star thistle, and mustard for a surplus.

The San Francisco Bay region includes, either in whole or in part, Marin, Sonoma, Napa, Solano, San Mateo, Contra Costa, Alameda, and Santa Clara counties, which border on Suisun, San Pablo, and San Francisco bays. The honey plants are essentially the same as those of the Great Valley. Early brood-rearing is stimulated by Eucalyptus, mustard, and wild radish, and by large areas of deciduous fruits which bloom from March to May. A great variety of garden truck, such as asparagus, celery, and parsnips is grown in the lower part of the Sacramento Valley; and celery and parsnip honey, says E. R. Root, who made a journey up the river in 1919, is stored in five and six story hives. Garden truck honey is a mixture from many plants, and is inferior in quality. A beekeeper at Walnut Grove operates 500 colonies scattered along the Sacramento River in yards of 150 colonies each. Bur clover, alfilerilla, white clover, horehound, and tarweed are of value. George A. Coleman, of the University of California, writes that in Alameda County there are about 50 apiaries which range from 25 to 150 colonies, besides several hundred small yards. The county contains between 8000 and 10,000 colonies of bees. "At Berkeley I have had very good success, both in raising bees and in the production of honey."

The San Joaquin Valley, especially the northern part, is a much better region for beekeeping than the Sacramento Valley. A company of beekeepers at Stockton reports that it operates 2000 colonies in twenty yards, and during ten years has averaged as high as 240 pounds per colony and as low as 80 pounds. A beekeeper at Dos Palos, Merced County, writes: "I have lived here 20 years and own about 2000 colonies of bees. In 1920 my crop was about sixty tons. I have never had a total failure, as my bees are scattered over a range 45 miles long; but not all of our apiaries give a surplus every year. Alfalfa is our main dependence, and occasionally we move to the orange districts." A half-century ago Fresno County was an arid plain on which only wheat and barley were grown. If there was sufficient rain, the crop matured; if not, everybody lost; but to-day the crops are watered by 900 miles of canals from the San Joaquin and King rivers. This county is the chief center of raisin-growing in the world; but the bloom of the vines is of very little benefit to the beekeeper. In 1919 the value of the fruit crop of all kinds was $70,000,000.

The county inspector of Fresno County writes that there are more than 200 commercial apiaries in the county, which range from 25 to 200 colonies; but many of them migrate from one section to another. At Selma the honey plants are alfalfa, yellow thistle, blue curls, and jackass clover. The crop is never an entire failure. Both Fresno and Tulare counties contain over 9000 colonies of bees. Tulare County possesses a great variety of climates and soils, and the land ranges in altitude from 268 to 15,000 feet on the summit of Mount Whitney, the highest mountain in the United States. The lower foothills of the mountains are covered with beautiful citrus groves, of which there are 45,000 acres in this county. There is also a large acreage of peach and prune trees. Cotton and cantaloupes are extensively planted. Kings is a small level county, which has every acre of its productive soil under cultivation. The crops of most value are wheat and fruits. Kern County is situated at the extreme southern end of the San Joaquin Valley. The acreage of apples, pears, oranges, alfalfa, and cotton is constantly increasing. In the vicinity of Bakersfield it is estimated that there are 5000 colonies of bees. The honey plants are Eucalyptus, willows, fruit bloom, Phacelia, alfilerilla, and alfalfa. In a good year the average surplus is 60 pounds, but the crop never fails wholly. Owing to the careless methods of many small farmers, European foul brood in some seasons destroys half the bees. In 1919 the upper San Joaquin region experienced one of the worst seasons in the history of bee culture in this section; and with the exception of a few acres of orange trees the honey plants failed almost entirely to yield nectar. Northward in the valley alfalfa yielded half a crop.

THE FOOTHILLS

The Coast Range on the western side of the Great Central Valley consists of numerous broken parallel chains of moderate height. The Sierra Nevada on the eastern side is a single continuous rugged range with a precipitous descent of 1000 feet to the mile, but on the western flank the slope is more gradual. On the foothills, or "the steps up to the high ranges," valley conditions prevail up to an elevation of about 1000 feet. Above this elevation the winter temperature is lower, snow flurries begin, and above 4000 feet the land is used principally for summer pasturage. The mountain winter is like that of the eastern states, with a very heavy snowfall; and the only streams which flow in summer in the Central Valley are fed by the snows of the Sierra. Unlike the forest of Oregon, the Sierra woodland is open, and the trees stand far apart, admitting the sunlight. The characteristic vegetation of the foothills is the chaparral, spiny impenetrable thickets densely covering large areas of the higher slopes of the Coast Ranges and the foothills and middle altitudes of the Sierra. The more common shrubs of the chaparral are manzanita, chamise, mountain lilacs or buckbush (*Ceanothus*), coffeeberry, pea chaparral, Christmas berry or toyon, poison oak (*Rhus*), and several species of scrub oak which are nectarless. Chamise, or greasewood, often excludes all other shrubs, when the thicket is known as a "chamisal." Pea chaparral (*Pickeringia montana*) is restricted chiefly to the Coast Range. A number of herbaceous plants found on the plains are also of value. In the valley many counties report only a few colonies of bees in the foothills of the mountains. A beekeeper at Coalings, Fresno County, in the Diablo Mountains of the Coast Ranges, writes: "Like every other mountain location in California, there is a crop about one year in four. When there is a good flow they all crowd in; when there is nothing they move away." "There are ten million acres of chaparral in the state suitable for bee range," writes George A. Coleman. "The chaparral extends along the inner and outer Coast Ranges from Siskiyou County to the Mexican border, and on the inner Sierra foothills for a strip 50 miles wide and 450 miles long."

NORTHERN CALIFORNIA

North of the Great Central Valley the Coast Ranges and the Sierra Nevada unite to form a rugged country, which is thinly populated. During the Tertiary Period a great flood of lava flowed over Lassen, Modoc, and Siskiyou counties, converting this section into a volcanic plain, which is largely barren because of the dryness of the climate. But the soil is fertile and a large area is already under irrigation. Many acres of alfalfa are grown and white clover is rapidly becoming more common; and as both of these plants are dependable sources of nectar here this section should in the near future offer excellent locations for the production of honey.

The northwest coast region is occupied by many disconnected chains of moderate elevation, between which are many valleys, low hills, and grassy slopes. The climate is mild, and there is a heavy annual rainfall of 40 to 100 inches. Fogs and high winds are very prevalent along the coast. The land is well suited to dairying, and both white and red clover succeed well; but the acreage of alfalfa is small. This region reports few colonies of bees and only a small amount of honey. The redwood belt with its heavy fogs is considered by Coleman as one of the poorest locations for beekeeping in the state.

SOUTHERN CALIFORNIA

Southern California, or that portion of the state south of the Tehachapi Mountains, has an area of over 50,000 square miles. The western section of this region is mountainous, consisting of a series of chains with a general trend from northwest to southeast. The Death Valley, Mohave, and Colorado deserts are in the southeastern portion of the state, and a large area is depressed below sea level. Fruit-growing is the principal industry of southern California, and with its favorable climate and the introduction of irrigation it is destined to supply the markets of North America with sub-tropical and many tropical fruits.

Beekeepers report excellent results in Monterey and San Luis Obispo counties; and as these two counties resemble in climate and vegetation the southwestern counties of California they may be briefly considered here. "In my Forest Apiary," says Coleman, "in the Monterey National Forest I get on an average 120 pounds of extracted honey and 100 sections of comb honey per colony, year after year, of very fine quality from wild alfalfa, madroña, manzanita, and black sage." A beekeeper at San Luis Obispo writes that there are about fifteen commercial apiaries in that locality, which contain from 25 to 200 colonies of bees. A crop of honey is obtained every year, and the surplus in a good season ranges from 100 to 200 pounds. The honey plants are willows, Eucalyptus, manzanita, Christmas berry, mountain lilacs, mustard, poison oak, black sage, alfilerilla, wild alfalfa. wild buckwheat, broomweed, tarweed, and black sage.

Bee culture and fruit-growing have reached their highest development in California in the seven southwestern counties of Santa Barbara, Ventura, Los Angeles, Riverside, San Bernardino, San Diego, and Imperial. It is a rugged or mountainous region, where the rolling hills are covered with open groves of oaks and chaparral. On the foothills the characteristic plants are wild buckwheat, black, purple, and white sages, orange, buckthorn, chamise, wild alfalfa, and sage brush. In some instances the hillsides from base to apex are entirely covered with two species of cactus (*Opuntia Engelmanii* and *O. prolifera*). The chaparral consists chiefly of manzanita, mountain lilacs (*Ceanothus*), sage, and mountain mahogany. There are in the open areas many herbaceous plants, a part of which are valuable to the beekeeper. Under irrigation, which is extensively practiced, the valleys are very productive.

The Lima bean is adapted to a costal strip twenty miles in width which is subject to heavy ocean fogs, and is mostly grown without irrigation. More than 50 per cent. of the entire bean crop in this country comes from Ventura, Orange, Santa Barbara, and San Diego counties, Ventura County leading with over 60,000 tons. Only the Lima bean is of value as a honey plant, although the black-eyed bean has been reported to yield an amber-colored honey, but this is rightly questioned by L. L. Andrews. The same authority says that in recent years the great San Fernando Valley in Los Angeles County has been planted extensively to bush Lima beans, which under irrigation have proven to be a reliable source of honey, and that many beekeepers find it profitable to move there after the orange flow.

In 1920 there were under cultivation in southern California 149,800 acres of Lima beans, California being credited with 85 per cent. of all the Lima beans grown in the world. Between Los Angeles and the Palms there are several bean ranches which contain from 2000 to 3500 acres in a single block. The vines bloom in July and August, when many beekeepers move their bees to the bean fields.

At Summerland, Santa Barbara County, every available acre was devoted, in 1919, to growing Lima beans, and the entire crop in the county was valued at $800,000. Cool sea fogs and the absence of protracted hot spells are required for the maturing of this plant, as otherwise it is apt to blight; but the heavy fogs retard the flight of the bees. At Santa Barbara there is one apiary which reports 500 colonies of bees, and two report 100 colonies each. The honey is mild-flavored, water-white, and candies quickly into a pastelike solid.

A few years ago the broad expanses of Ventura County were considered only fit for grazing; but with the introduction of irrigation the land has been converted into profitable fruit and truck farms. Owing to competition from the Orient the acreage of beans has been greatly reduced and large tracts have been planted with vegetables. At the present time, according to the bee inspector, there are some 18,000 colonies of bees in this county. A beekeeper writes from Fillmore that there are 20 apiaries in that locality which range from 20 to 300 colonies. "I began beekeeping in 1880 with 47 hives, and operated at one time 700 colonies and produced 30 tons of honey. I now have 250 colonies, and will secure about six tons this season (1920)." At Ventura, M. H. Mendleson had at one time 2000 colonies of bees, and he has kept as many as 900 colonies in one yard. The main crop is gathered from the sages, and about July 1 he moves his co'onies to the bean fields. The honey flow from bean is usually reliable, and his average surplus from this source is about 50 pounds per colony annually. In dry years the sage crop is a failure, but in a good season following ample winter rains he has had as high as two supers on every hive filled with honey in three days. During an exceptional year (there have been only two such seasons in his 38 years of experience) he secured an average of 300 pounds per colony from sage. Previous to 1880 enormous yields were obtained from the sages. In 1878, R. Wilkin is reported to have obtained a surplus that seems unbelievable; but at that time the mountain slopes were densely covered with the sages and there were splendid rainfalls.

Formerly at his Piru range Mendleson secured tons of wild sunflower honey, but to-day none is gathered as the range is overstocked with cattle. A decrease in the rainfall also has occurred, due to the burning and clearing off of large areas of trees and brush that protected the water-sheds. On the stubble fields after the grain has been harvested turkey mullein, or woolly white drought weed (*Eremocarpus setigera*), springs up very abundantly, and secretes nectar well after a wet winter. It is an annual, whitish-green plant, covered with rough spreading hairs, and forming close mats upon the ground. The strong-flavored honey is similar to a thick dough and is difficult to extract. There are in Ventura County thousands of oranges, lemons, and apricots. A few years ago there was a great difference of opinion among orange-growers as to the value of honeybees as pollinators of the bloom; but Mendleson states that one large orange producer refused to take any rent, as he had found that a much larger quantity of fruit was set when the bees were present.

Four-fifths of Los Angeles County, which has an area of 3800 square miles, are capable of cultivation. At the close of the Great War it was estimated that the county contained 65,000 colonies of bees, and the crop of honey was valued at $1,000,000. Since then, owing to a succession of poor years, the number of colonies has been greatly reduced.

Migratory beekeeping is very largely practiced by progressive beekeepers in this section. The home apiary is located in the foothills near the orange groves. After the honey flow from orange bloom, which is usually of short duration, is over, the apiary is moved to the sage ranges, then to the bean fields, and perhaps later to wild buckwheat. In some instances from 1000 to 2000 colonies are operated by one apiarist with hired help, and enormous yields are obtained. Bees are also moved in the spring into southern California by the carload from Nevada and Utah. Many of the colonies are run for increase, and from the larger colonies four or five nuclei are made. After a crop has been harvested from orange bloom and the colonies have become strong they are moved back to these states in time for the flow from alfalfa. There is danger from forest fires in the wild sage ranges, and in the San Fernando forest fire of 1906 there were burned in Los Angeles County 16,000 colonies of bees.

Near Los Angeles there are numerous apiaries of various sizes. The main crop comes from orange bloom; sage follows, first black, then purple and white, if there has been sufficient rain. Each variety of sage is adapted to certain localities, black sage being more widely distributed. In the fall winter stores are gathered from wild buckwheat, sumac, and blue curls if there has been an average rainfall; if not, the bees must be fed. The average surplus per colony is from 50 to 60 pounds; but by moving the apiary it is possible to secure 200 pounds. At Owensmouth, Los Angeles County, there are over 2500 colonies in fifteen apiaries. A beekeeper writes: "I have made bees pay well here. I now have 300 colonies, and have kept 400 to 500 in some seasons. All my surplus is sold at home or by parcel post. In a good year the average surplus per colony is from 100 to 150 pounds."

A beekeeper living at Lankershim, Los Angeles County, writes (1920): "At present my bees are in the bean district about five miles from home. They have been there three weeks, but have stored very little surplus. This spring in the orange district, 35 miles from here, they gathered a fair crop. Then I moved to the sage region, seven miles from here, where a fine crop was obtained. In the Lima-bean fields there are two or three thousand colonies in a radius of three miles for about two months, then they are moved away." In the vicinity of Pasadena there are a great number of small yards. A beekeeper at Hollywood writes: "I have an apiary of 300 colonies at Moorpark. In 1918 I obtained 4800 pounds of sage honey. An apiary of 300 colonies in the orange groves made six tons of orange honey; but when moved to the bean fields in June they secured hardly enough surplus for winter. The average profit per colony in Ventura and Los Angeles counties is about five dollars."

The southwestern part of San Bernardino and the western portion of Riverside counties are typical orange and sage locations, quite like those of Los Angeles County. San Bernardino County is the largest county in the United States. It has more than 200,000 acres under irrigation or included in projects, and many thousand acres of alfalfa and fruit trees under cultivation. This county produces more than one-third of the citrus fruits of California, one of the largest citrus districts being located at Redlands. East of the San Bernardino Mountains the land is largely desert, which is valueless for beekeeping. L. L. Andrews, of Corona, Riverside County, who operates 1000 colonies of bees and has been in the business for twenty-five years, writes that there are in this locality about a dozen apiaries which range from 50 to 200 colonies. A crop of honey is usually obtained every year, chiefly from orange and sage. In 1920 strong colonies averaged four pounds of honey per day for fifteen days from black sage. At an elevation of 3000 feet in the San Jacinto Mountains there is more rain, and white sage and wild buckwheat bloom later than in the lowlands.

In San Diego County, in the southwestern corner of the state, in the vicinity of Fall Brook there are more than 2000 colonies in yards which range from 25 to 300 colonies. In a period of five years there is usually one large crop, one failure, and three fair crops. In a good year 30 pounds of comb honey and 60 pounds of extracted honey may be obtained. Eight years ago a yard of 90 colonies averaged 300 pounds per colony, and made a large increase in the number of colonies. At Descanso, in the south-central part of the county, there are 13 apiaries which range from 50 to 300 colonies. "I have lived in this locality for 25 years," writes a beekeeper, "and have had fairly good success. European foul brood was unknown here until four years ago, and till then we gave the bees very little attention and had good yields." At Santa Ysabel in the central part of the county the best record of a beekeeper who has lived there since 1887 is 24,000 pounds of honey from 180 colonies. There has been only one complete failure in 13 years. The number of colonies in this locality is about 600. Near the coast from sea level to an altitude of perhaps 500 feet the sages are not very abundant, and are not especially valuable as honey plants—due probably to lower elevation and lighter rainfall. The great sage ranges, which yield the surplus of sage honey, are above 500 feet and below 3500 feet. Above this elevation the plants are much less common, although reported on the mountains at a height of over 5000 feet.

The more important early-blooming plants in the southern coast counties are Eucalyptus, willows, Ceanothus, manzanita, buckthorn, bur clover, pepper tree, sumac, and a spring-blooming species of tarweed. "After traveling over all of California, I am convinced," writes E. R. Root, in 1919, "that there is as much black sage as there ever was. There is five times as much sage honey produced to-day as there was forty years ago—not that sage is more abundant, but beekeepers are more numerous. To-day within fifty miles of Los Angeles there are more bees and beekeepers than in any other part of the United States, and fully 50 per cent. of them are located in the sage ranges." In San Diego County white sage is far more abundant than black sage, and is the source of a large part of the sage honey produced in this county. It extends northward to Santa Barbara County. South of Los Angeles County black sage is relatively not very common. It is reported northward by Jepson as far as Mt. Diablo. Very little purple sage is found in San Diego and Orange counties. It is abundant in the Angeles National Forest, and is the dominant sage in the northern part of Los Angeles and Ventura counties. It is rare north of San Luis Obispo. Ventura County is the heart of the purple-sage territory, Orange County of the black sage, and San Diego County of the white sage.

"The citrus region of southern California is well stocked or overstocked," according to E. F. Phillips, "as are parts of the sage areas; but there are vast areas of fine dependable sage region insufficiently covered by bees, and some of the best sage areas that I ever saw were hardly touched by bees. The future development of the sage areas depends chiefly on the building of suitable roads back into the mountains, and this is being done rapidly. Some of the best sage region will probably always remain inaccessible, for this vast area can not be reached by bees placed in the canyons of the foothills."

The conditions described above apply to southern California for the years immediately preceding 1921. The year 1920 was one of the best known in this state for the production of honey. In the past fifty years there have been only two or three seasons in which equally good conditions prevailed. For illustration, an apiary of 280 colonies, spring count, produced 37 tons of honey. Orange bloom, sage, and wild buckwheat all gave a big flow. But the four years following have been most unsatisfactory. There has been much cold, cloudy, and windy weather with very little rain, or with rain at the wrong time. Beekeepers often failed to secure a surplus, and frequently were obliged to feed in order to carry their bees through the winter. Many colonies were neglected, and were found dead in the spring. But conditions vary very widely in California, and in some localities fair to good crops of honey were produced. This was generally true in localities where there were large areas of irrigated alfalfa. But beekeepers dependent on orange bloom, the sages, and wild buckwheat very widely met with severe disappointment and loss.

"The year 1921," declared a well-known beekeeper, "in southern California from the beekeeper's viewpoint will go down in history in a class by itself." For the entire state it was estimated that only a quarter of a normal crop was secured. During orange bloom many colonies on scales showed a daily loss in weight. In Los Angeles County field bees perished by thousands. But toward the close of the season sufficient winter stores were obtained from wild buckwheat in a great number of bee yards. While the crop of 1922 was better than that of 1921, Ventura County produced only about a third of a normal crop; but Riverside County obtained a surplus from black sage and wild buckwheat. Only in rare instances, as a rule, was a satisfactory amount of honey gathered, and many colonies did not obtain enough to carry them through the winter. The season of 1923 was described by a beekeeper in Ventura County as a "terrible" one for the production of honey. The orange flow was one of the poorest on record. In San Diego County the surplus extracted was the lightest for many years. A convention of beekeepers in Riverside County declared the crop the poorest ever known in that section. These reports apply to orange, sage, and the other wild ranges; but in the Imperial, Palo Verde, and other irrigated tracts a part of a crop was secured from alfalfa. A beekeeper writes: "The four years previous to 1925 were so dry that the honey crops from beans were almost entire failures."

The year 1924 opened with a large number of apiaries in a very neglected condition. In many of them 75 per cent. of the colonies were dead. Only a few had sufficient stores to carry them to a honey flow in the spring. While during this season orange, sage, and wild buckwheat did not give an abundant flow, most beekeepers found their hives well filled with honey for winter, and a small surplus was generally reported. In certain sections alfalfa gave a remarkable flow. In most localities the flow was never heavy, but was a little more than the daily needs of the colonies. These four unfavorable years have not, in the opinion of beekeepers of long experience, been at all discouraging. Strong colonies nearly always secured some surplus; it was the weak colonies which failed With better seasons, this industry will pay as well as or better than the average business representing the same investment and labor.

THE SOUTHEASTERN DESERTS

Imperial County was established in 1907 by cutting off the eastern desert portion of San Diego County. Imperial Valley, which was once an arm of the sea, is about 140 miles long and 20 to 50 miles wide, and has an area larger than the state of Rhode Island. The land is very level, and the silty soil is of great depth and of almost inexhaustible fertility. There are more than a million acres of irrigable land, of which 600,000 acres are now supplied with water by canals leading from the Colorado River. The annual rainfall is less than two inches. There is no snow, and only a few light frosts. In August the temperature often rises to 120 degrees F., but owing to the dry atmosphere the heat is less oppressive than the same temperature would be in the eastern states. In no part of the world has agriculture developed more rapidly or more profitably. In 1900 the land was a hot sandy desert almost destitute of vegetation. In 1920 there were 100,000 acres of alfalfa and 90,000 acres of cotton. But there will probably be a large decrease in the acreage of cotton, as the result of the great decline in price, and an increase in the area of alfalfa.

In February the first honey comes from wild hollyhock; and arrow-weed (*Pluchea sericea*), which grows along the ditches, a little later yields a fine-flavored honey. Cotton is the source of a light-colored, very sweet honey; but the surplus comes almost wholly from alfalfa. The color of the alfalfa honey is light amber to amber, and the flavor is a little inferior to the alfalfa honey of Nevada. The secretion of nectar is very reliable, and a crop failure is almost unknown. A fair average per colony is 75 pounds. It is estimated that there are 15,000 colonies of bees in the valley, and the total annual surplus has been estimated to range from 12 to 40 carloads. Few bee ranges are unoccupied, and no one should take up beekeeping in the valley without first investigating. Imperial Valley may be a bee paradise, but it is not a paradise for a man to live in. The valley was formerly known as the Colorado Desert.

In the southeastern part of the state are the barren desert wastes occupied by the counties of Riverside, San Bernardino, and Inyo. In the San Bernardino

Desert a whole year has passed without any measurable rainfall; and, exposed to the intense heat and aridity, no vegetation can live except thorny cacti, the creosote bush, mesquite, palo verde, and arrow-weed. In the Mohave Desert are weird groves of tree yuccas covering many square miles. Along the Colorado River, in the eastern portion of Imperial and Riverside counties, there is a large early flow from mesquite. In elevation the land ranges from 250 feet below sea level to 10,978 feet above the sea at San Jacinto Park. On the east border of Riverside County lies the Palo Verde Valley, where 90,000 acres can be watered from the Colorado River, and 37,000 are under intensive cultivation. In 1920 the principal crop of the valley was short-staple cotton, of which 20,000 acres were grown. The acreage of alfalfa in the county was 35,000 acres; and of oranges, lemons, and apricots there were many thousand acres.

Inyo County, which comprises 10,294 square miles of rugged mountains, deserts, and fertile valleys, lies east of the Sierra Nevada Range. Owens River Valley, which is 199 miles long and 6 to 20 miles wide, is the agricultural section of the county. The rainfall is about 6 inches, and there are no fogs. Wild buckwheat grows all along the foothills of the valley. About five miles from Independence there is a wild-buckwheat range where large crops of honey have been secured. "A crop of honey," says Richter, "is practically always assured at high altitudes where there are honey-producing plants present." A surplus is usually obtained every year from this plant; but sheep pastured in the national-forest areas are very destructive to the flower buds. The area around Bishop is the largest in the county available for alfalfa, and consequently for beekeeping. Alfalfa honey is as white here as in Nevada, where conditions are very similar. Sweet clover yields nectar until frost, while in the lower valleys it usually dries up in July and August. This county is the great comb-honey district of California, for, owing to the dryness of the atmosphere, the honey is so thick that extracting is almost impossible. The valley is, moreover, at an elevation of 4000 feet, and the cold nights and cool days also tend to thicken the honey. On the other hand, in the Imperial Valley it is so hot that comb honey can not be successfully produced. Inyo County has reported 6000 colonies of bees, and an annual production of 300,000 pounds of honey. The Muth-Rasmussen apiary a few years ago contained 300 colonies.

INDEX

LIST OF ILLUSTRATIONS